ATLAS OF
HUMAN CENTRAL NERVOUS SYSTEM DEVELOPMENT
Series

Volume 1: The Human Brain during the First Trimester 3.5- to 4.5-mm Crown-Rump Lengths

Volume 2: The Human Brain during the First Trimester 6.3- to 10.5-mm Crown-Rump Lengths

Volume 3: The Human Brain during the First Trimester 15- to 18-mm Crown-Rump Lengths

Volume 4: The Human Brain during the First Trimester 21- to 23-mm Crown-Rump Lengths

Volume 5: The Human Brain during the First Trimester 31- to 33-mm Crown-Rump Lengths

Volume 6: The Human Brain during the First Trimester 40- to 42-mm Crown-Rump Lengths

Volume 7: The Human Brain during the First Trimester 57- to 60-mm Crown-Rump Lengths

Volume 8: The Human Brain during the Second Trimester 96- to 150-mm Crown-Rump Lengths

Volume 9: The Human Brain during the Second Trimester 160- to 170-mm Crown-Rump Lengths

Volume 10: The Human Brain during the Second Trimester 190- to 210-mm Crown-Rump Lengths

Volume 11: The Human Brain during the Third Trimester 225- to 235-mm Crown-Rump Lengths

Volume 12: The Human Brain during the Third Trimester 260- to 270-mm Crown-Rump Lengths

Volume 13: The Human Brain during the Third Trimester 310- to 350-mm Crown-Rump Lengths

Volume 14: The Spinal Cord during the First Trimester

Volume 15: The Spinal Cord during the Second and Third Trimesters and the Early Postnatal Period

The Human Brain during the Third Trimester 310- to 350-mm Crown-Rump Lengths

This thirteenth of 15 short atlases reimagines the classic 5 volume *Atlas of Human Central Nervous System Development*. This volume presents serial sections from specimens between 310 mm and 350 mm with detailed annotations. An introduction summarizes human CNS developmental highlights around 9 months of gestation. The Glossary (available separately) gives definitions for all the terms used in this volume and all the others in the *Atlas*.

Key Features

- Classic anatomical atlases
- Detailed labeling of structures in the developing brain offers updated terminology and the identification of unique developmental features, such as germinal matrices of specific neuronal populations and migratory streams of young neurons
- Appeals to neuroanatomists, developmental biologists, and clinical practitioners
- A valuable reference work on brain development that will be relevant for decades

The Human Brain during the Third Trimester 310- to 350-mm Crown-Rump Lengths

Atlas of Human Central Nervous System Development, Volume 13

Shirley A. Bayer
Joseph Altman

CRC Press
Taylor & Francis Group
Boca Raton London New York

CRC Press is an imprint of the
Taylor & Francis Group, an **informa** business

Designed cover: Shirley A. Bayer and Joseph Altman

First edition published 2024
by CRC Press
2385 NW Executive Center Drive, Suite 320, Boca Raton FL 33431

and by CRC Press
4 Park Square, Milton Park, Abingdon, Oxon, OX14 4RN

CRC Press is an imprint of Taylor & Francis Group, LLC

LCCN no. 2022008216

ISBN: 978-1-032-22885-3 (hbk)
ISBN: 978-1-032-22880-8 (pbk)
ISBN: 978-1-003-27464-3 (ebk)

DOI: 10.1201/9781003274643

Typeset in Times Roman
by KnowledgeWorks Global Ltd.

Publisher's note: This book has been prepared from camera-ready copy provided by the authors.
Access the Support Material: www.routledge.com/9781032228853

CONTENTS

ACKNOWLEDGMENTS

We thank the late Dr. William DeMyer, pediatric neurologist at Indiana University Medical Center, for access to his personal library on human CNS development. We also thank the staff of the National Museum of Health and Medicine that was at the Armed Forces Institute of Pathology, Walter Reed Hospital, Washington, D.C. when we collected data in 1995 and 1996: Dr. Adrianne Noe, Director; Archibald J. Fobbs, Curator of the Yakovlev Collection; Elizabeth C. Lockett; and William Discher. We are most grateful to the late Dr. James M. Petras at the Walter Reed Institute of Research who made his darkroom facilities available so that we could develop all the photomicrographs on location rather than in our laboratory in Indiana. Finally, we thank Chuck Crumly, Neha Bhatt, Kara Roberts, Michele Dimont, and Rebecca Condit for expert help during production of the manuscript.

AUTHORS

Shirley A. Bayer received her PhD from Purdue University in 1974 and spent most of her scientific career working with Joseph Altman. She was a professor of biology at Indiana-Purdue University in Indianapolis for several years, where she taught courses in human anatomy and developmental neurobiology while continuing to do research in brain development. Her lengthy publication record of dozens of peer-reviewed, scientific journal articles extends back to the mid 1970s. She has co-authored several books and many articles with her late spouse, Joseph Altman. It was her research (published in *Science* in 1982) that proved that new neurons are added to granule cells in the dentate gyrus during adult life, a unique neuronal population that grows. That paper stimulated interest in the dormant field of adult neurogenesis.

Joseph Altman, now deceased, was born in Hungary and migrated with his family via Germany and Australia to the US. In New York, he became a graduate student in psychology in the laboratory of Hans-Lukas Teuber, earning a PhD in 1959 from New York University. He was a postdoctoral fellow at Columbia University, and later joined the faculty at the Massachusetts Institute of Technology. In 1968, he accepted a position as a professor of biology at Purdue University. During his career, he collaborated closely with Shirley A. Bayer. From the early 1960s-2016, he published many articles in peer-reviewed journals, books, monographs, and free online books that emphasized developmental processes in brain anatomy and function. His most important discovery was adult neurogenesis, the creation of new neurons in the adult brain. This discovery was made in the early 1960s while he was based at MIT, but was largely ignored in favor of the prevailing dogma that neurogenesis is limited to prenatal development. After Dr. Bayer's paper proved new neurons are added to granule cells in the hippocampus, Dr. Altman's monumental discovery became more accepted. During the 1990s, new researchers "rediscovered" and confirmed his original finding. Adult neurogenesis has recently been proven to occur in the dentate gyrus, olfactory bulb, and striatum through the measurement of Carbon-14—the levels of which changed during nuclear bomb testing throughout the 20th century—in postmortem human brains. Today, many laboratories around the world are continuing to study the importance of adult neurogenesis in brain function. In 2011, Dr. Altman was awarded the Prince of Asturias Award, an annual prize given in Spain by the Prince of Asturias Foundation to individuals, entities, or organizations globally who make notable achievements in the sciences, humanities, and public affairs. In 2012, he received the International Prize for Biology - an annual award from the Japan Society for the Promotion of Science (JSPS) for "outstanding contribution to the advancement of research in fundamental biology." This Prize is one of the most prestigious honors a scientist can receive. When Dr. Altman died in 2016, Dr. Bayer continued the work they started over 50 years ago. In her late husband's honor, she created the Altman Prize, awarded each year by JSPS to an outstanding young researcher in developmental neuroscience.

INTRODUCTION

A. Specimens and Organization

Volume 13 in the Atlas Series presents the human brain in three normal specimens from the Yakovlev Collection[1] at 9 months at the end of the third trimester. These specimens were analyzed in Volume 2 of the original *Atlas of Human Central Nervous System Development* (Bayer and Altman, 2004). These fetuses are viable *ex utero*. Nearly all the structures present in the adult brain are recognizable and are maturing from the diencephalon to the medulla. But dwindling remnants of the embryonic nervous system remain in the cerebral cortex and the cerebellar cortex.

This volume contains serial grayscale photographs of Nissl-stained sections of a sagittal specimen (Y180-61, **Part II**), a horizontal specimen (Y301-62, **Part III**) and a frontal specimen (Y217-65, **Part IV**) with crown-rump (CR) lengths of 310-mm to 350-mm. All specimens are in the 37th gestational week (GW). **Sagittal plates** are ordered from medial to lateral; the anterior part of each photographed section is facing left, posterior right. **Horizontal plates** are presented from dorsal (first) to ventral (last); the anterior part of each section faces left, the posterior part faces right; the midline is in the horizontal center. **Frontal plates** are presented from anterior (first) to posterior (last); the dorsal part of each section is at the top, the ventral part at the bottom; the midline is in the vertical center. Each **Plate** is in two parts: **A**, on the left, shows the full-contrast photograph without labels; **B,** on the right, shows a low-contrast copy of each photograph with unabbreviated labels. For each specimen, a series of serially spaced **low-magnification plates** show the entire section to identify large structures. The brain core is shown in many **high-magnification plates** to identify smaller structures. In addition, several **very-high-magnification plates** show the cerebral and cerebellar cortices. Because our emphasis is on development, transient structures that appear only in immature brains are labeled in ***italics***, either directly in some of the high-magnification plates or in **bold numbers** that refer to labels in a list. During dissection, embedding, cutting, and staining, some of the sections illustrated were torn; that damage is sometimes surrounded by *dashed lines*.

1. The *Yakovlev Collection* (designated by a **Y** prefix in the specimen number) is the work of Dr. Paul Ivan Yakovlev (1894–1983), a neurologist affiliated with Harvard University. Throughout his career, Yakovlev collected many diseased and normal human brains. He invented a giant microtome that was capable of sectioning entire human brains. Later, he became interested in the developing brain and collected many during the second and third trimesters. The normal brains in the developmental group were cataloged by Haleem (1990) and were examined by us during 1996 and 1997. The collection was moved to the National Museum of Health and Medicine when the Armed Forces Institute of Pathology (AFIP) closed at Walter Reed Hospital and is still available for research.

B. Developmental Highlights

Figures 1-2 compare early, middle, and late third trimester brains in horizontal sections of the dorsal telencephalon. The overall growth of the brain is obvious when the sections are stacked in a "layer cake" arrangement (**Fig. 1**). The side-by- side comparison (**Fig. 2**) shows that by the end of the third trimester, the cortical plate slightly decreases its thickness and rapidly grows lengthwise. The white matter (*pale yellow*) greatly increases its thickness in most cortical areas except the insula. On the other hand, layers 3-6 of the *stratified transitional field* slow their growth and eventually will disappear. Cortical gyrification continues to elaborate as more fibers enter, and the white matter increases its relative volume. Fibers come from the thalamus (via the enlarging internal capsule) and from the opposite cortex (via the expanding corpus callosum). By now, many associational fibers are accumulating in ipsilateral sides of the cortex to establish connections with neurons in different cortical areas (Altman and Bayer, 2015). Cortical circuitry is continuing to elaborate and the early gyri are more difficult to identify as many secondary and tertiary gyri appear.

Figure 3 provides some insight concerning secondary and tertiary gyrification. The sequence of postnatal myelination gives us a clue to the reason the

Superimposed Third Trimester Brains in the Horizontal Plane

Bottom layer: GW 37
Middle layer: GW 30
Top layer: GW 26

NEP- neuroepithelium
STF-stratified transitional field
SVZ-subventricular zone

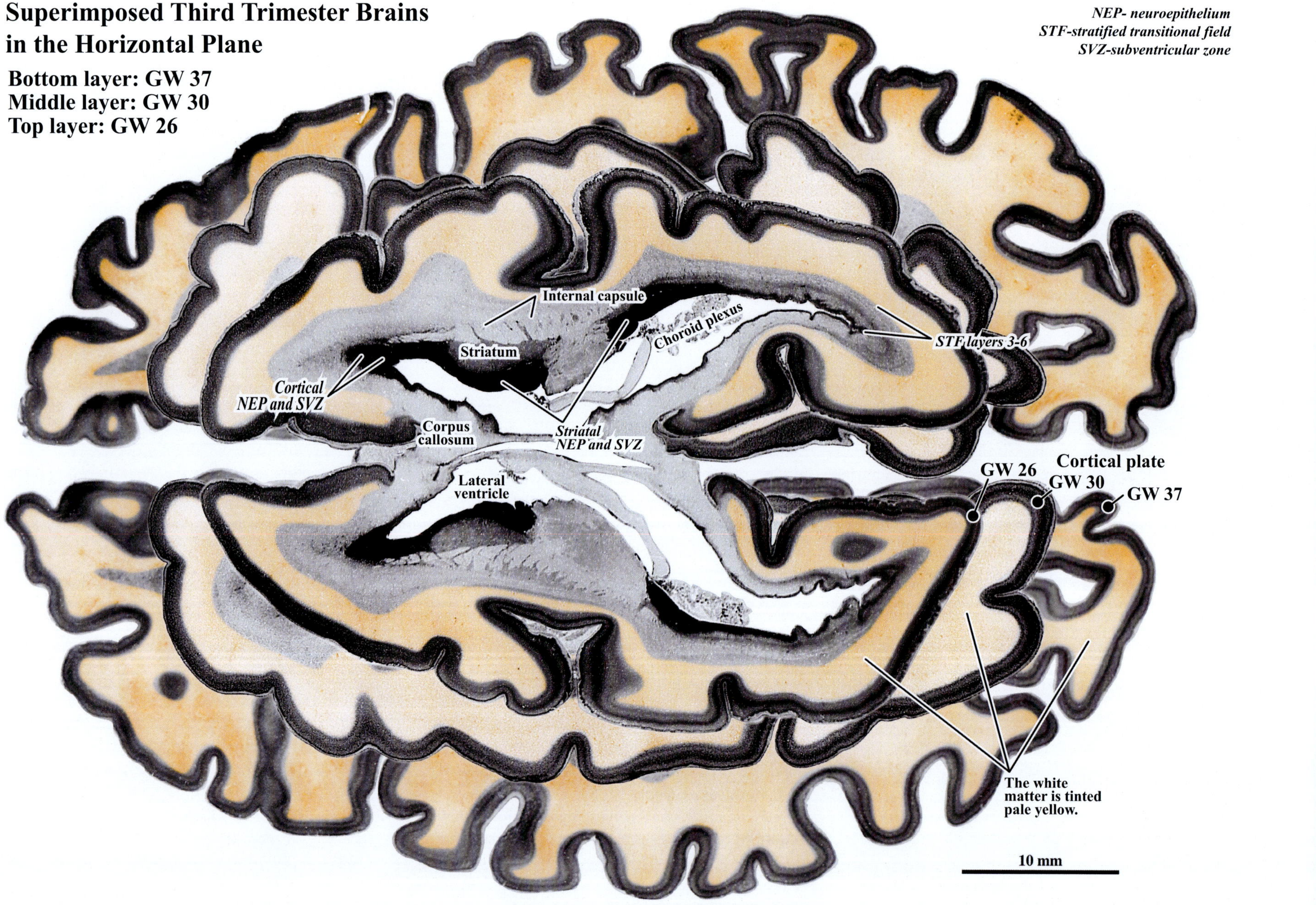

Side-by-Side Comparison of Third Trimester Brains in the Horizontal Plane

Growth related to increased gyrification:

Cortical plate *slightly decreases* thickness and *greatly increases* linear surface
White matter increases depth in most areas (insula excepted)
STF layers 3-6 are less prominent and decrease in relative size

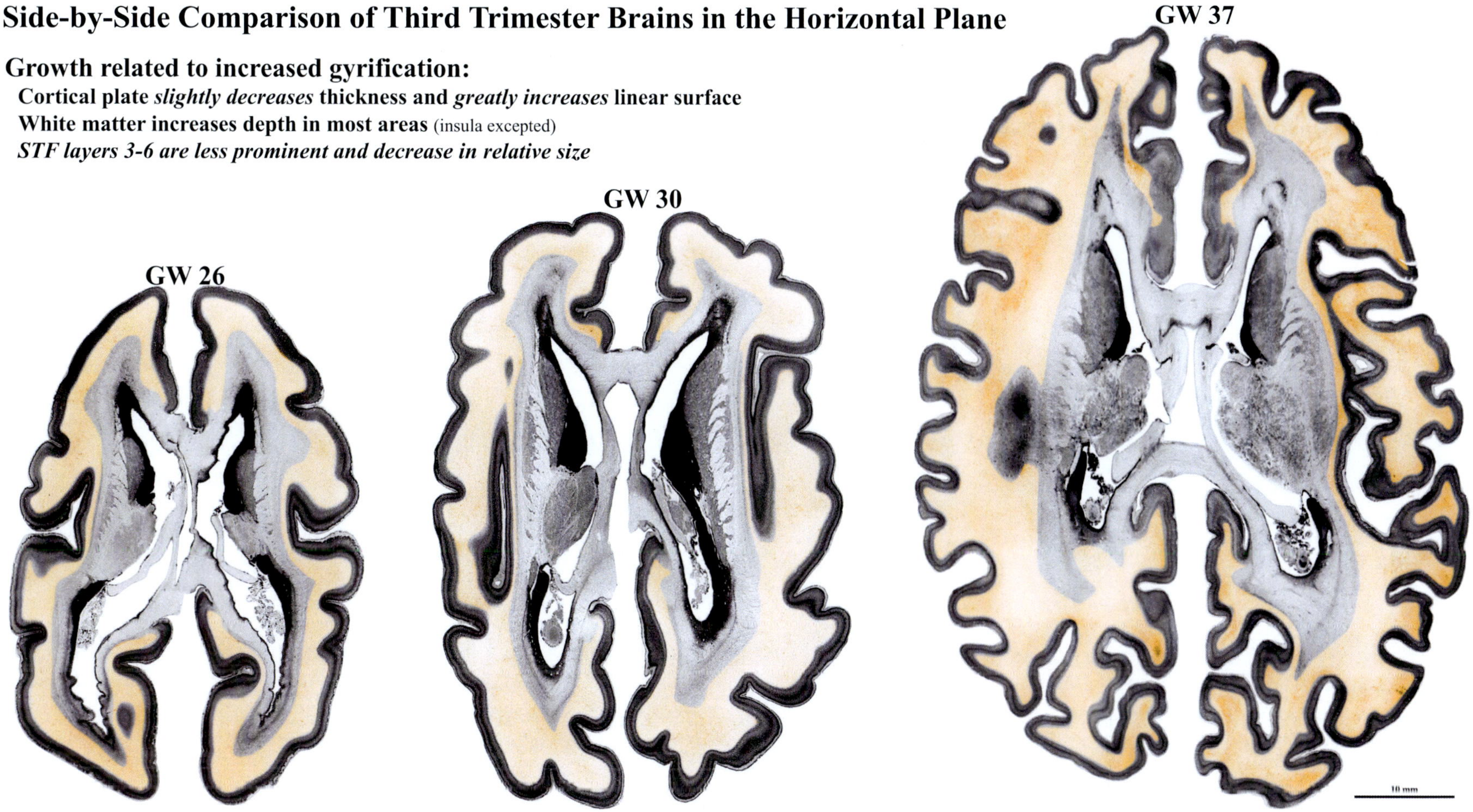

Figures 1 and 2 (*facing pages*). A comparison early (GW 26), middle (GW 30), and late (GW 37) trimester brains cut the horizontal plane. All brains are shown at the same scale to emphasize the relative growth of different components of the cerebral cortex. Note that the internal capsule and corpus callosum continue to thicken because many new axons enter. Massive numbers of axons invade the white matter (*pale yellow*). These axons pile up in the white matter and their terminal branches grow into the cortical plate, causing it to thicken and lengthen as synaptic contacts are beginning to be established. That growth results in increased gyrification to the point that the early sulci and gyri are difficult to distinguish. (GW 26 is shown in Plate 28, Volume 11, Bayer and Altman, in press; GW 30 in Plate 58, Volume 12, Bayer and Altman, in press; GW 37 in Plate 21, this volume.)

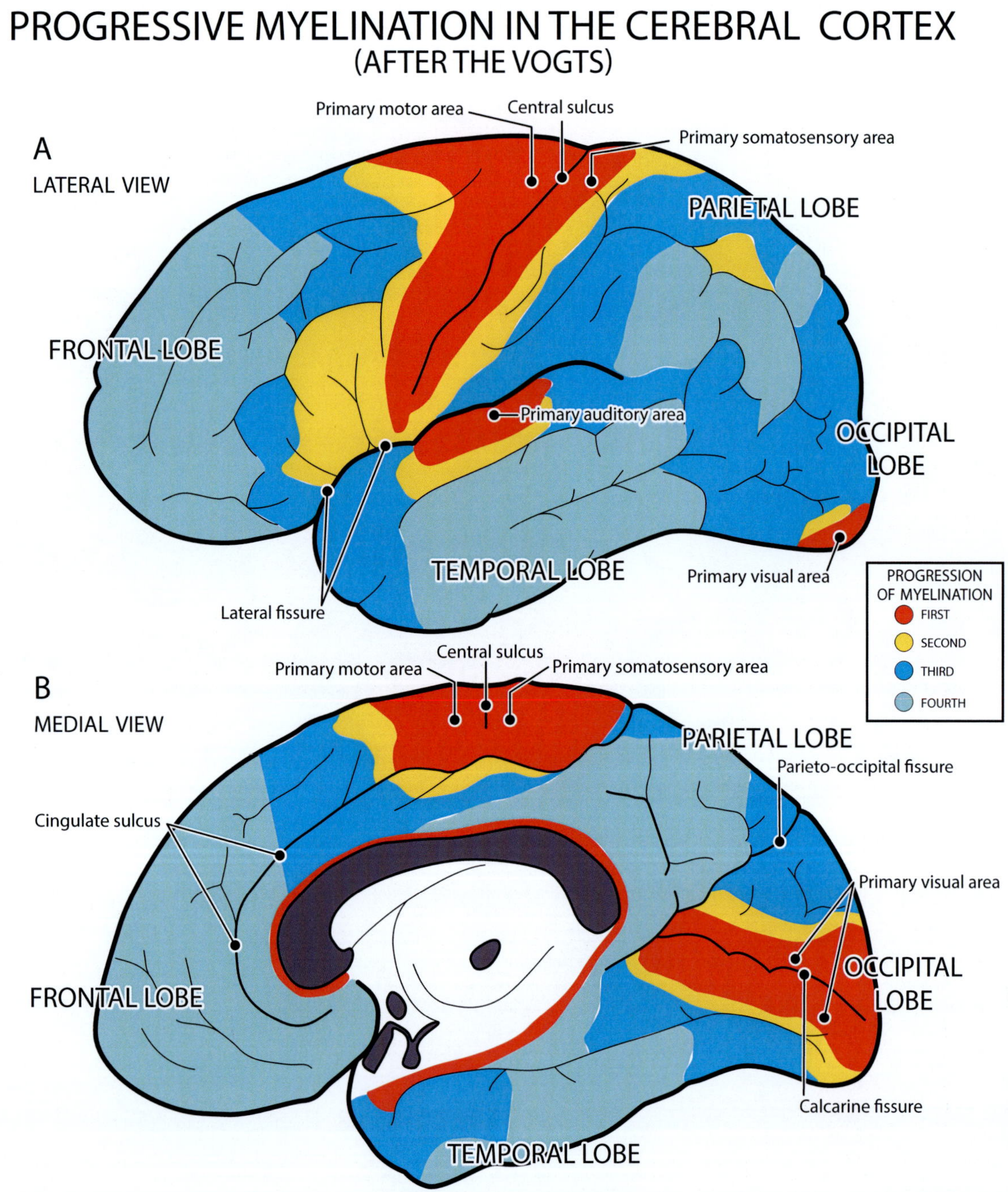

primary gyri are "hidden" by secondary and tertiary gyrification. The early gyri and sulci are the primary cortical areas (*red*); note they are a relatively small part of the cortex. Primary areas show little individual variability and are highly conserved in cortical development. They begin to myelinate early in infancy. Secondary areas (*yellow*) surrounding the primary areas myelinate slightly later, adjacent to the primary gyri. Processing here is closely related to processes in the primary areas. Third-order (*dark blue*) and fourth-order (*light blue*) areas mature much later and myelinate last. The late maturing parts of the cortex are association areas where high-order integrative and cognitive processes occur. These areas produce new gyri that mask the prominent appearance of the early gyri in the early third trimester. One can already see the proliferation of new gyri and sulci in the brains presented in this volume, that are near birth. Note that these areas take up the most space in the cortex, and it is here that there is an increase in individual variability. Some parts of the association areas in the frontal lobe (the prefrontal cortex) do not myelinate until the early adult period.

Figure 3. The sequence of myelination based on extensive studies by Vogt and Vogt (1902, 1904). **A**, the lateral cortex, **B**, the medial cortex. Colored areas indicate the myelination sequence: first (*red*), second (*yellow*), third (*dark blue*), fourth (*light blue*). Modified Figure 51 from Altman and Bayer (2015). The original composite figure was first published by Ariëns Kappers et al. (1936).

REFERENCES

Altman J, Bayer SA. (2015) *Development of the Human Neocortex.* Ocala, FL, Laboratory of Developmental Neurobiology, neurondevelopment.org.

Ariéns Kappers CU, Huber GC, Crosby EC (1936) *The Comparative Anatomy of the Nervous System of Vertebrates, Including Man.* 2 volumes. New York, Macmillan (Reprinted in 3 volumes, New York, Hafner, 1960)

Bayer SA, Altman J (2004) *Atlas of Human Central Nervous System Development*, Volume 2: *The Human Bran during the Third Trimester.* Boca Raton, FL, CRC Press.

Bayer SA, Altman J (in press) *The Human Brain during the Second Trimester 225- to 235-mm Crown-Rump Lengths, Atlas of Human Central Nervous System Development*, Volume 11. Taylor and Francis, CRC Press.

Bayer SA, Altman J (in press) *The Human Brain during the Second Trimester 260- to 270-mm Crown-Rump Lengths, Atlas of Human Central Nervous System Development*, Volume 12. Taylor and Francis, CRC Press.

Curtis BA, Jacobson S, Marcus EM (1972) *An Introduction to the Neurosciences*, Philadelphia: W. B. Saunders.

Haleem M (1990) *Diagnostic Categories of the Yakovlev Collection of Normal and Pathological Anatomy and Development of the Brain.* Washington, D.C. Armed Forces Institute of Pathology.

Larroche JC (1966) The development of the central nervous system during intrauterine life. In: *Human Development*, F. Falkner (ed.), Philadelphia: W. B. Saunders, pages 257-276.

Vogt C, Vogt O (1902) *Bieträge zur Hirnfasernlehre* Volume 1, *Zur Erforschung der Hirnfaserung*. Jena, Fischer.

Vogt C, Vogt O (1904) *Die Markretfung des Kindergehirns während der ersten vier Lebensmonate und thre metholologische Bedeutung*, Volume 2, Jena, Fischer.

PART II: Y180-61 CR 310 mm (GW 37) Sagittal

This specimen is case number B-180-61 (Perinatal RPSL) in the Yakovlev Collection. A female infant survived for one hour after birth. Death occurred because a hyaline membrane obstructed the airway to the lungs. The brain was cut in the sagittal plane in 35-µm thick sections and is classified as a Normative Control in the Yakovlev Collection (Haleem, 1990). There is no photograph of this brain before it was embedded and cut, so the photograph of another GW 37 brain that Larroche published in 1966 (**Figure 4**) is used to show medial surface features.

Photographs of 8 Nissl-stained sections are shown in **Plates 1-8**. The core of the brain and the cerebellum are shown at higher magnification in **Plates 9–16**. Very-high-magnification views of different regions of the cerebellar cortex are shown in **Plates 17** and **18**. Because the section numbers decrease from **Plate 1** (most medial) to **Plate 8** (most lateral), they are from the left side of the brain; the right side has higher section numbers proceeding medial to lateral. The cutting plane of this brain is nearly parallel to the midline. However, the posterior part of each section is angled more toward the right than the anterior part. For example, in **Plate 1** the paraflocculus from the right side of the brain is beneath the cerebellar vermis. The paraflocculus is very small in **Plate 2** and the left paraflocculus appears in **Plate 3**. The sections chosen for illustration are spaced closer together near the midline to show small structures in the diencephalon, midbrain, pons, and medulla.

Y180-61 contains remnants of the germinal matrices in all lobes of the cerebral cortex where the ***neuroepithelium/subventricular zone*** is still generating neocortical interneurons. Migrating and sojourning neurons and/or glia are in the dwindling ***stratified transitional fields*** in all lobes of the cerebral cortex, more prominent in sensory rather than motor cortical areas.

Many neurons, glia, and their mitotic precursor cells are still migrating through the olfactory peduncle toward the olfactory bulb in the ***rostral migratory stream*** from a presumed source area in the germinal matrix at the junction between the cerebral cortex, striatum, and nucleus accumbens. Lateral parts of the cerebral cortex has streams of neurons and glia in the ***lateral migratory stream*** that percolates through the claustrum, endopiriform nucleus, external capsule, and uncinate fasciculus. These cells appear to be heading toward the insular cortex, primary olfactory cortex, temporal cortex, and basolateral parts of the amygdaloid complex.

In the basal ganglia, a prominent ***neuroepithelium/subventricular zone*** covers the striatum and nucleus accumbens where neurons are being generated. The ***subgranular zone*** in the hilus of the dentate gyrus is generating granule cells. Other structures in the telencephalon, such as the septum, fornix, and Ammon's horn part of the hippocampus, have only a densely stained layer at the ventricle; presumably generating glia, cells of the choroid plexus, and the ependymal lining of the ventricle.

Most of the structures in the diencephalon appear to be settled and maturing; the third ventricle is lined by a thin ***glioepithelium/ependyma***. In the midbrain and anterior pons, there is a slightly thicker and more convoluted ***glioepithelium/ependyma*** lining the posterior cerebral aqueduct and anterior fourth ventricle. The posterior pons and entire medulla have a thin ***glioepithelium/ependyma*** lining the rest of the fourth ventricle.

The ***external germinal layer*** is prominent over the entire surface of the cerebellar cortex and is producing basket, stellate, and granule cells. The ***germinal trigone*** is still visible at the base of the nodulus and along the floccular peduncle; choroid plexus cells and glia are originating here.

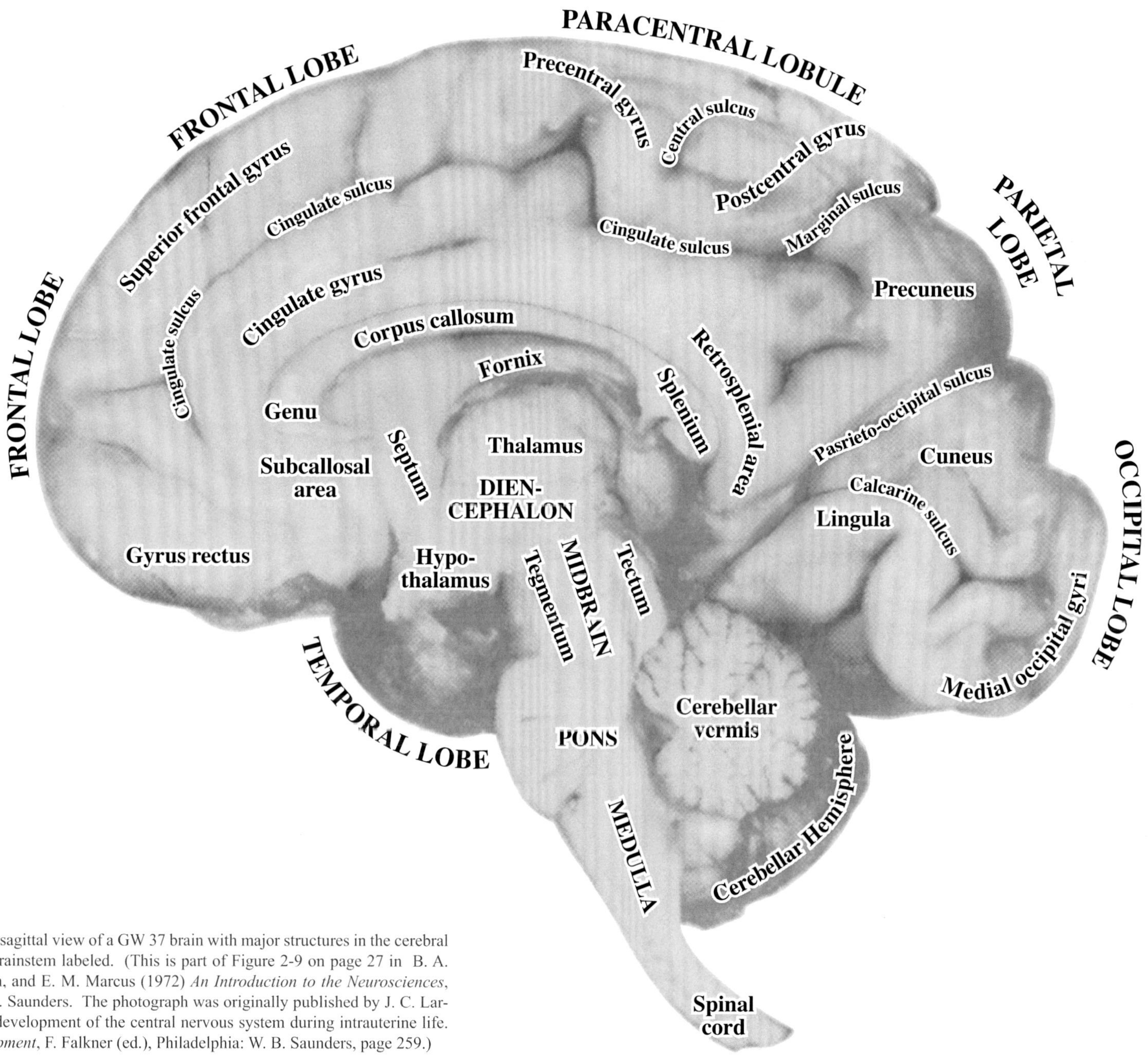

Figure 4. Midline sagittal view of a GW 37 brain with major structures in the cerebral hemispheres and brainstem labeled. (This is part of Figure 2-9 on page 27 in B. A. Curtis, S. Jacobson, and E. M. Marcus (1972) *An Introduction to the Neurosciences*, Philadelphia: W. B. Saunders. The photograph was originally published by J. C. Larroche (1966) The development of the central nervous system during intrauterine life. In: *Human Development*, F. Falkner (ed.), Philadelphia: W. B. Saunders, page 259.)

PLATE 1A
CR 310 mm
GW 37, Y180-61
Sagittal
Section 961

See detail of the brain core and cerebellum in Plates 9A and B.

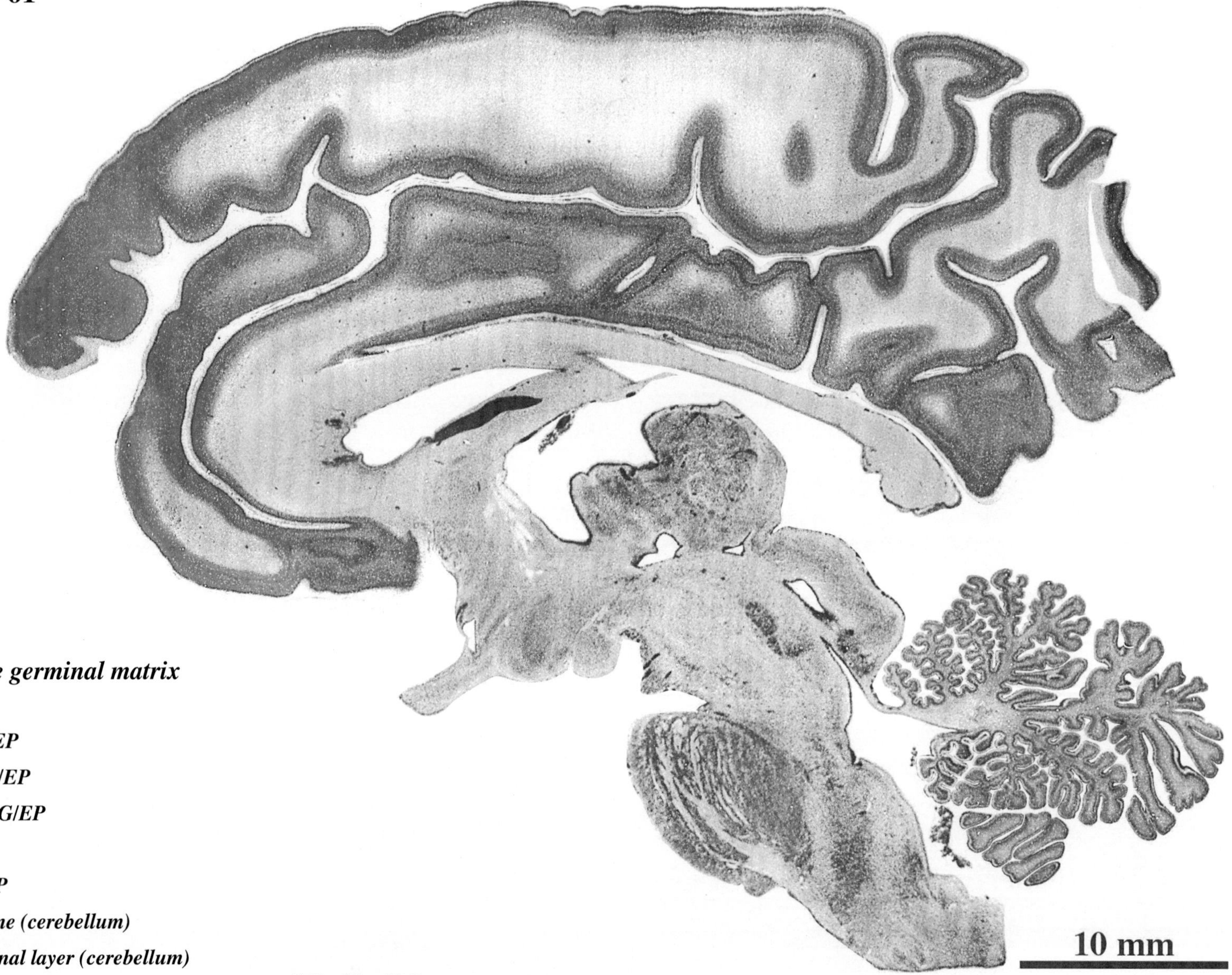

Remnants of the germinal matrix

1 ***Callosal sling***
2 ***Strionuclear GEP***
3 ***Diencephalic G/EP***
4 ***Mesencephalic G/EP***
5 ***Pontine G/EP***
6 ***Medullary G/EP***
7 ***Germinal trigone (cerebellum)***
8 ***External germinal layer (cerebellum)***
9 ***Subpial granular layer (cortical)***

GEP - Glioepithelium
G/EP - Glioepithelium/ependyma

PLATE 1B

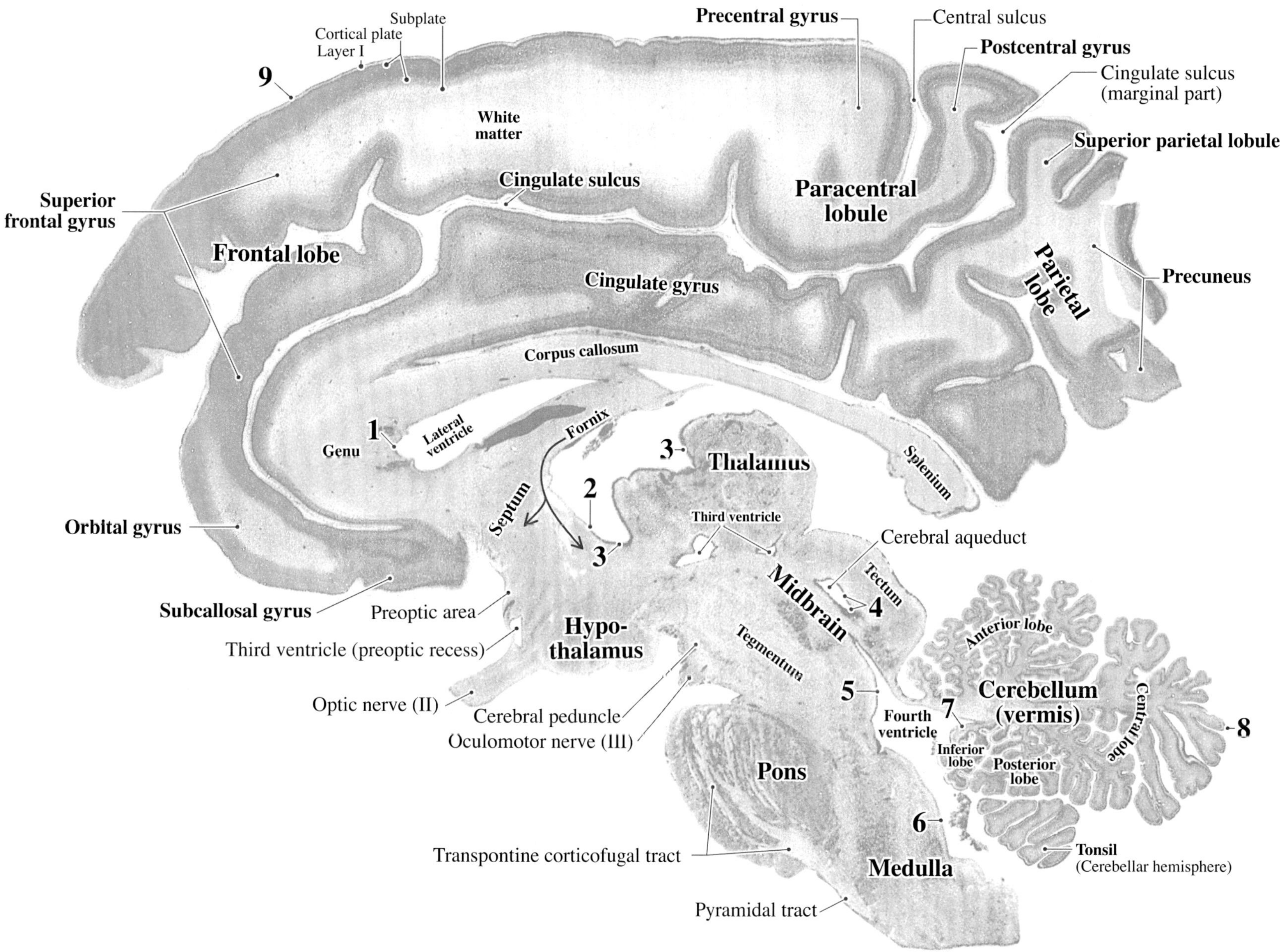
Subplate
Cortical plate
Layer I
9
White matter
Precentral gyrus
Central sulcus
Postcentral gyrus
Cingulate sulcus (marginal part)
Superior parietal lobule
Cingulate sulcus
Paracentral lobule
Superior frontal gyrus
Frontal lobe
Cingulate gyrus
Parietal lobe
Precuneus
Corpus callosum
Lateral ventricle
Fornix
1
Genu
3
Thalamus
Splenium
2
Septum
Orbital gyrus
Third ventricle
3
Cerebral aqueduct
Tectum
Midbrain
4
Subcallosal gyrus
Preoptic area
Hypo-thalamus
Anterior lobe
Third ventricle (preoptic recess)
Tegmentum
5
Cerebellum (vermis)
Central lobe
Optic nerve (II)
Cerebral peduncle
Fourth ventricle
7
8
Oculomotor nerve (III)
Inferior lobe
Posterior lobe
Pons
6
Tonsil (Cerebellar hemisphere)
Transpontine corticofugal tract
Medulla
Pyramidal tract

PLATE 2A
CR 310 mm
GW 37, Y180-61
Sagittal
Section 941

See detail of the brain core and cerebellum in Plates 10A and B.

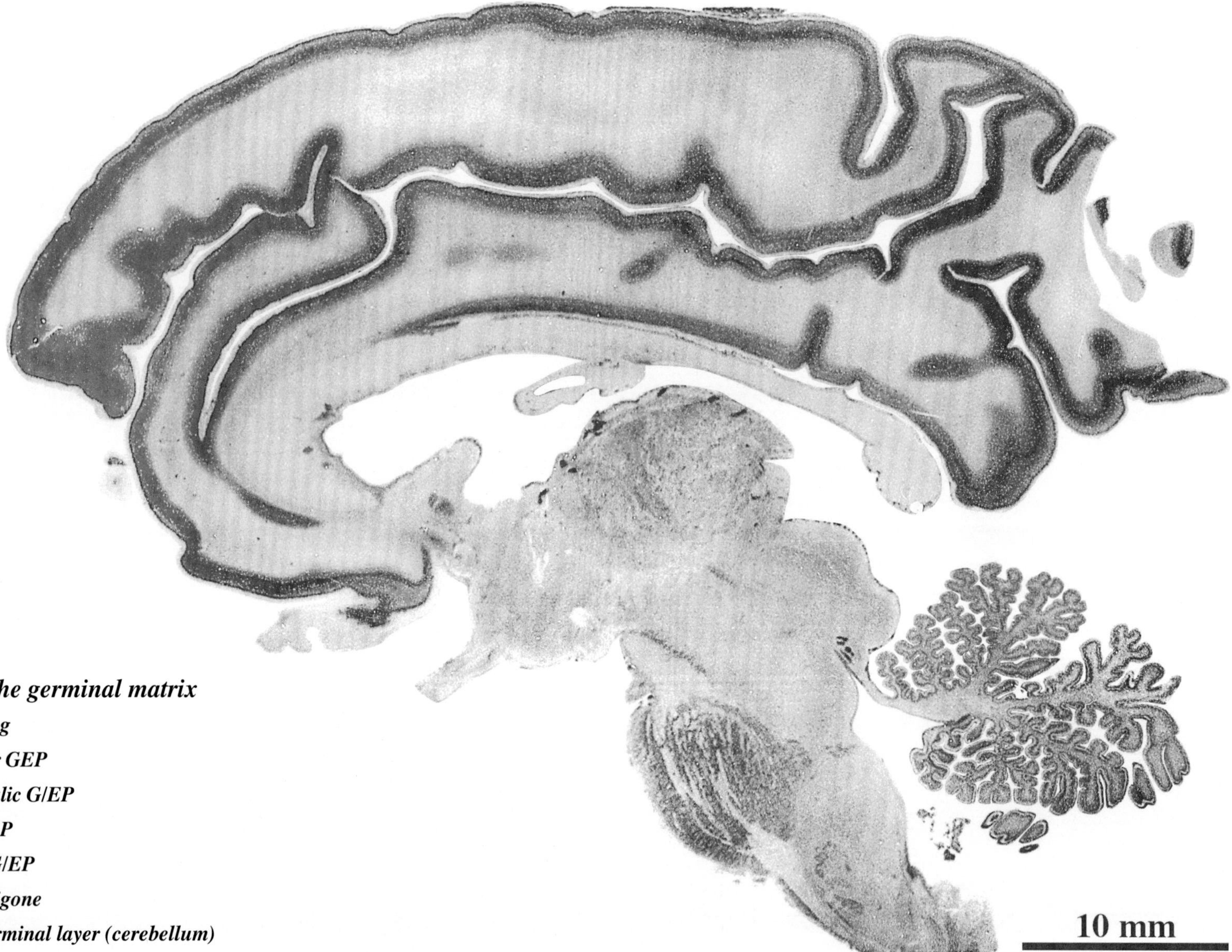

Remnants of the germinal matrix

1 ***Callosal sling***
2 ***Strionuclear GEP***
3 ***Mesencephalic G/EP***
4 ***Pontine G/EP***
5 ***Medullary G/EP***
6 ***Germinal trigone***
7 ***External germinal layer (cerebellum)***
8 ***Subpial granular layer (cortical)***

GEP - Glioepithelium
G/EP - Glioepithelium/ependyma

PLATE 2B

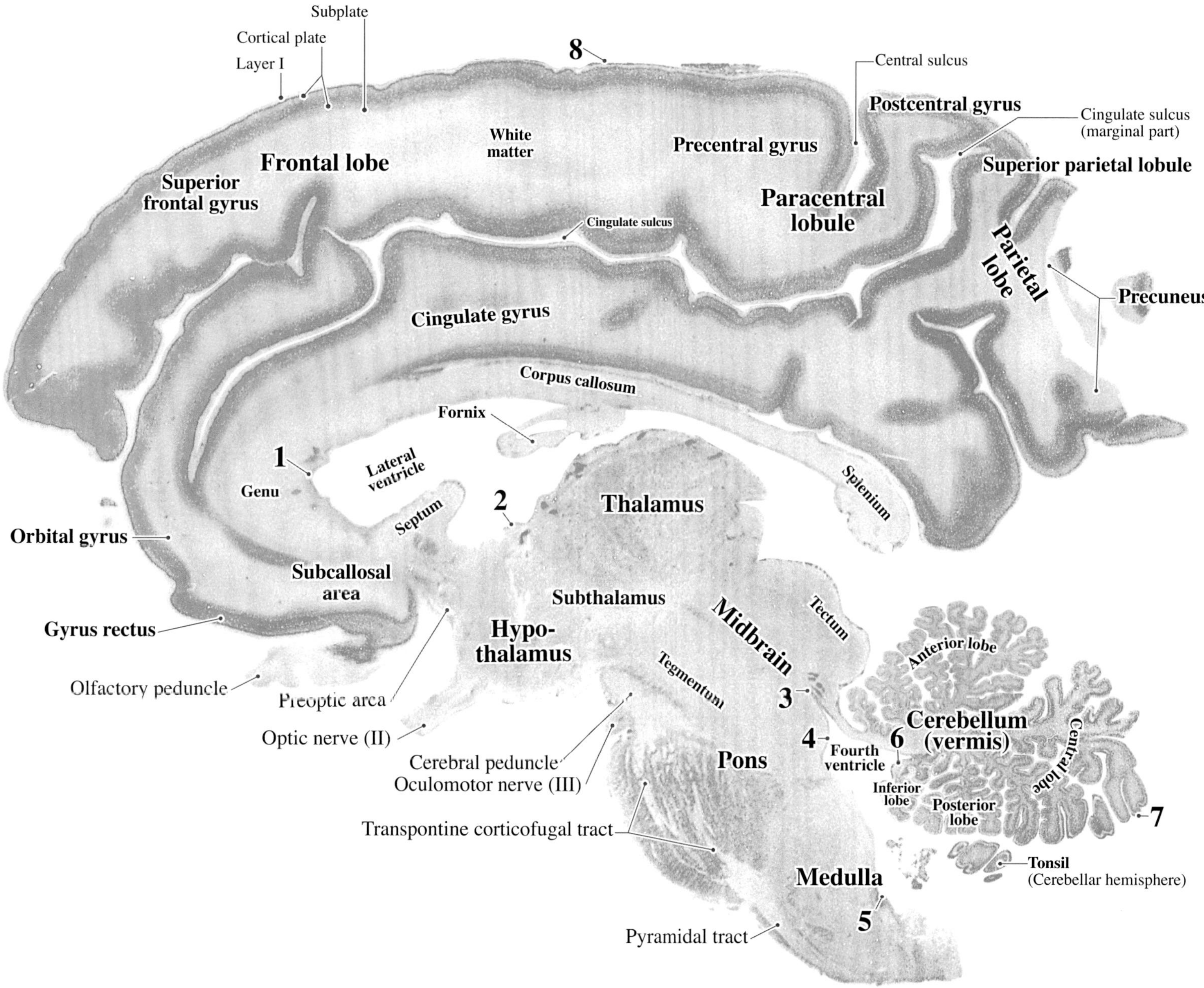
Subplate
Cortical plate
Layer I
8
Central sulcus
Postcentral gyrus
Cingulate sulcus
(marginal part)
White
matter
Precentral gyrus
Frontal lobe
Superior parietal lobule
Superior
frontal gyrus
Paracentral
lobule
Cingulate sulcus
Parietal
lobe
Precuneus
Cingulate gyrus
Corpus callosum
Fornix
1
Lateral
ventricle
Genu
2
Thalamus
Splenium
Septum
Orbital gyrus
Subcallosal
area
Subthalamus
Tectum
Midbrain
Gyrus rectus
Hypo-
thalamus
Anterior lobe
Olfactory peduncle
Preoptic area
Tegmentum
3
Cerebellum
(vermis)
Central lobe
Optic nerve (II)
4
Fourth
ventricle
6
Pons
Cerebral peduncle
Oculomotor nerve (III)
Inferior
lobe
Posterior
lobe
7
Transpontine corticofugal tract
Tonsil
(Cerebellar hemisphere)
Medulla
5
Pyramidal tract

PLATE 3A
CR 310 mm
GW 37, Y180-61
Sagittal
Section 841

See detail of the brain core and cerebellum in Plates 11A and B.

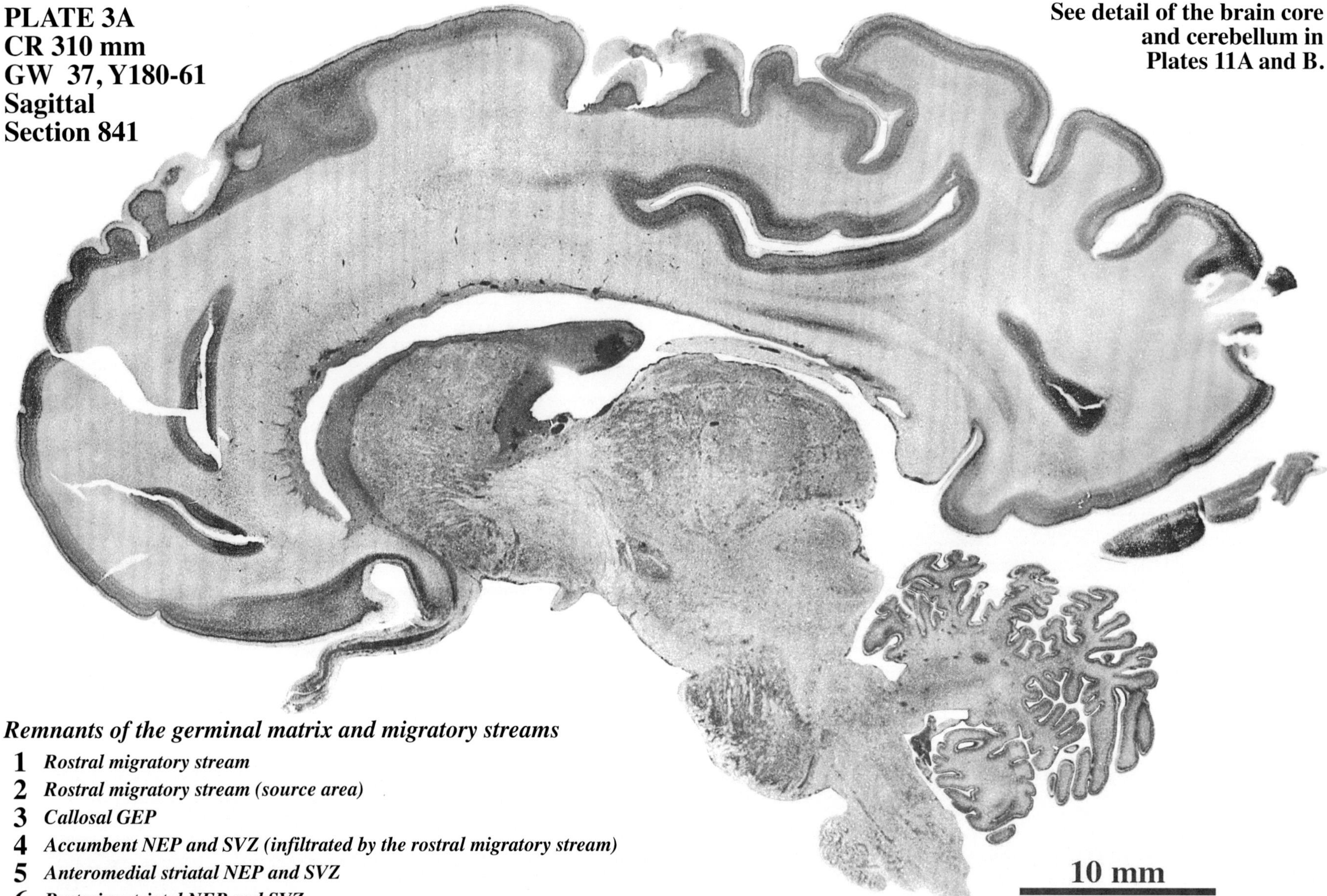

Remnants of the germinal matrix and migratory streams

1 *Rostral migratory stream*
2 *Rostral migratory stream (source area)*
3 *Callosal GEP*
4 *Accumbent NEP and SVZ (infiltrated by the rostral migratory stream)*
5 *Anteromedial striatal NEP and SVZ*
6 *Posterior striatal NEP and SVZ*
7 *Strionuclear GEP*
8 *Pontine and medullary G/EP*
9 *External germinal layer (cerebellum)*
10 *Subpial granular layer (cortical)*

GEP - Glioepithelium
G/EP - Glioepithelium/ependyma
NEP - Neuroepithelium
SVZ - Subventricular zone

PLATE 3B

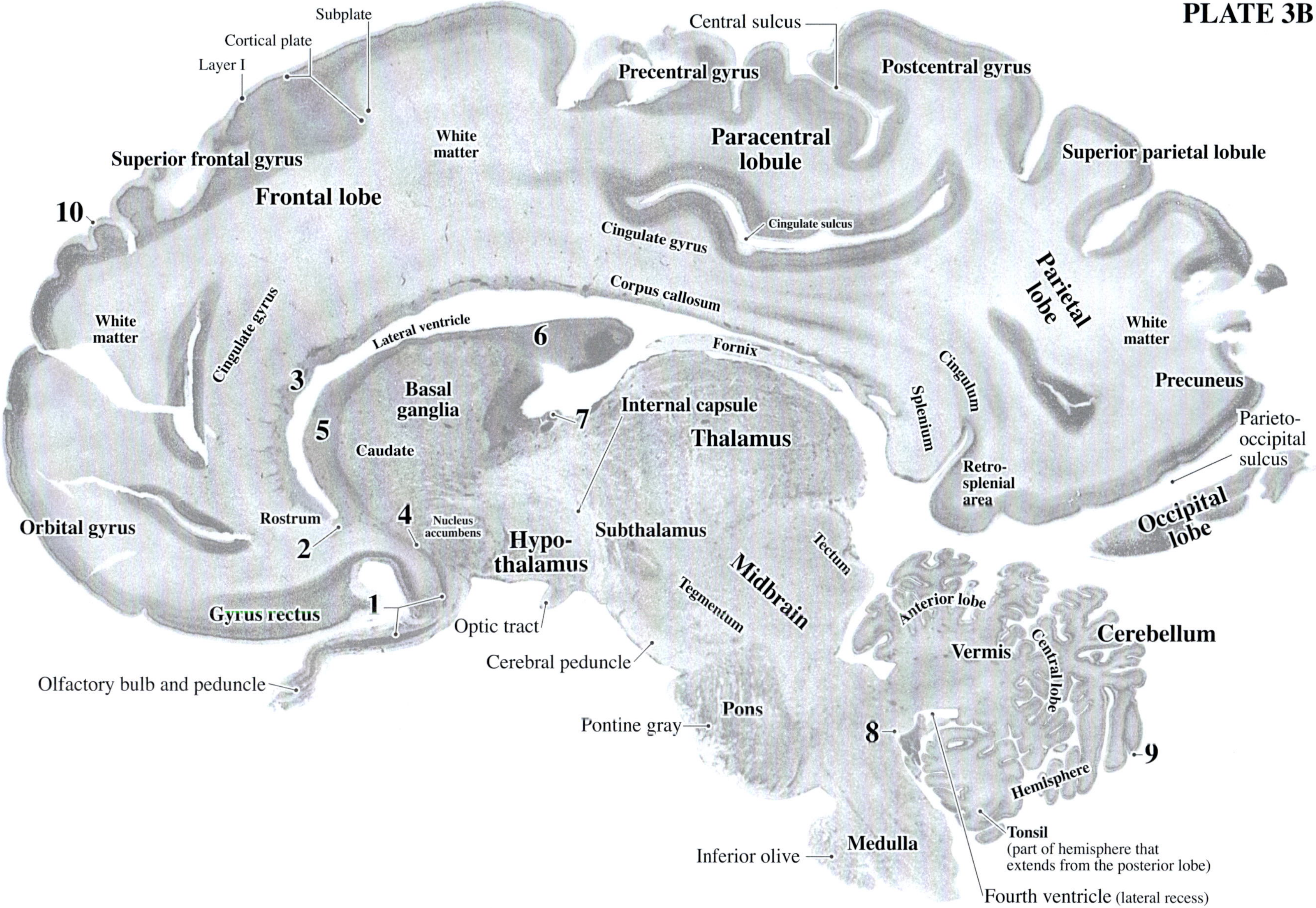

Subplate
Cortical plate
Layer I
Central sulcus
Precentral gyrus
Postcentral gyrus
White matter
Paracentral lobule
Superior frontal gyrus
Superior parietal lobule
Frontal lobe
10
Cingulate sulcus
Cingulate gyrus
Parietal lobe
Corpus callosum
White matter
Cingulate gyrus
Lateral ventricle
6
White matter
Fornix
Cingulum
3
Basal ganglia
Precuneus
Internal capsule
7
Splenium
5
Parieto-occipital sulcus
Caudate
Thalamus
Retro-splenial area
Rostrum
4
Nucleus accumbens
Orbital gyrus
Subthalamus
Hypo-thalamus
Occipital lobe
2
Tectum
Midbrain
Tegmentum
Anterior lobe
1
Gyrus rectus
Optic tract
Cerebellum
Vermis
Central lobe
Cerebral peduncle
Olfactory bulb and peduncle
Pons
Pontine gray
8
9
Hemisphere
Tonsil
(part of hemisphere that extends from the posterior lobe)
Medulla
Inferior olive
Fourth ventricle (lateral recess)

PLATE 4A
CR 310 mm
GW 37, Y180-61
Sagittal
Section 781

See detail of the brain core and cerebellum in Plates 12A and B.

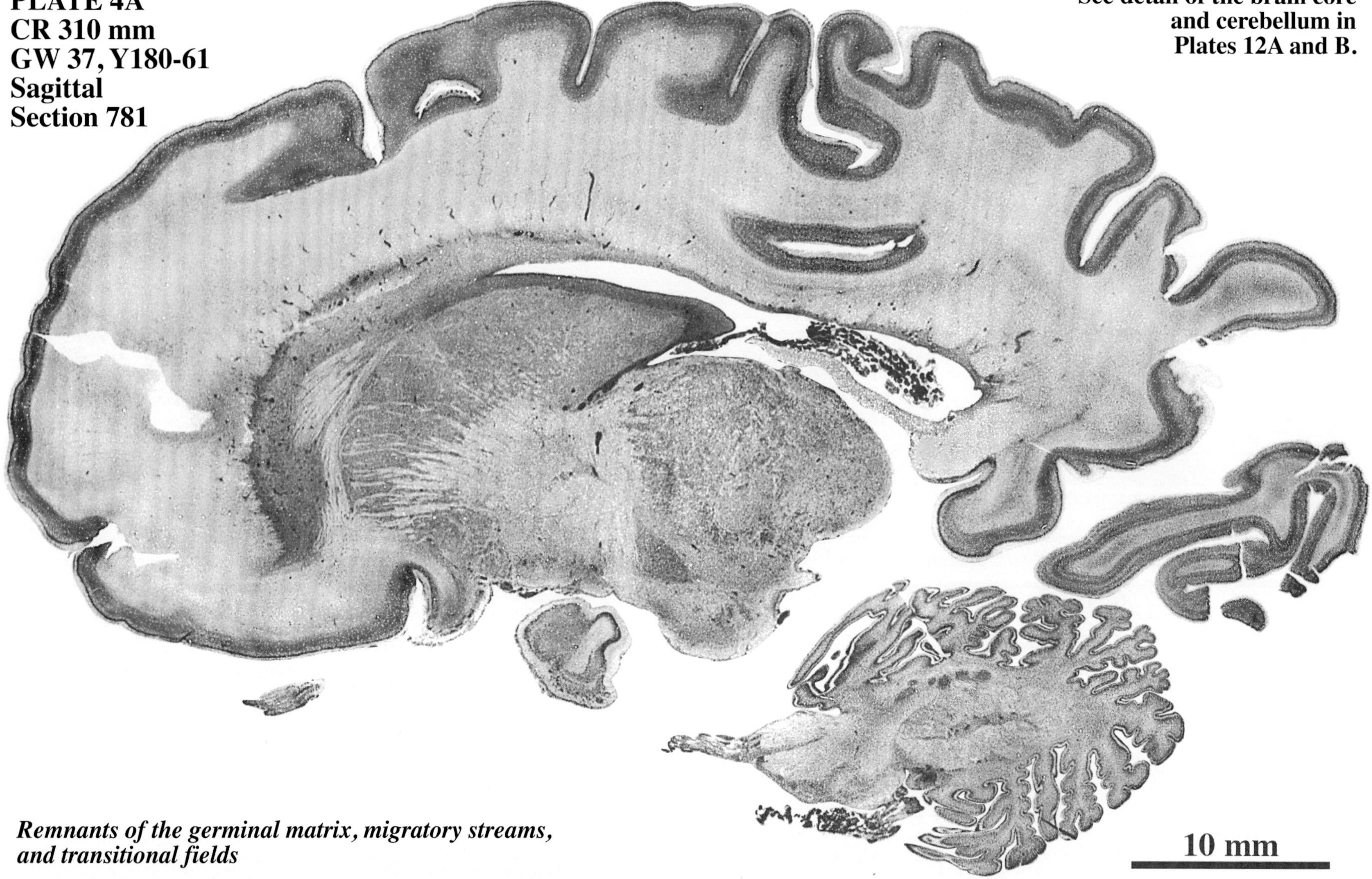

Remnants of the germinal matrix, migratory streams, and transitional fields

1 *Rostral migratory stream*
2 *Frontal NEP and SVZ*
3 *Frontal STF*
4 *Callosal GEP*
5 *Fornical GEP*
6 *Anterolateral striatal NEP and SVZ*
7 *Posterior striatal NEP and SVZ*
8 *Strionuclear GEP*
9 *External germinal layer (cerebellum)*
10 *Subpial granular layer (cortical)*

GEP - Glioepithelium
NEP - Neuroepithelium
STF - Stratified transitional field
SVZ - Subventricular zone

PLATE 4B

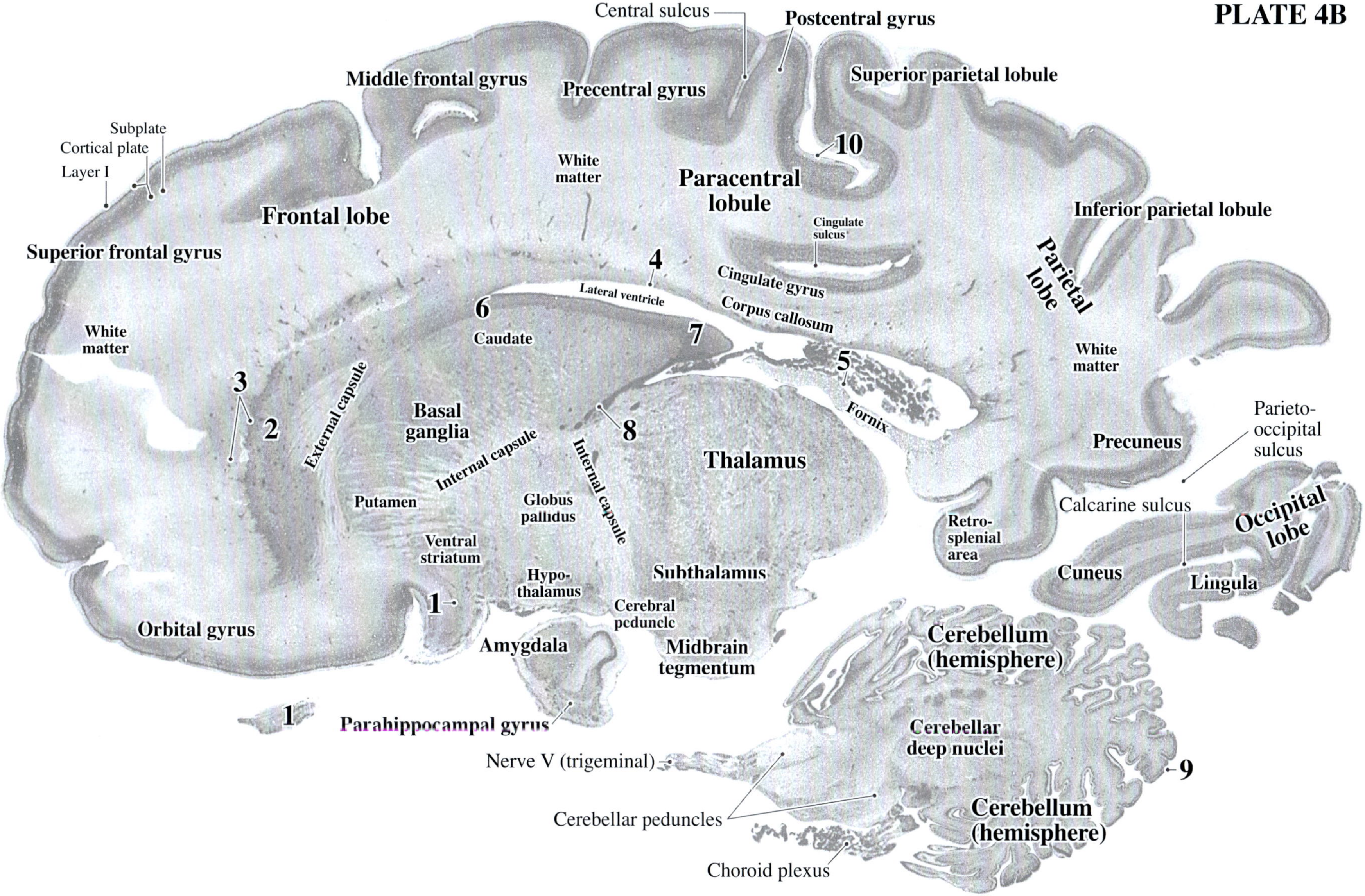
Central sulcus
Postcentral gyrus
Middle frontal gyrus
Precentral gyrus
Superior parietal lobule
Subplate
Cortical plate
Layer I
10
White matter
Paracentral lobule
Frontal lobe
Inferior parietal lobule
Cingulate sulcus
Superior frontal gyrus
4
Cingulate gyrus
Parietal lobe
Lateral ventricle
6
Corpus callosum
White matter
7
Caudate
White matter
5
3
External capsule
Basal ganglia
Fornix
2
8
Internal capsule
Internal capsule
Thalamus
Precuneus
Parieto-occipital sulcus
Putamen
Globus pallidus
Calcarine sulcus
Occipital lobe
Retro-splenial area
Ventral striatum
Subthalamus
Cuneus
Lingula
Hypo-thalamus
1
Cerebral peduncle
Orbital gyrus
Amygdala
Midbrain tegmentum
Cerebellum (hemisphere)
1
Parahippocampal gyrus
Cerebellar deep nuclei
Nerve V (trigeminal)
9
Cerebellar peduncles
Cerebellum (hemisphere)
Choroid plexus

PLATE 5A
CR 310 mm
GW 37, Y180-61
Sagittal
Section 721

See detail of the brain core and cerebellum in Plates 13A and B.

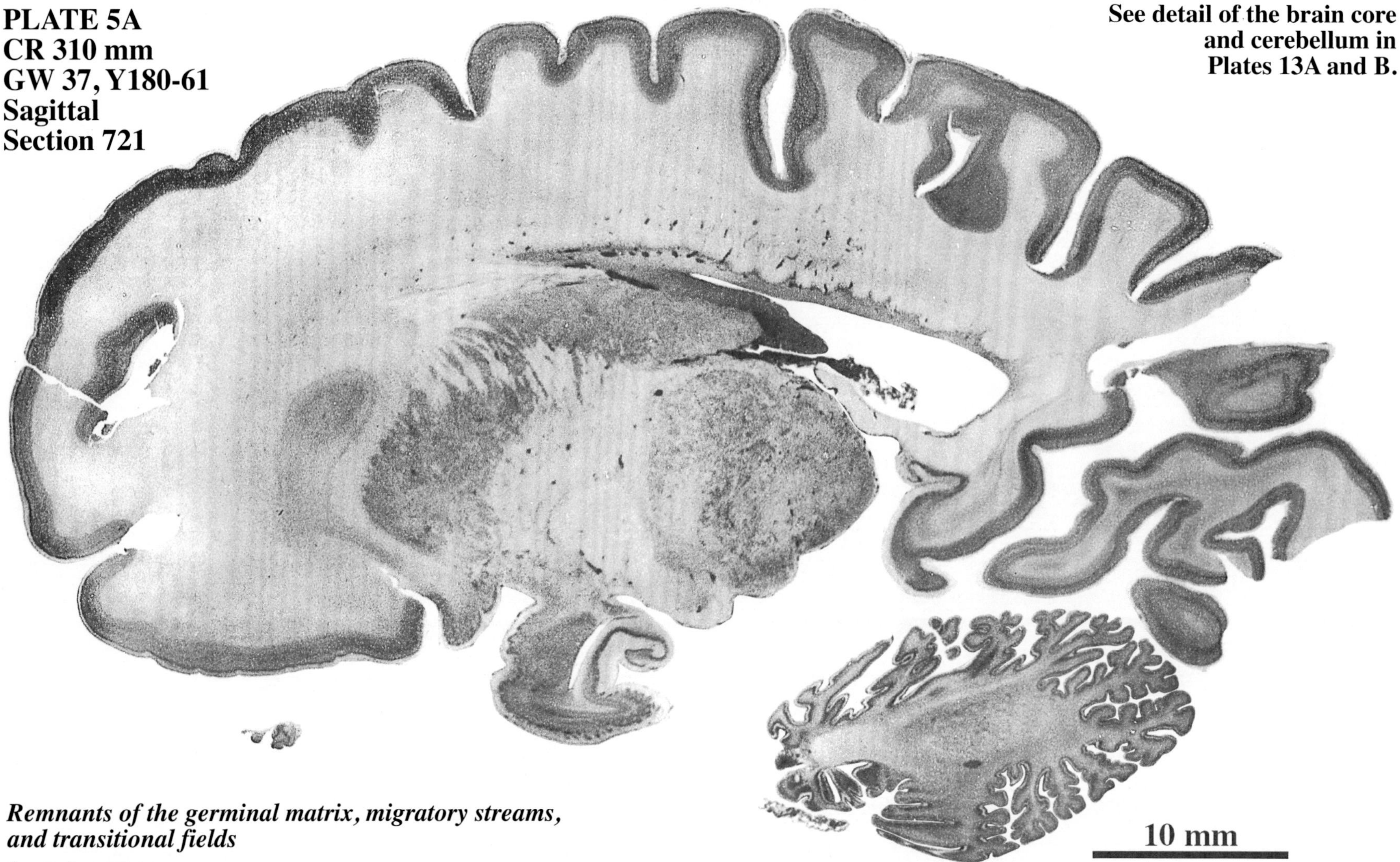

Remnants of the germinal matrix, migratory streams, and transitional fields

1 *Callosal GEP*
2 *Fornical GEP*
3 *Alvear GEP*
4 *Subgranular zone (dentate)*
5 *Lateral migratory stream (cortical)*
6 *Amygdaloid G/EP*
7 *Anterolateral striatal NEP and SVZ*
8 *Posterior striatal NEP and SVZ*
9 *Strionuclear GEP*
10 *External germinal layer (cerebellum)*
11 *Subpial granular layer (cortical)*

GEP - Glioepithelium
G/EP - Glioepithelium/ependyma
NEP - Neuroepithelium
SVZ - Subventricular zone

PLATE 5B

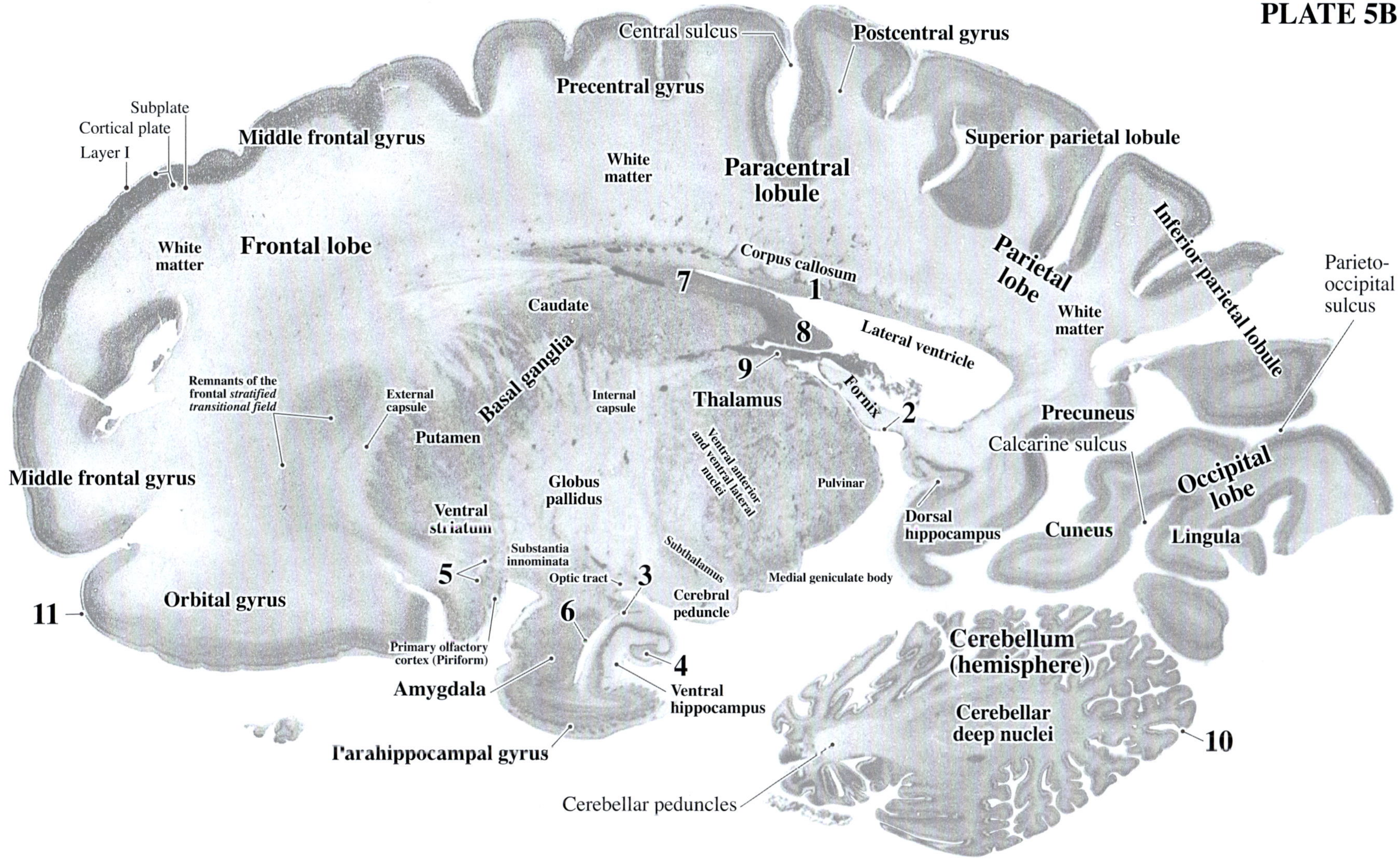

Central sulcus
Postcentral gyrus
Precentral gyrus
Subplate
Cortical plate
Layer I
Middle frontal gyrus
Superior parietal lobule
White matter
Paracentral lobule
White matter
Frontal lobe
Corpus callosum
Parietal lobe
Inferior parietal lobule
Parieto-occipital sulcus
7
1
White matter
Caudate
8
Lateral ventricle
Basal ganglia
9
Fornix
Remnants of the frontal *stratified transitional field*
External capsule
Internal capsule
Thalamus
2
Precuneus
Putamen
Calcarine sulcus
Ventral anterior and ventral lateral nuclei
Middle frontal gyrus
Globus pallidus
Pulvinar
Occipital lobe
Ventral striatum
Dorsal hippocampus
Cuneus
Lingula
Substantia innominata
Subthalamus
5
Optic tract
3
Medial geniculate body
Orbital gyrus
Cerebral peduncle
11
6
Cerebellum (hemisphere)
Primary olfactory cortex (Piriform)
4
Amygdala
Ventral hippocampus
Cerebellar deep nuclei
10
Parahippocampal gyrus
Cerebellar peduncles

PLATE 6A
CR 310 mm
GW 37, Y180-61
Sagittal
Section 661

See detail of the brain core and cerebellum in Plates 14A and B.

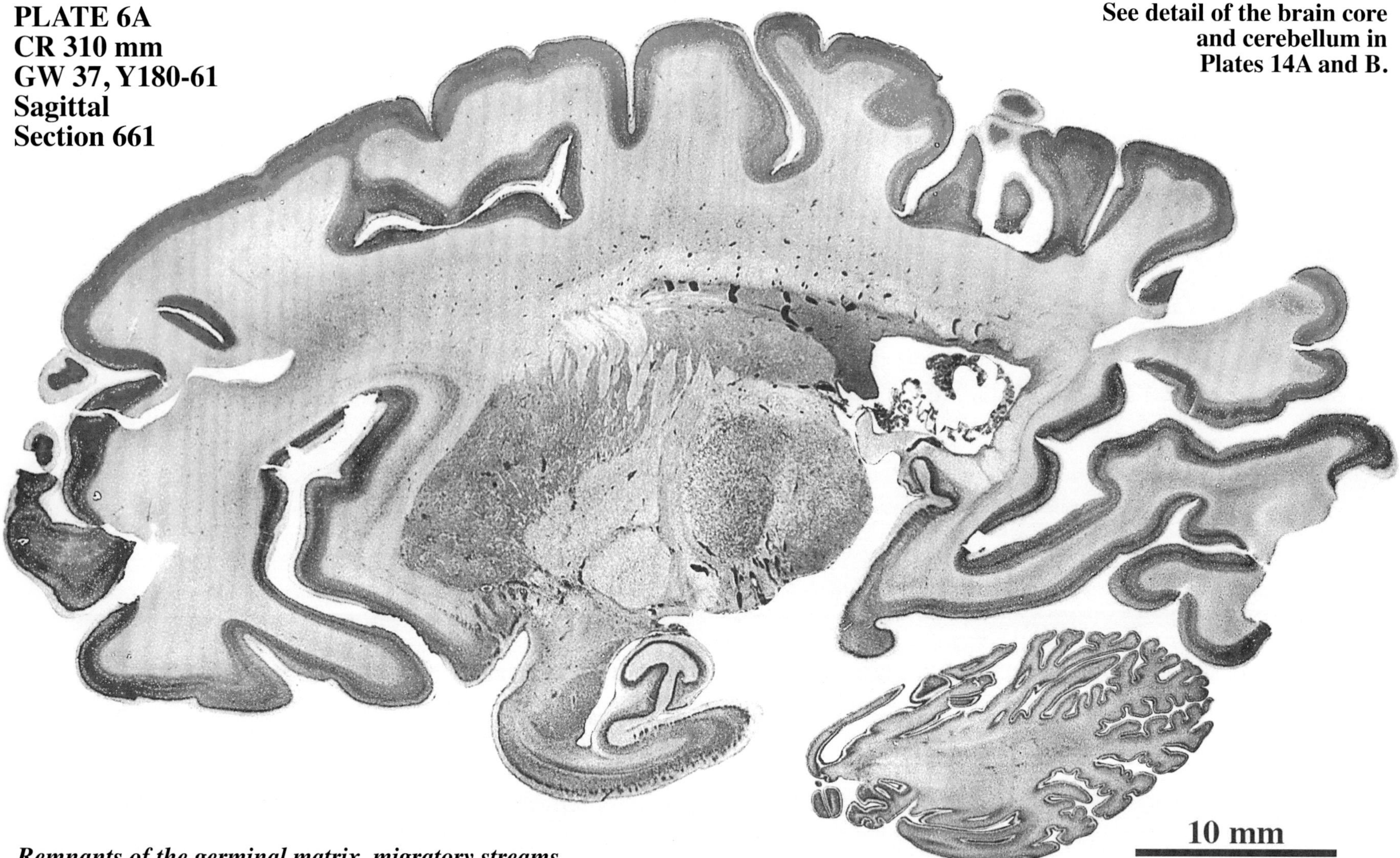

Remnants of the germinal matrix, migratory streams, and transitional fields

1 *Callosal GEP*
2 *Fornical GEP*
3 *Alvear GEP*
4 *Subgranular zone (dentate)*
5 *Lateral migratory stream (cortical)*
6 *Amygdaloid G/EP*
7 *Posterior striatal NEP and SVZ*
8 *Strionuclear GEP*
9 *External germinal layer (cerebellum)*
10 *Subpial granular layer (cortical)*

GEP - Glioepithelium
G/EP - Glioepithelium/ependyma
NEP - Neuroepithelium
SVZ - Subventricular zone

PLATE 6B

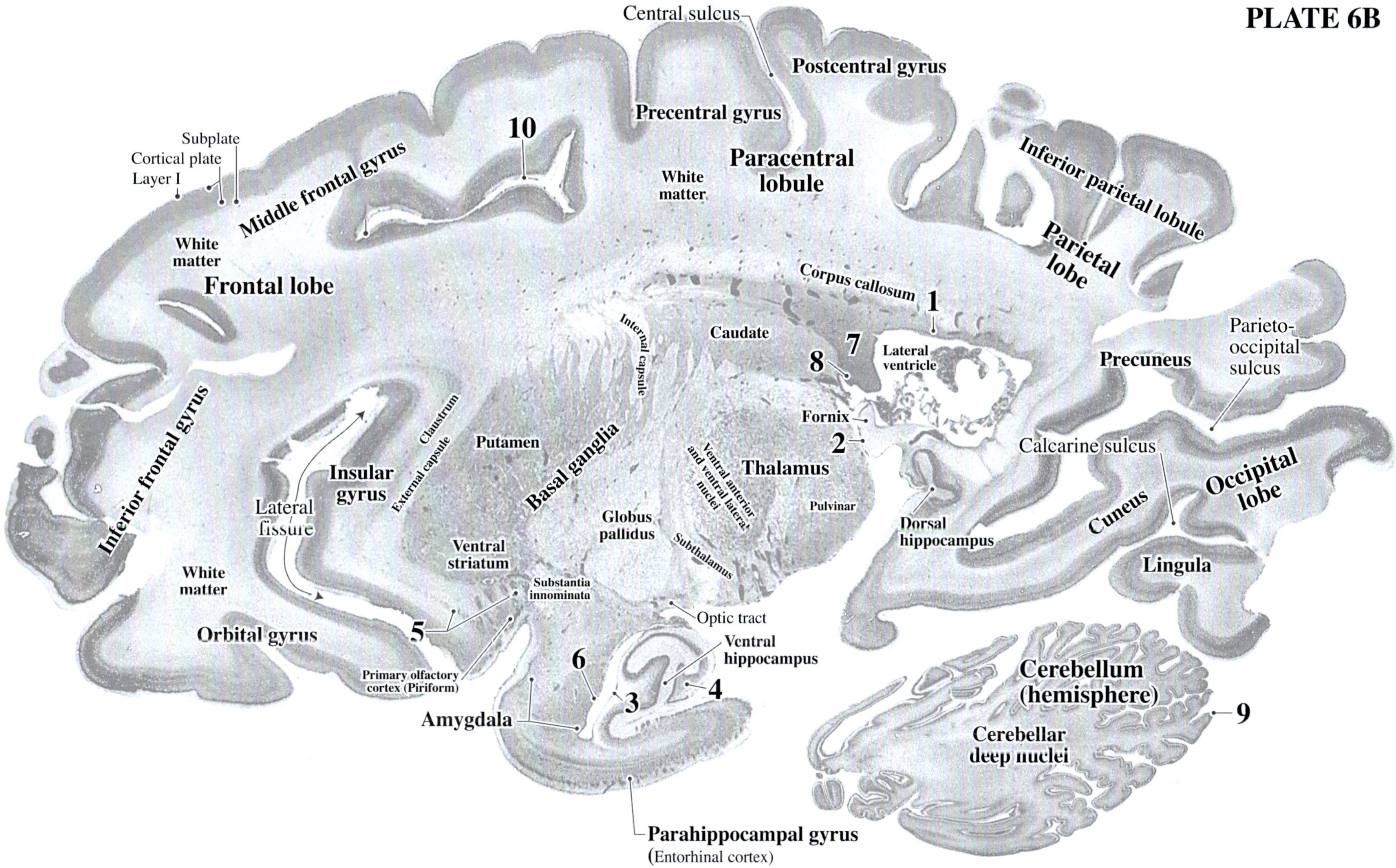

Central sulcus
Postcentral gyrus
Precentral gyrus
10
Subplate
Cortical plate
Layer I
Middle frontal gyrus
White matter
Paracentral lobule
Inferior parietal lobule
Parietal lobe
White matter
Frontal lobe
Corpus callosum
1
Internal capsule
Caudate
7
8
Lateral ventricle
Precuneus
Parieto-occipital sulcus
Claustrum
Putamen
Fornix
2
Calcarine sulcus
Inferior frontal gyrus
Insular gyrus
External capsule
Basal ganglia
Ventral anterior and ventral lateral nuclei
Thalamus
Pulvinar
Dorsal hippocampus
Occipital lobe
Lateral fissure
Globus pallidus
Cuneus
Ventral striatum
Subthalamus
Lingula
White matter
Substantia innominata
Optic tract
Orbital gyrus
5
Ventral hippocampus
6
Cerebellum (hemisphere)
Primary olfactory cortex (Piriform)
3
4
9
Amygdala
Cerebellar deep nuclei
Parahippocampal gyrus
(Entorhinal cortex)

PLATE 7A
CR 310 mm
GW 37, Y180-61
Sagittal
Section 621

See detail of the brain core and cerebellum in Plates 15A and B.

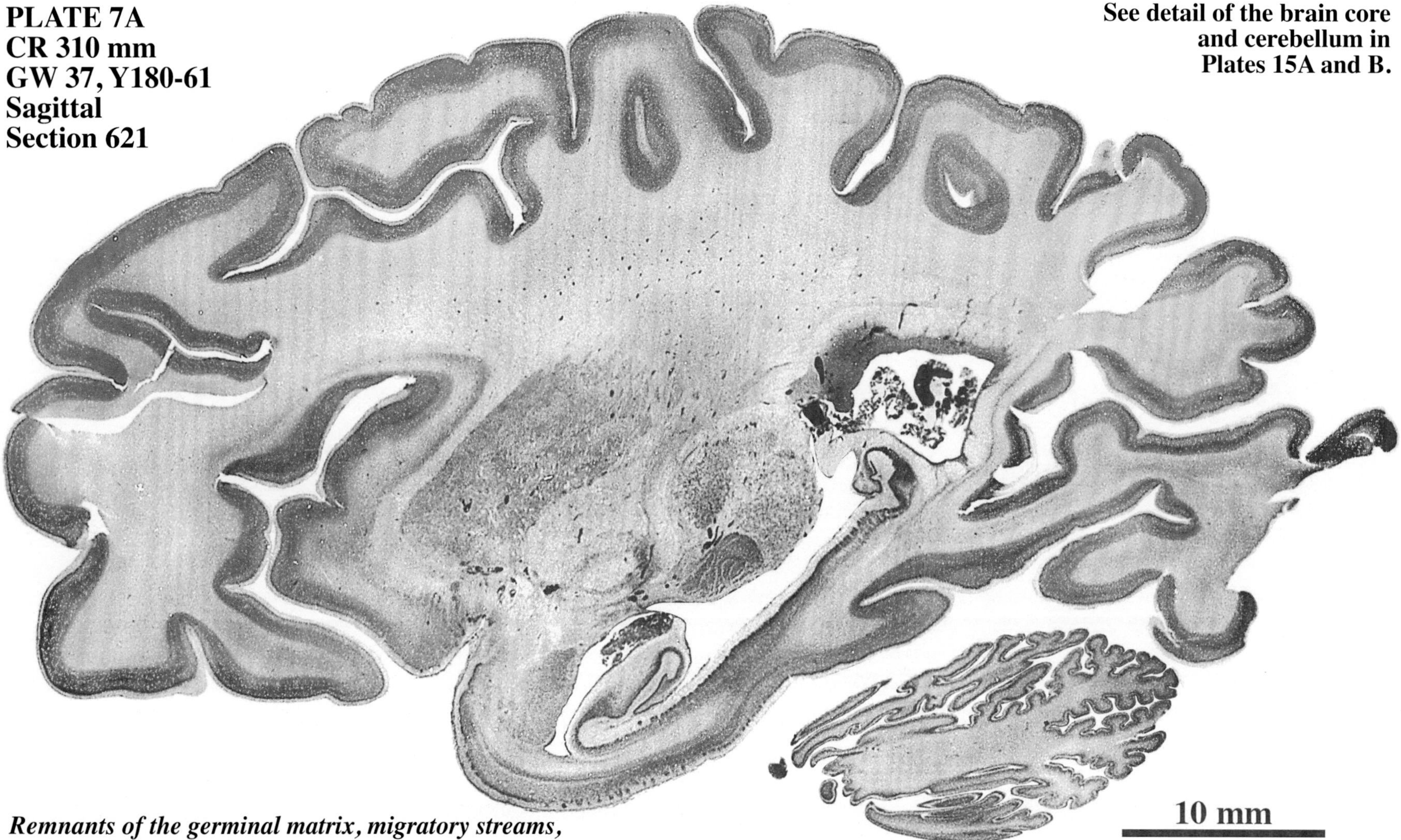

Remnants of the germinal matrix, migratory streams, and transitional fields

1 *Callosal GEP*

2 *Fornical GEP*

3 *Parahippocampal NEP, SVZ, and STF*

4 *Alvear GEP*

5 *Subgranular zone (dentate)*

6 *Lateral migratory stream (cortical)*

7 *Amygdaloid G/EP*

8 *Posterior striatal NEP and SVZ*

9 *Strionuclear GEP*

10 *External germinal layer (cerebellum)*

11 *Subpial granular layer (cortical)*

GEP - Glioepithelium
G/EP - Glioepithelium/ependyma
NEP - Neuroepithelium
STF - Stratified transitional field
SVZ - Subventricular zone

PLATE 7B

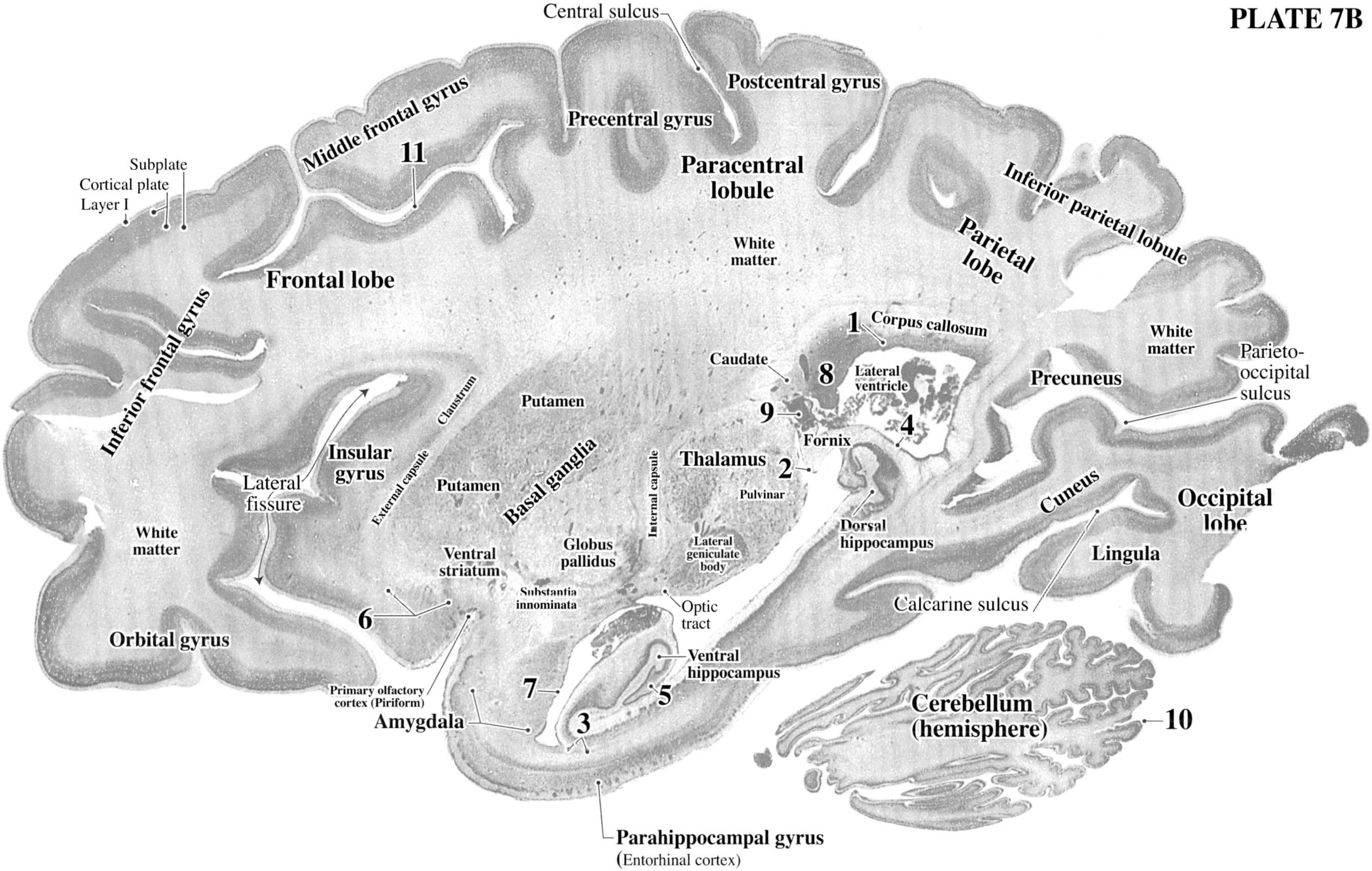

Central sulcus
Postcentral gyrus
Precentral gyrus
Middle frontal gyrus
11
Subplate
Cortical plate
Layer I
Paracentral lobule
Inferior parietal lobule
Parietal lobe
White matter
Frontal lobe
Inferior frontal gyrus
1
Corpus callosum
Caudate
8
Lateral ventricle
White matter
Parieto-occipital sulcus
Precuneus
Claustrum
Putamen
9
4
Fornix
Insular gyrus
External capsule
Basal ganglia
Internal capsule
Thalamus
2
Lateral fissure
Putamen
Pulvinar
Cuneus
Occipital lobe
White matter
Globus pallidus
Lateral geniculate body
Dorsal hippocampus
Lingula
Ventral striatum
Substantia innominata
Optic tract
Calcarine sulcus
6
Orbital gyrus
Ventral hippocampus
Primary olfactory cortex (Piriform)
7
5
Amygdala
3
Cerebellum (hemisphere)
10
Parahippocampal gyrus
(Entorhinal cortex)

PLATE 8A
CR 310 mm
GW 37, Y180-61
Sagittal
Section 501

See detail of the brain core and cerebellum in Plates 16A and B.

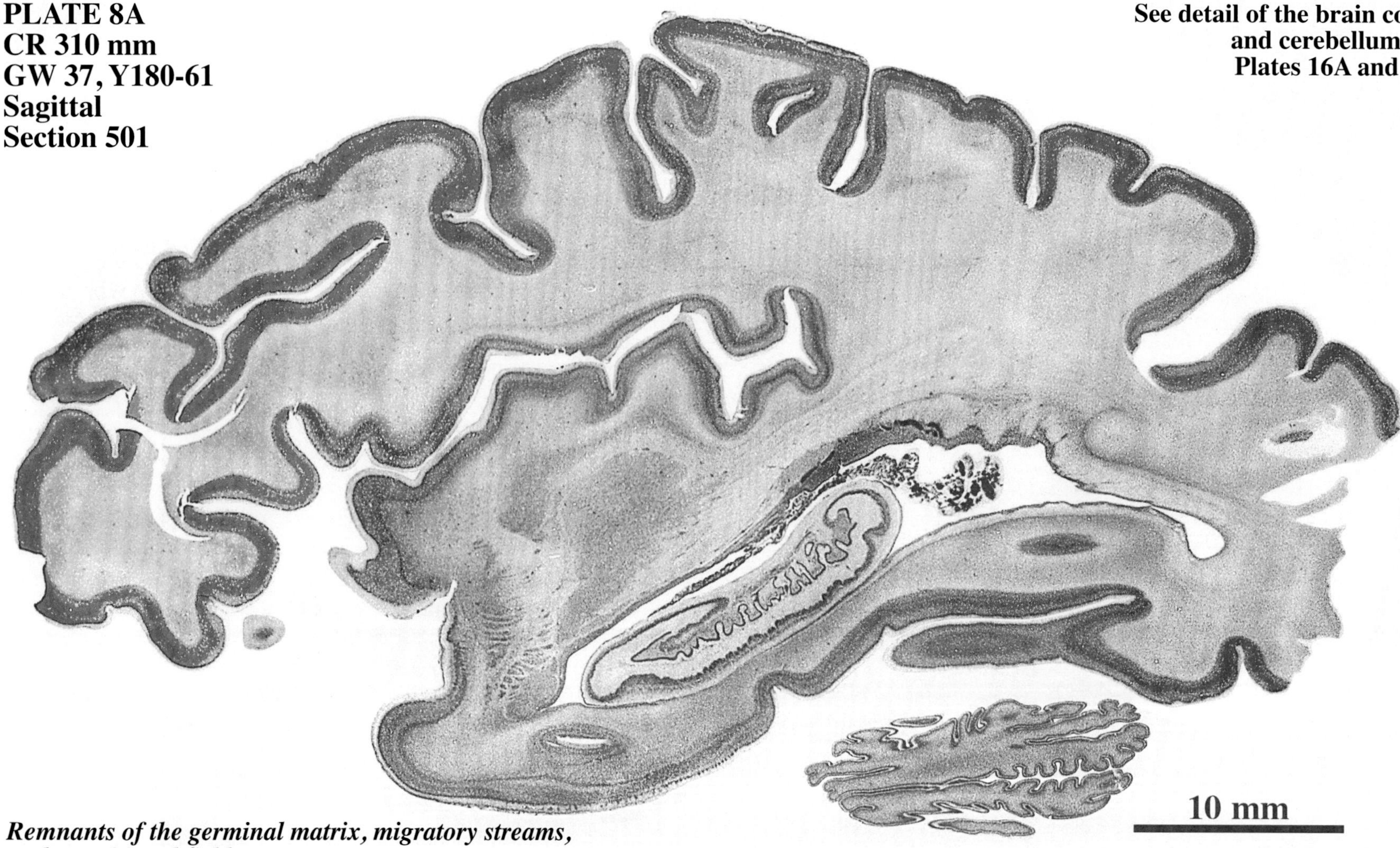

Remnants of the germinal matrix, migratory streams, and transitional fields

1 *Parietal NEP, SVZ, and STF*
2 *Occipital NEP, SVZ, and STF*
3 *Parahippocampal NEP, SVZ, and STF*
4 *Alvear GEP*
5 *Subgranular zone (dentate)*
6 *Lateral migratory stream (cortical)*
7 *Amygdaloid G/EP*
8 *Posterior striatal NEP and SVZ*
9 *Strionuclear GEP*
10 *External germinal layer (cerebellum)*
11 *Subpial granular layer (cortical)*

GEP - Glioepithelium
G/EP - Glioepithelium/ependyma
NEP - Neuroepithelium
STF - Stratified transitional field
SVZ - Subventricular zone

PLATE 8B

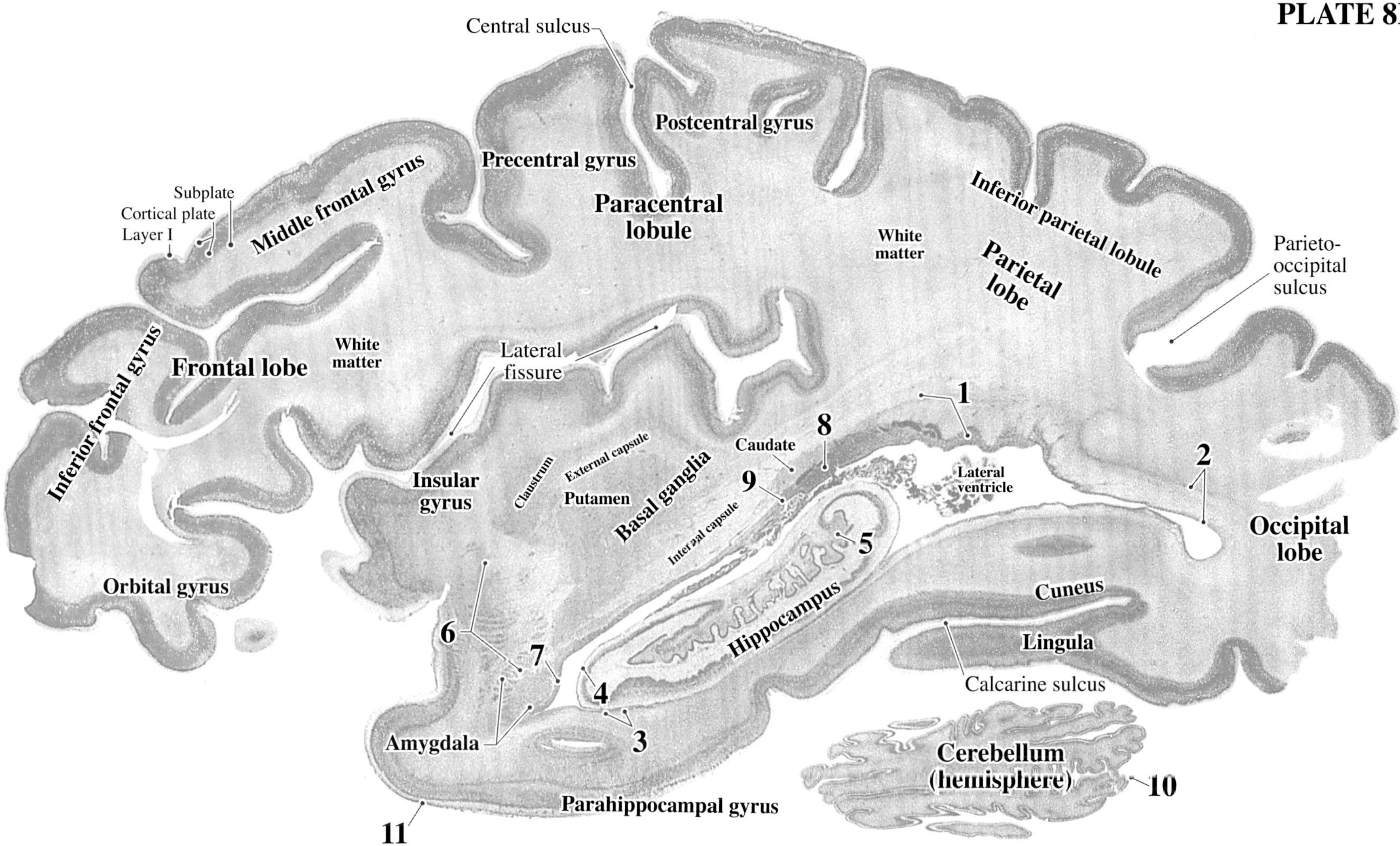
Central sulcus
Postcentral gyrus
Precentral gyrus
Paracentral lobule
Subplate
Cortical plate
Layer I
Middle frontal gyrus
Inferior parietal lobule
White matter
Parietal lobe
Parieto-occipital sulcus
Inferior frontal gyrus
Frontal lobe
White matter
Lateral fissure
1
8
Caudate
Lateral ventricle
2
Insular gyrus
Claustrum
External capsule
Putamen
Basal ganglia
9
Internal capsule
5
Occipital lobe
Orbital gyrus
Hippocampus
Cuneus
6
7
Lingula
4
Calcarine sulcus
Amygdala
3
Cerebellum (hemisphere)
10
Parahippocampal gyrus
11

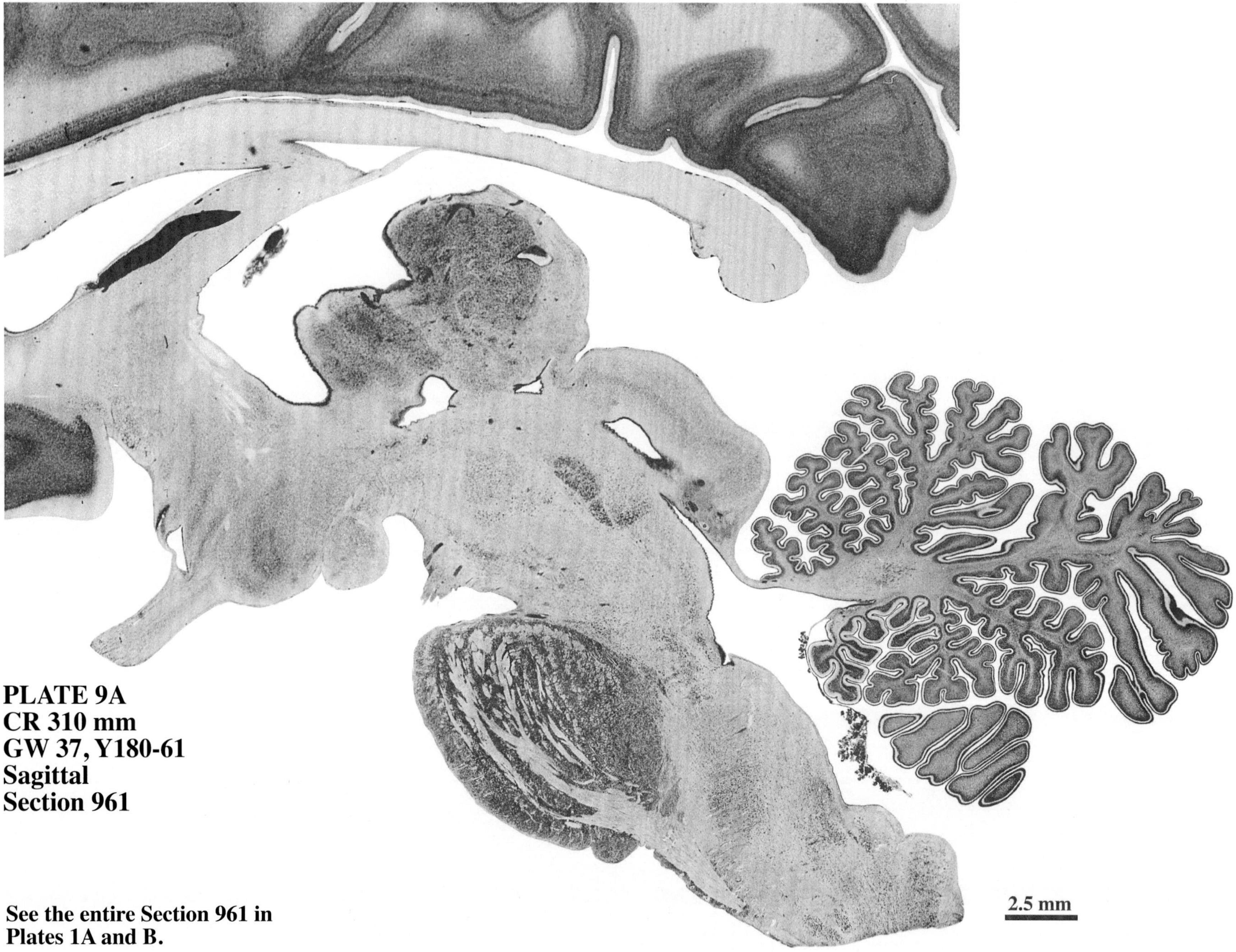

PLATE 9A
CR 310 mm
GW 37, Y180-61
Sagittal
Section 961

2.5 mm

See the entire Section 961 in Plates 1A and B.

PLATE 9B

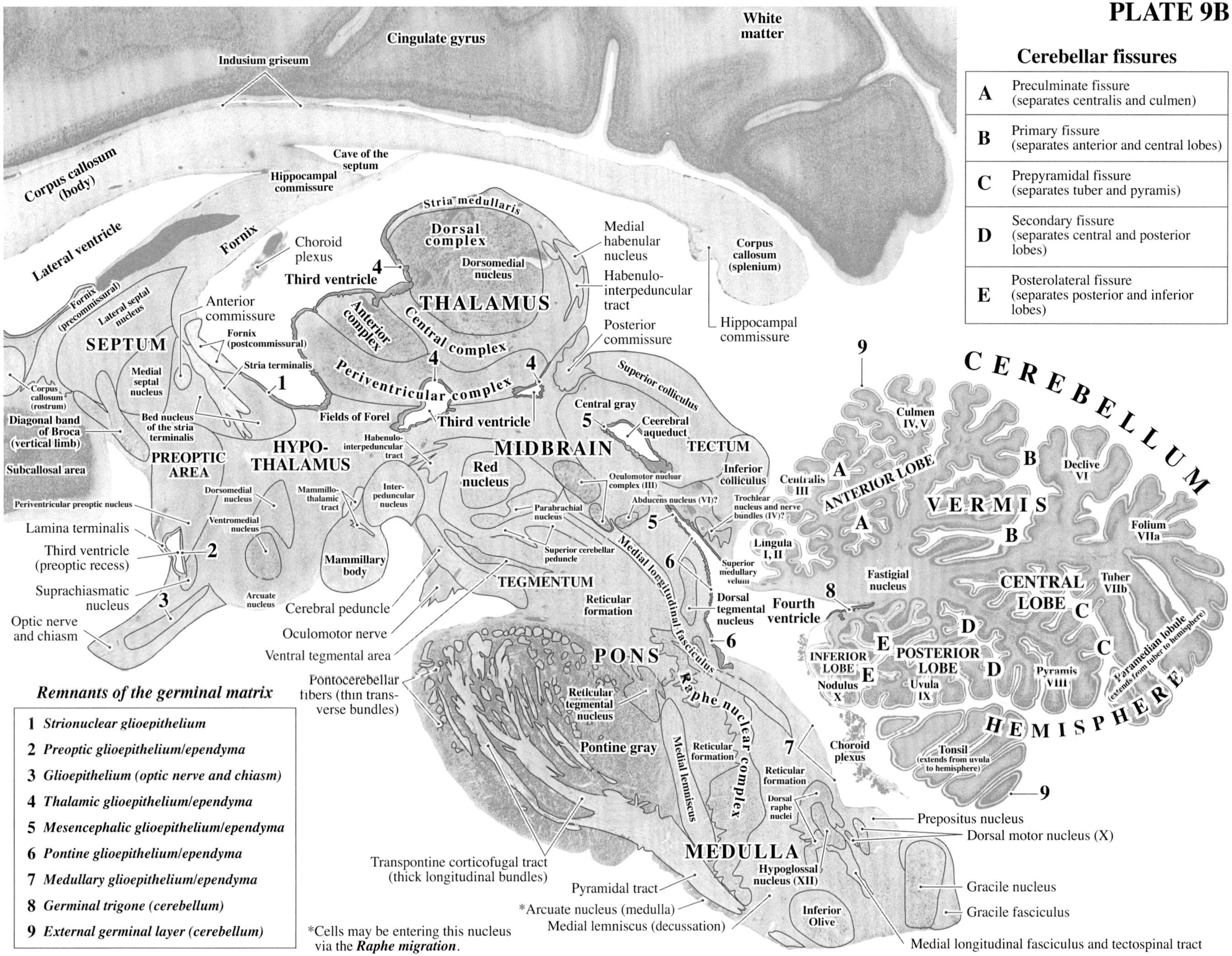

Cerebellar fissures

A	Preculminate fissure (separates centralis and culmen)
B	Primary fissure (separates anterior and central lobes)
C	Prepyramidal fissure (separates tuber and pyramis)
D	Secondary fissure (separates central and posterior lobes)
E	Posterolateral fissure (separates posterior and inferior lobes)

Remnants of the germinal matrix

1. *Strionuclear glioepithelium*
2. *Preoptic glioepithelium/ependyma*
3. *Glioepithelium (optic nerve and chiasm)*
4. *Thalamic glioepithelium/ependyma*
5. *Mesencephalic glioepithelium/ependyma*
6. *Pontine glioepithelium/ependyma*
7. *Medullary glioepithelium/ependyma*
8. *Germinal trigone (cerebellum)*
9. *External germinal layer (cerebellum)*

*Cells may be entering this nucleus via the ***Raphe migration***.

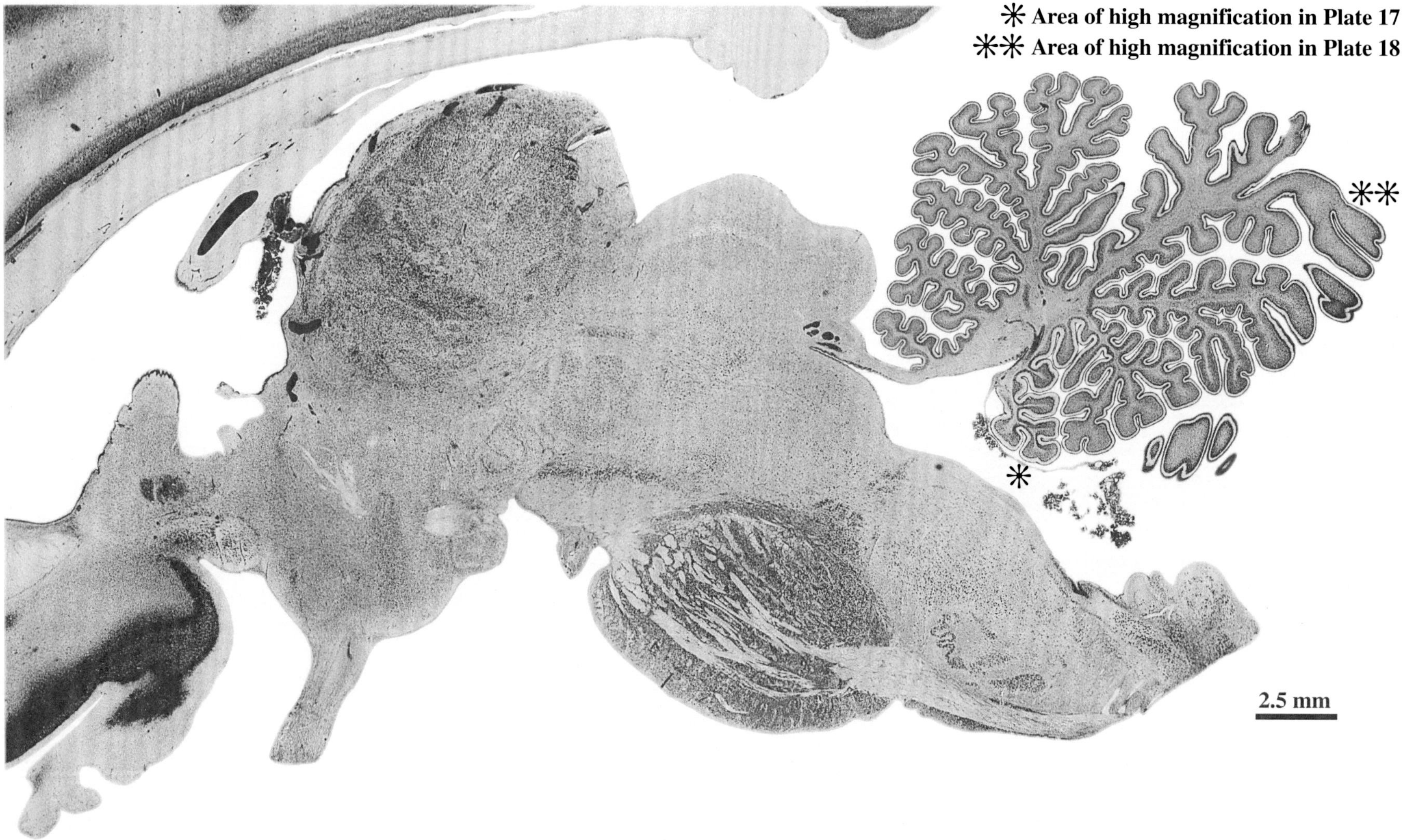

PLATE 10A
CR 310 mm, GW 37, Y180-61
Sagittal, Section 941

See the entire Section 941 in Plates 2A and B.

PLATE 10B

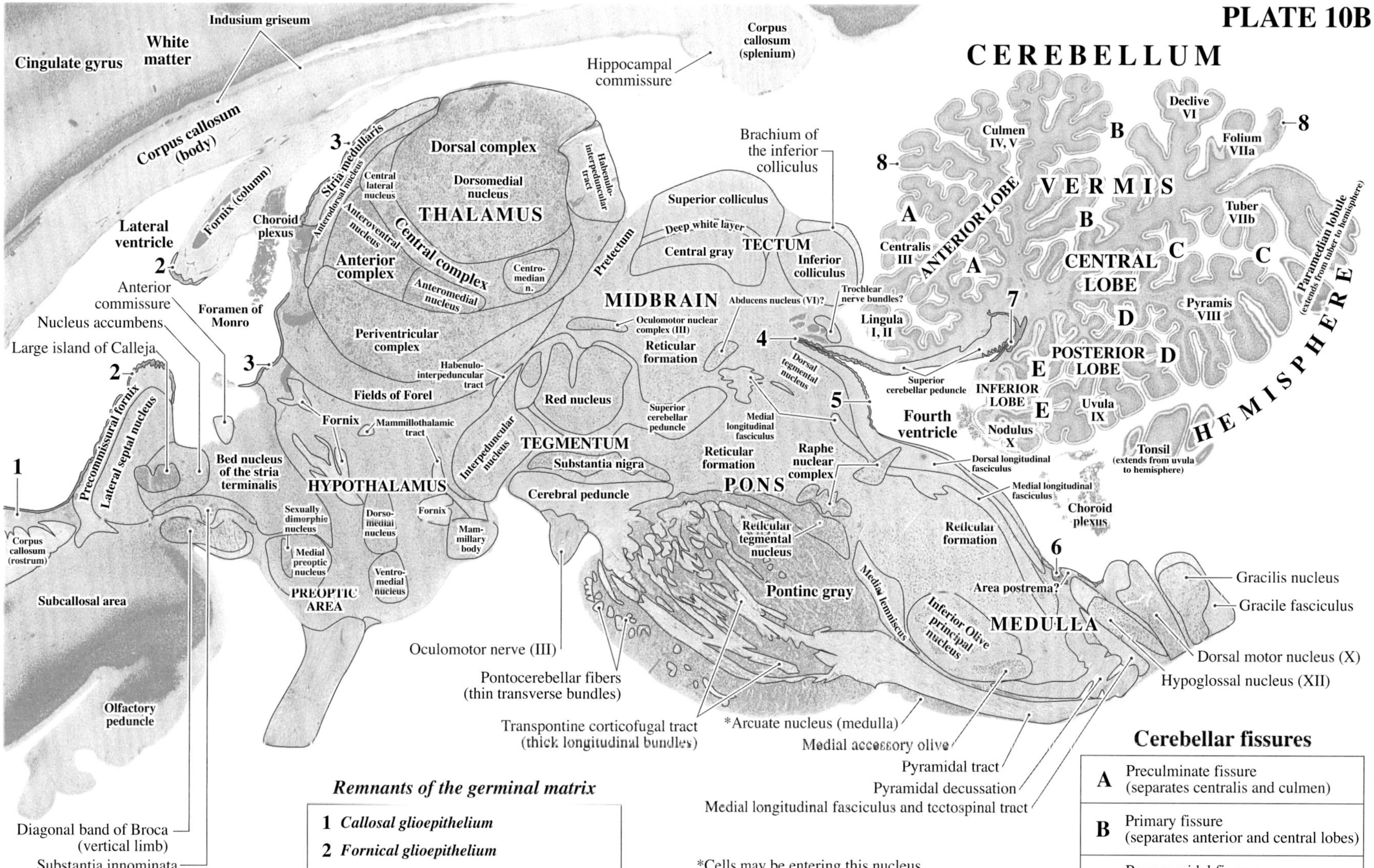

Remnants of the germinal matrix

1	***Callosal glioepithelium***
2	***Fornical glioepithelium***
3	***Strionuclear glioepithelium***
4	***Mesencephalic glioepithelium/ependyma***
5	***Pontine glioepithelium/ependyma***
6	***Medullary glioepithelium/ependyma***
7	***Germinal trigone (cerebellum)***
8	***External germinal layer (cerebellum)***

*Cells may be entering this nucleus via the ***Raphe migration***.

Cerebellar fissures

A	Preculminate fissure (separates centralis and culmen)
B	Primary fissure (separates anterior and central lobes)
C	Prepyramidal fissure (separates tuber and pyramis)
D	Secondary fissure (separates central and posterior lobes)
E	Posterolateral fissure (separates posterior and inferior lobes)

PLATE 11A
CR 310 mm, GW 37, Y180-61
Sagittal, Section 841

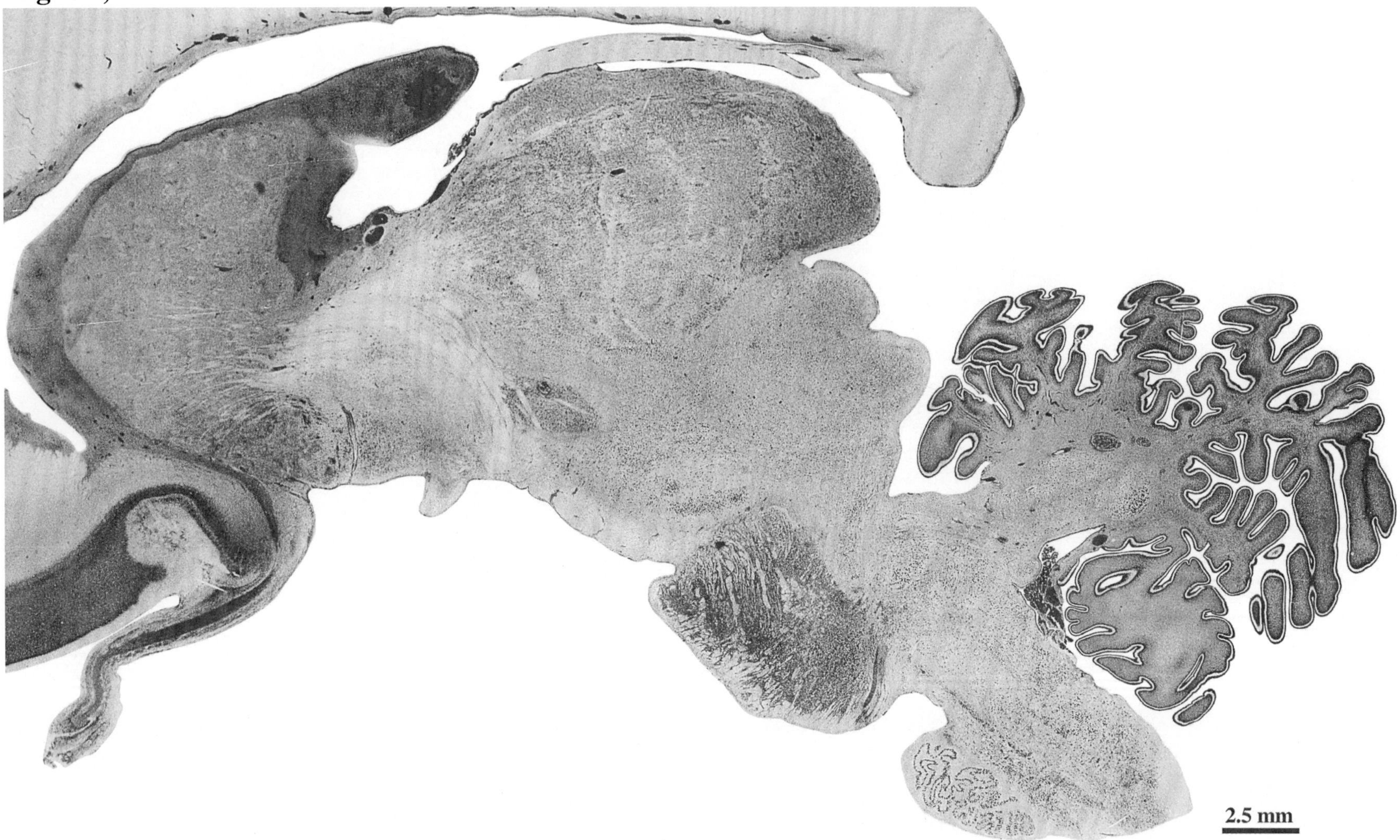

See the entire Section 841 in Plates 3A and B.

PLATE 11B

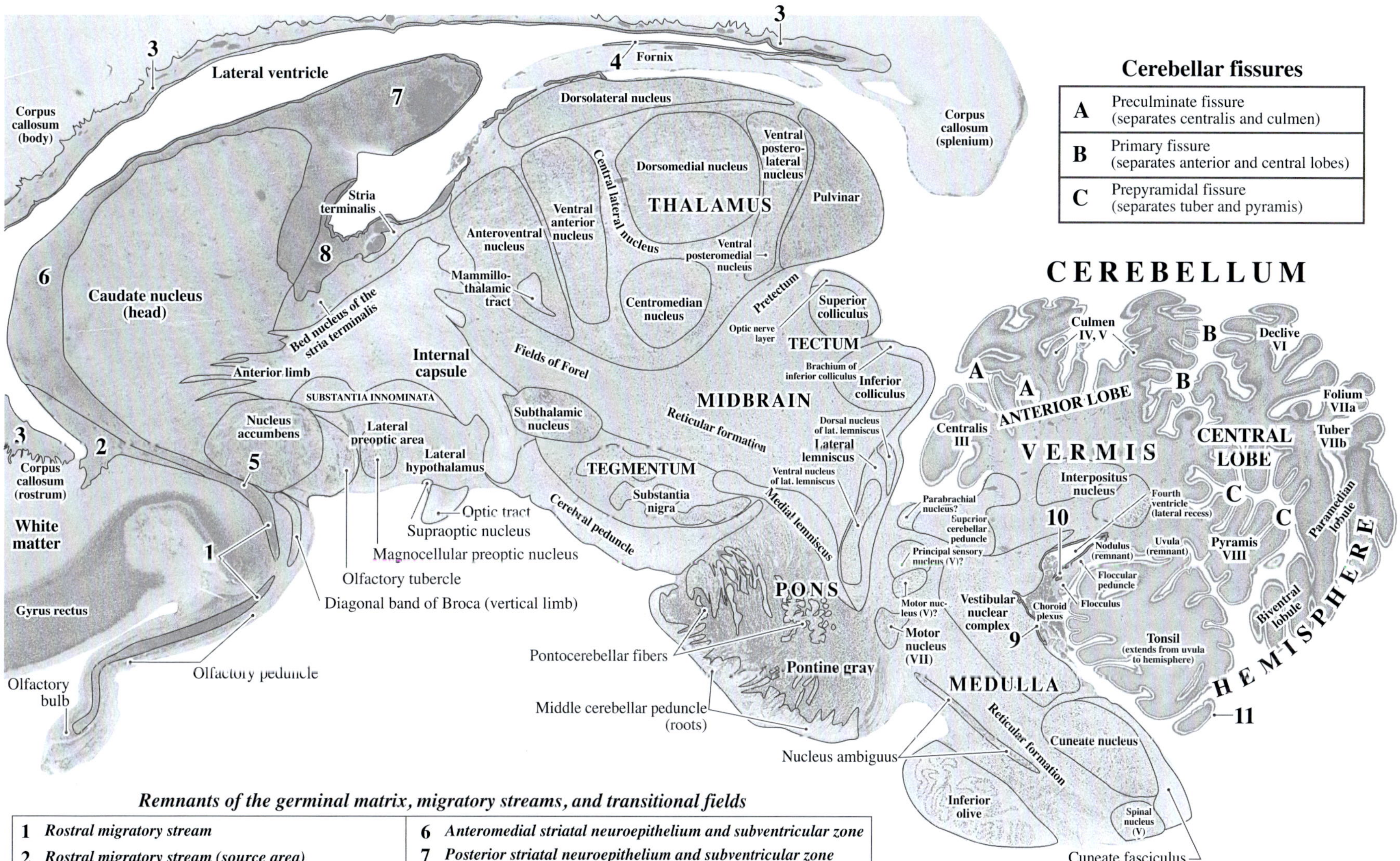

Cerebellar fissures

A	Preculminate fissure (separates centralis and culmen)
B	Primary fissure (separates anterior and central lobes)
C	Prepyramidal fissure (separates tuber and pyramis)

Remnants of the germinal matrix, migratory streams, and transitional fields

1	*Rostral migratory stream*	6	*Anteromedial striatal neuroepithelium and subventricular zone*
2	*Rostral migratory stream (source area)*	7	*Posterior striatal neuroepithelium and subventricular zone*
3	*Callosal glioepithelium*	8	*Strionuclear glioepithelium*
4	*Fornical glioepithelium*	9	*Pontine and medullary glioepithelium/ependyma*
5	*Accumbent neuroepithelium and subventricular zone (infiltrated by the rostral migratory stream)*	10	*Germinal trigone (cerebellum)*
		11	*External germinal layer (cerebellum)*

PLATE 12A
CR 310 mm, GW 37, Y180-61, Sagittal, Section 781

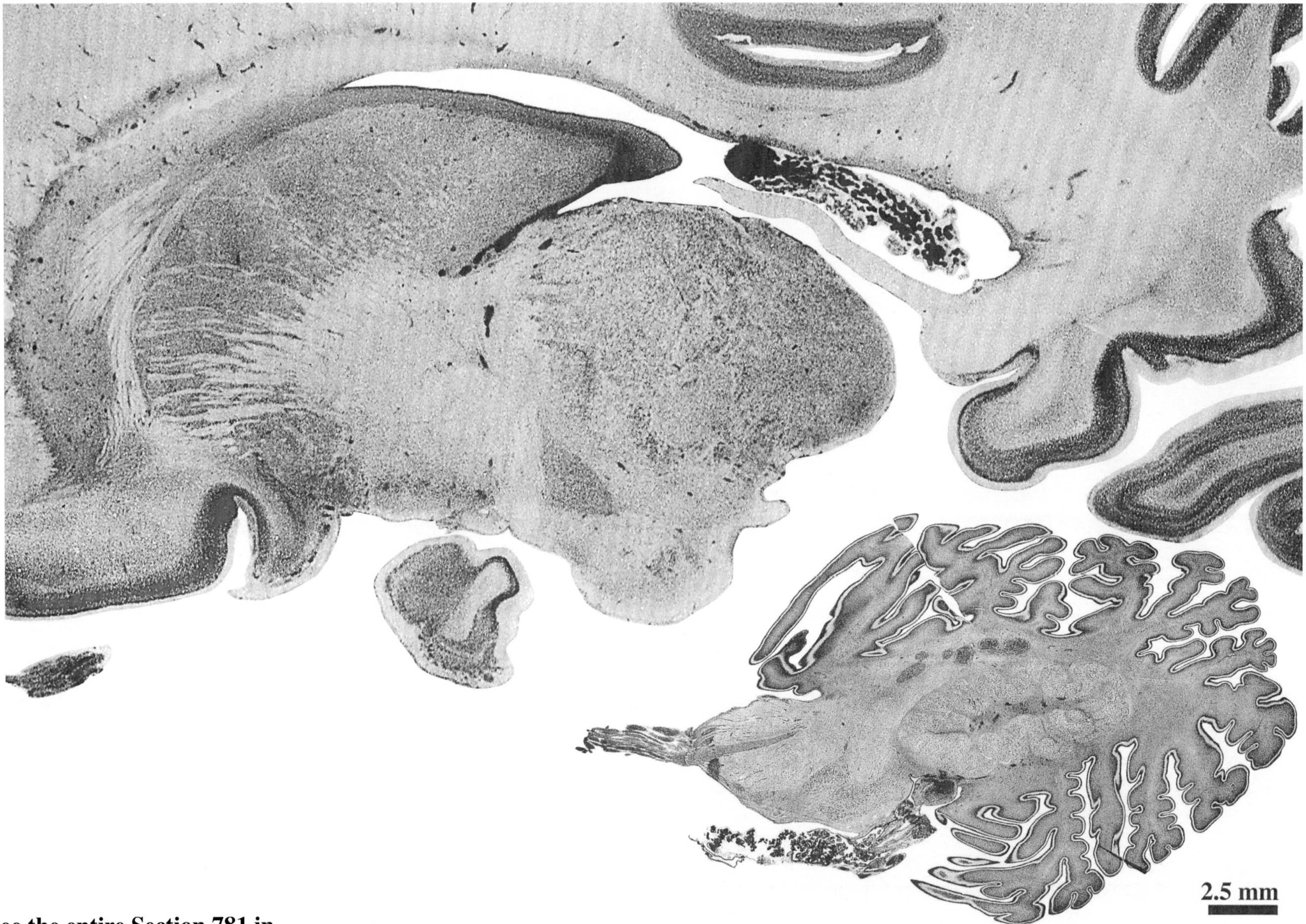

See the entire Section 781 in Plates 4A and B.

PLATE 12B

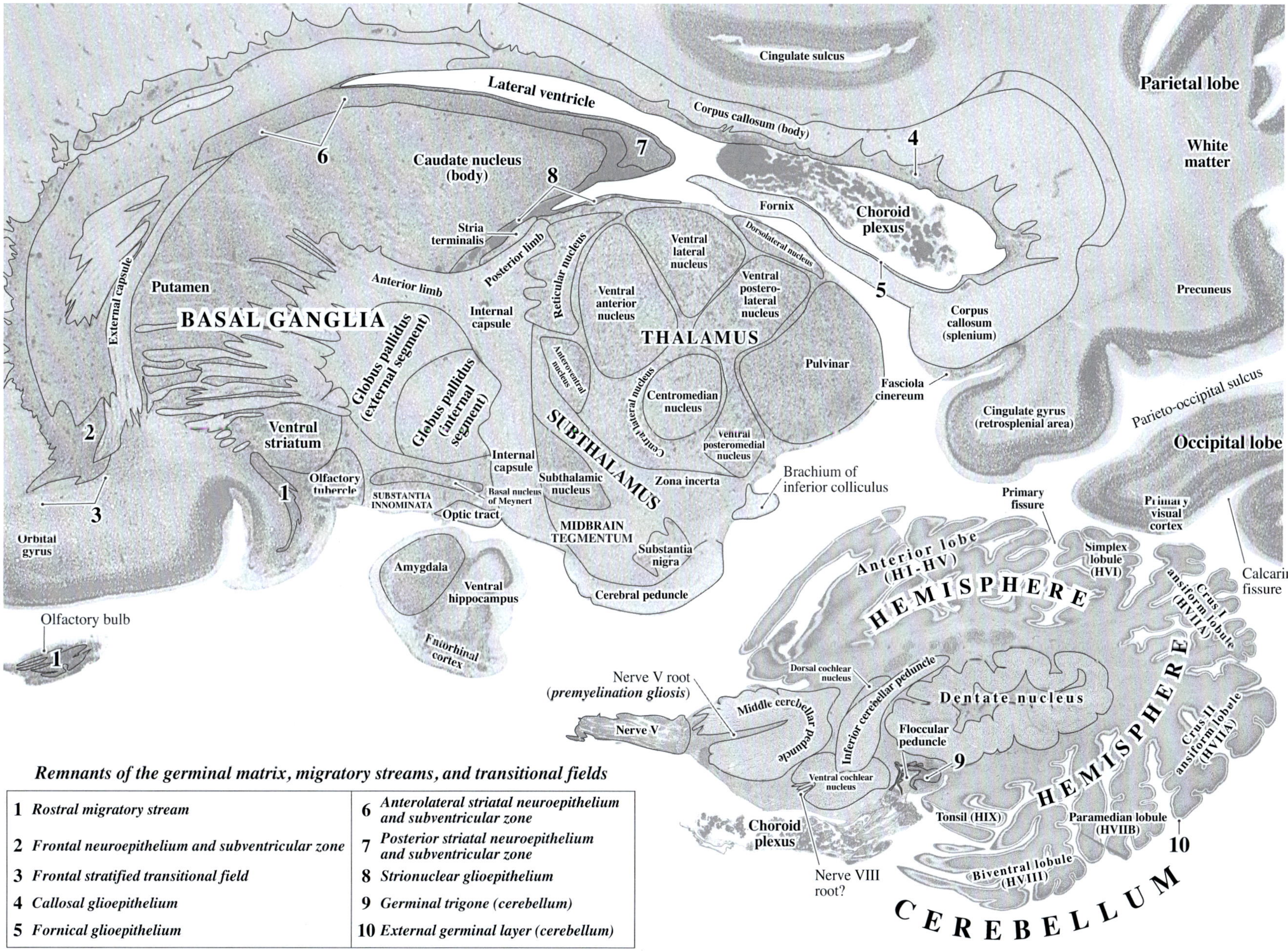

Remnants of the germinal matrix, migratory streams, and transitional fields

1 *Rostral migratory stream*	6 *Anterolateral striatal neuroepithelium and subventricular zone*
2 *Frontal neuroepithelium and subventricular zone*	7 *Posterior striatal neuroepithelium and subventricular zone*
3 *Frontal stratified transitional field*	8 *Strionuclear glioepithelium*
4 *Callosal glioepithelium*	9 *Germinal trigone (cerebellum)*
5 *Fornical glioepithelium*	10 *External germinal layer (cerebellum)*

PLATE 13A
CR 310 mm, GW m37, Y180-61, Sagittal, Section 721

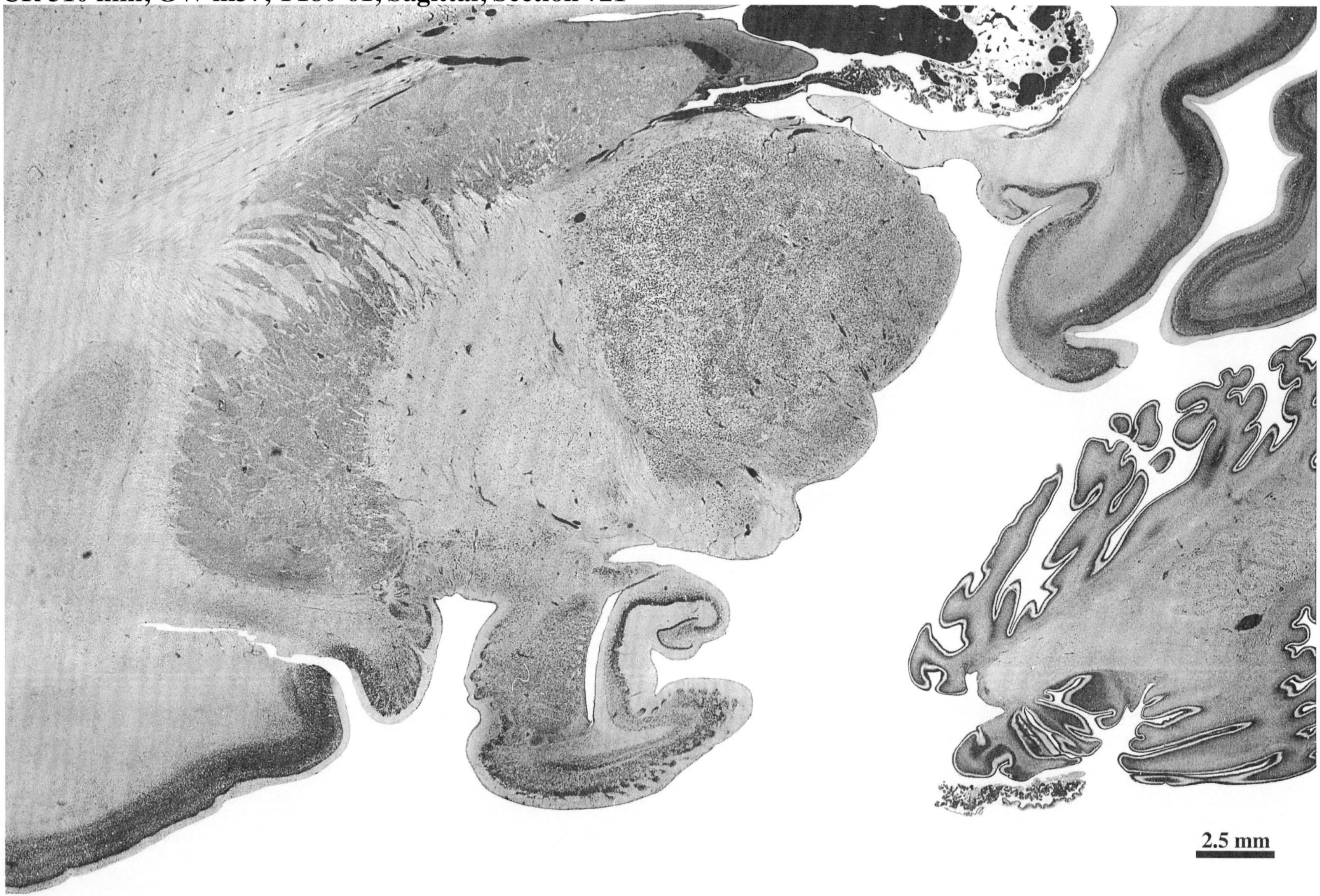

See the entire Section 721 in Plates 5A and B.

PLATE 13B

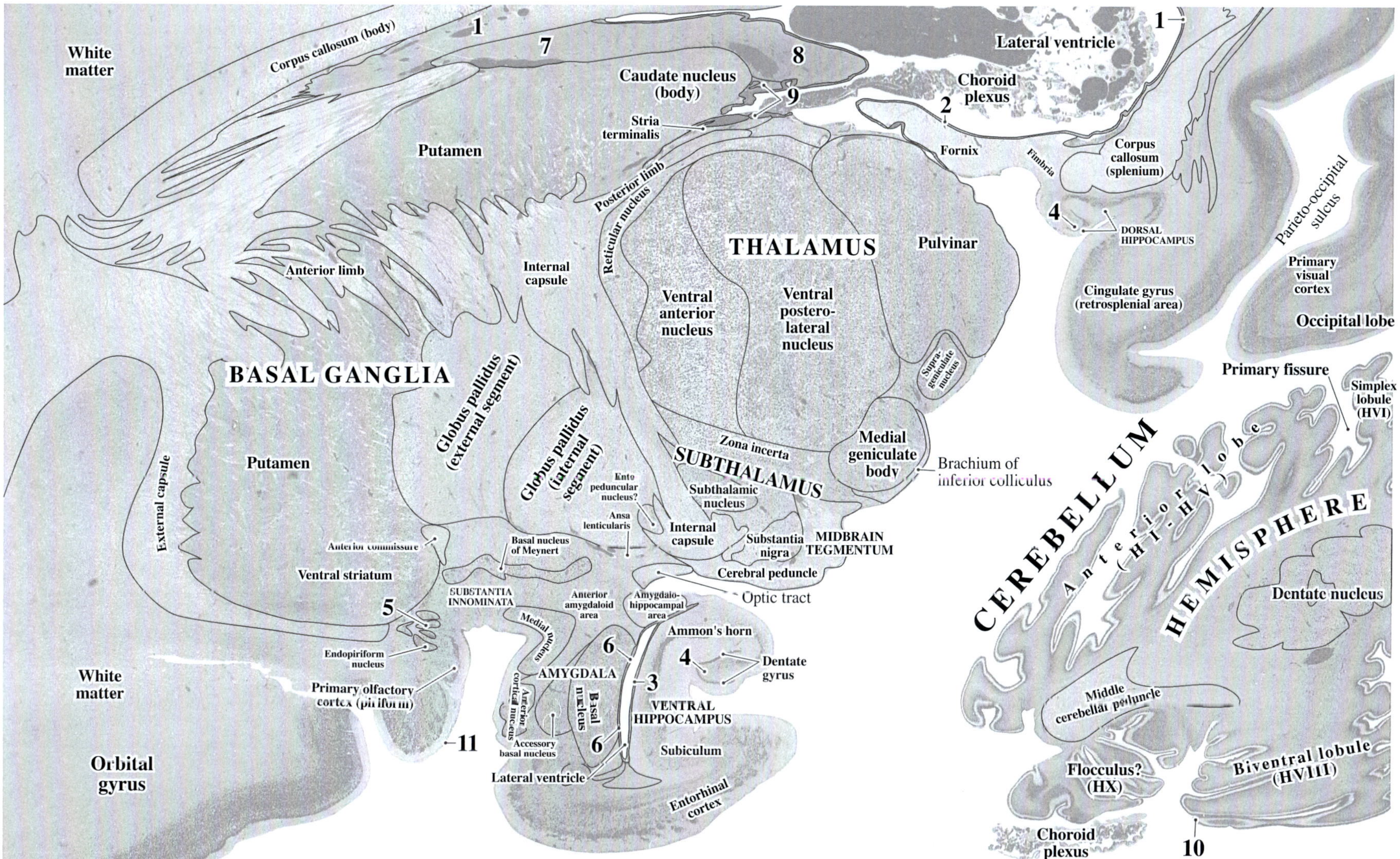

Remnants of the germinal matrix, migratory streams, and transitional fields

1	*Callosal glioepithelium*	6	*Amygdaloid glioepithelium/ependyma*
2	*Fornical glioepithelium*	7	*Anterolateral striatal neuroepithelium and subventricular zone*
3	*Alvear glioepithelium*	8	*Posterior striatal neuroepithelium and subventricular zone*
4	*Subgranular zone (dentate)*	9	*Strionuclear glioepithelium*
5	*Lateral migratory stream (cortical)*	10	*External germinal layer (cerebellum)*
		11	*Subpial granular layer (cortical)*

PLATE 14A
CR 310 mm, GW 37, Y180-61, Sagittal, Section 661

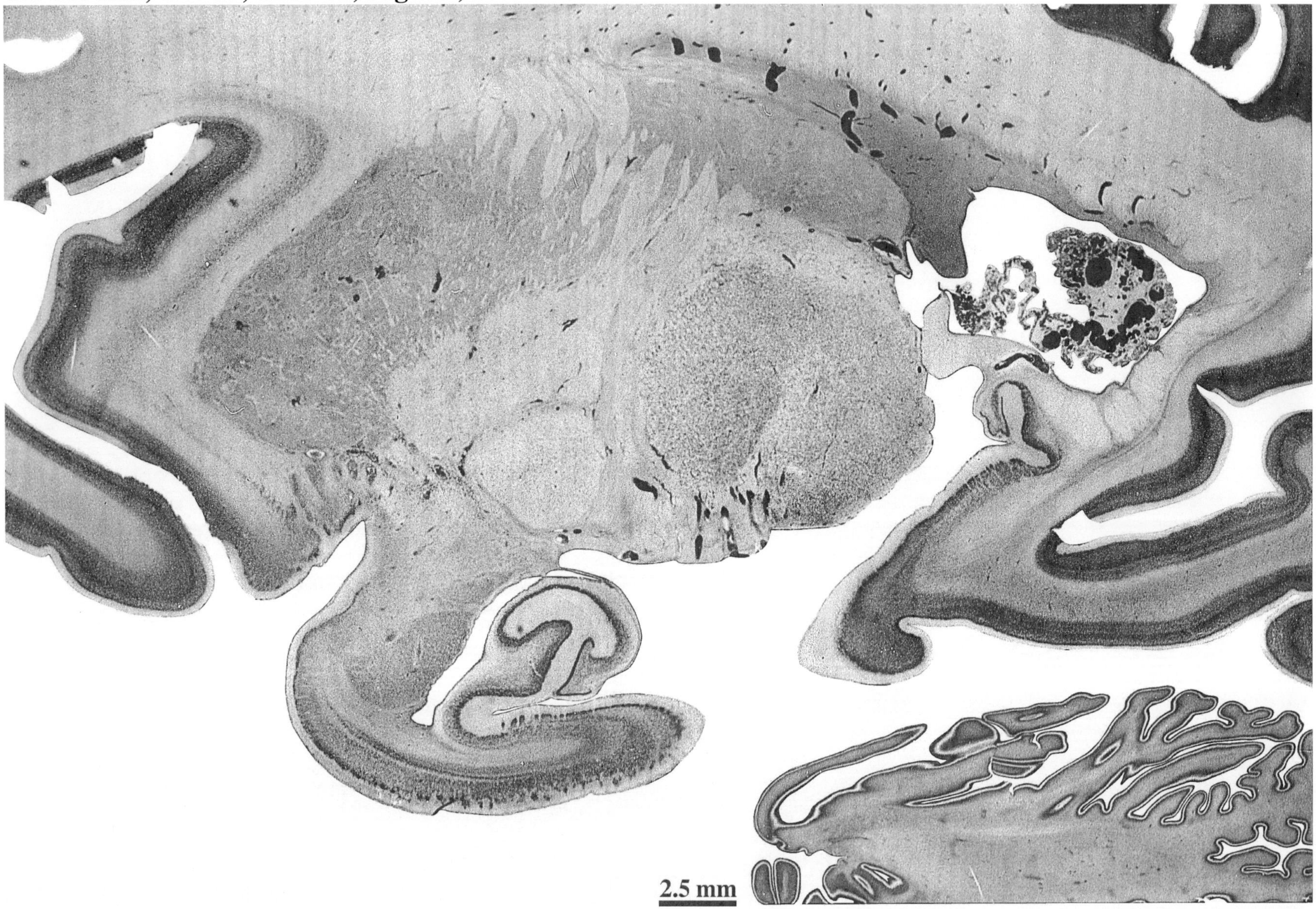

2.5 mm

See the entire Section 661 in Plates 6A and B.

PLATE 14B

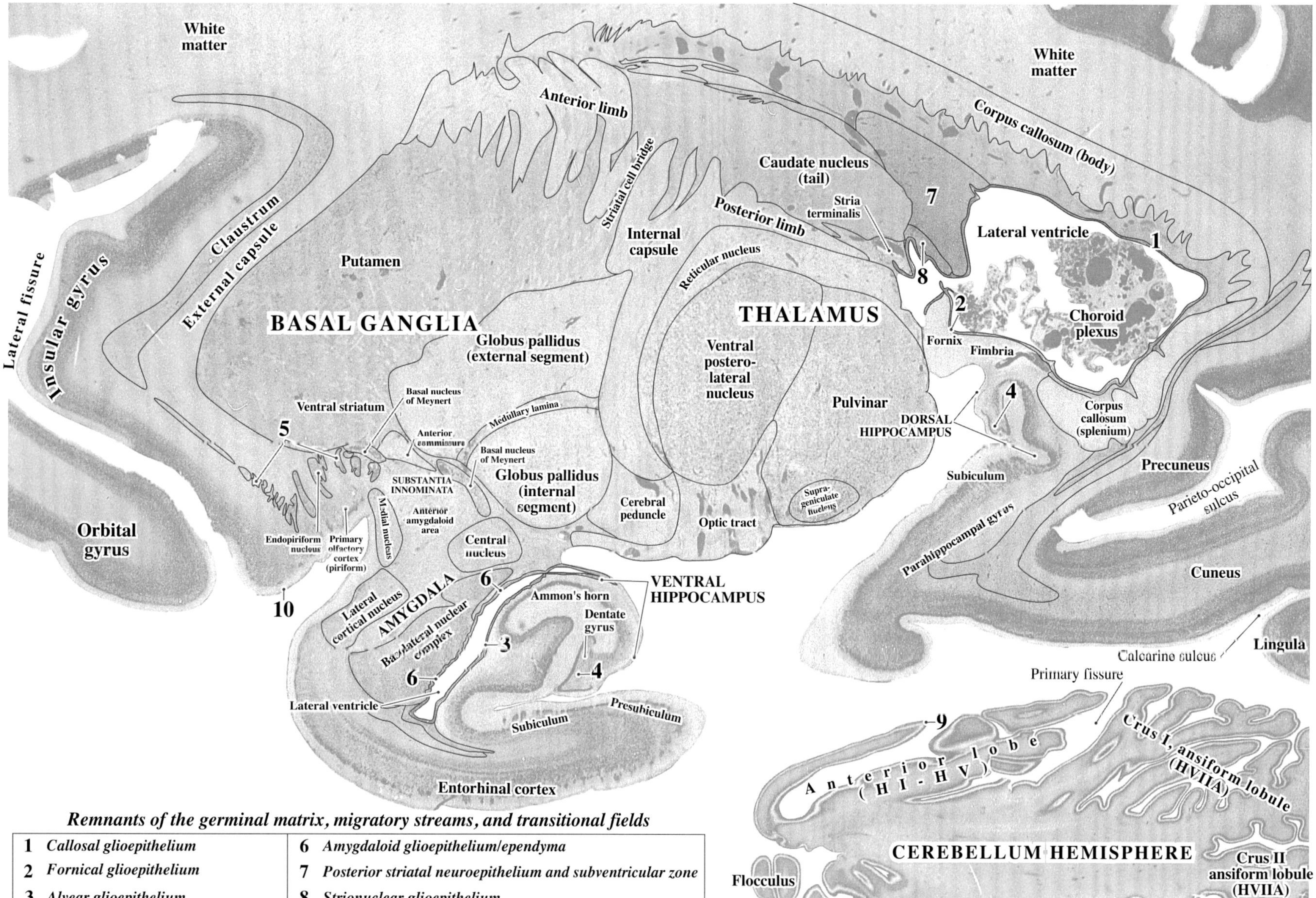

Remnants of the germinal matrix, migratory streams, and transitional fields

1	*Callosal glioepithelium*	6	*Amygdaloid glioepithelium/ependyma*
2	*Fornical glioepithelium*	7	*Posterior striatal neuroepithelium and subventricular zone*
3	*Alvear glioepithelium*	8	*Strionuclear glioepithelium*
4	*Subgranular zone (dentate)*	9	*External germinal layer (cerebellum)*
5	*Lateral migratory stream (cortical)*	10	*Subpial granular layer (cortical)*

PLATE 15A
CR 310 mm, GW 37, Y180-61, Sagittal, Section 621

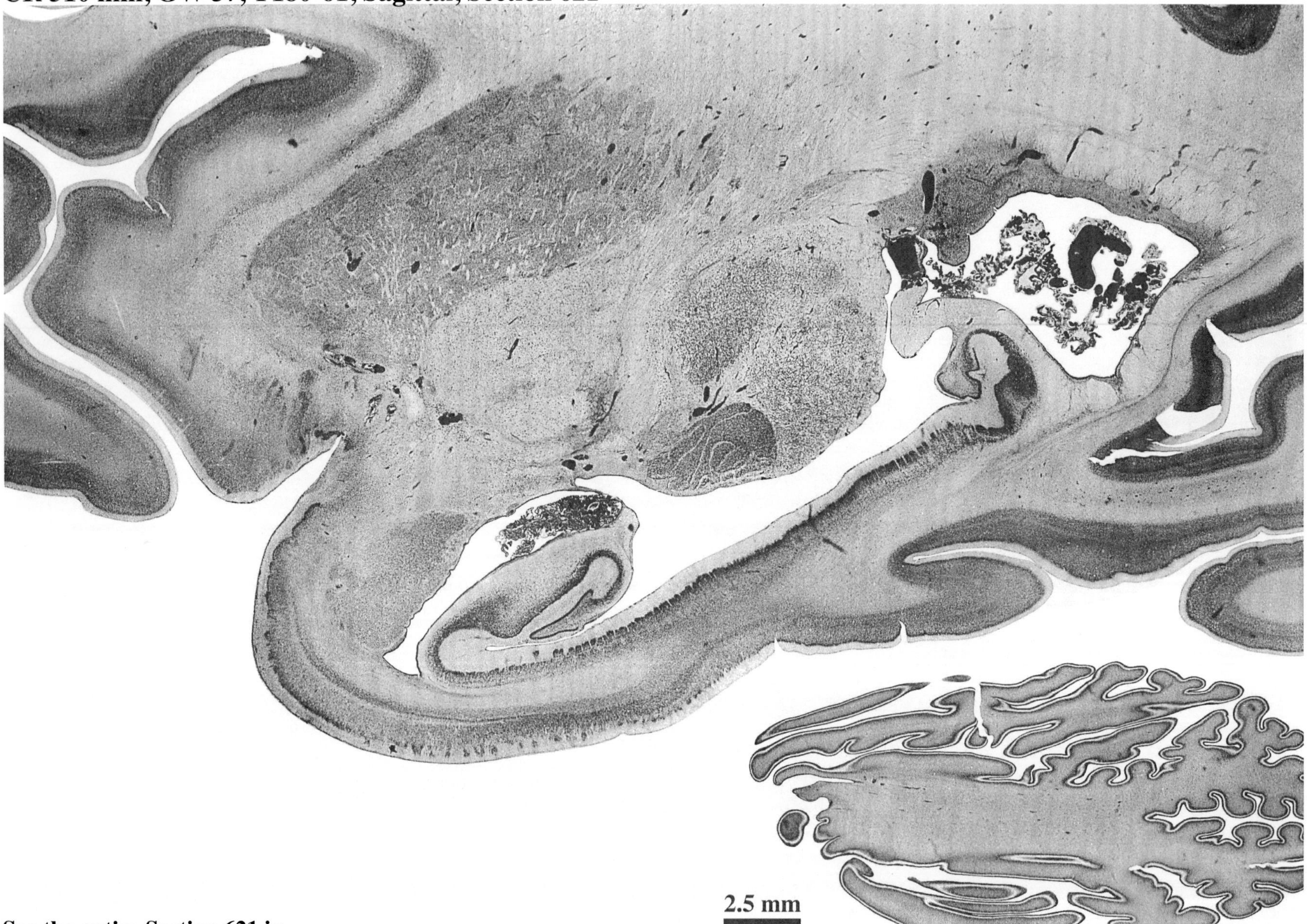

2.5 mm

See the entire Section 621 in Plates 7A and B.

PLATE 15B

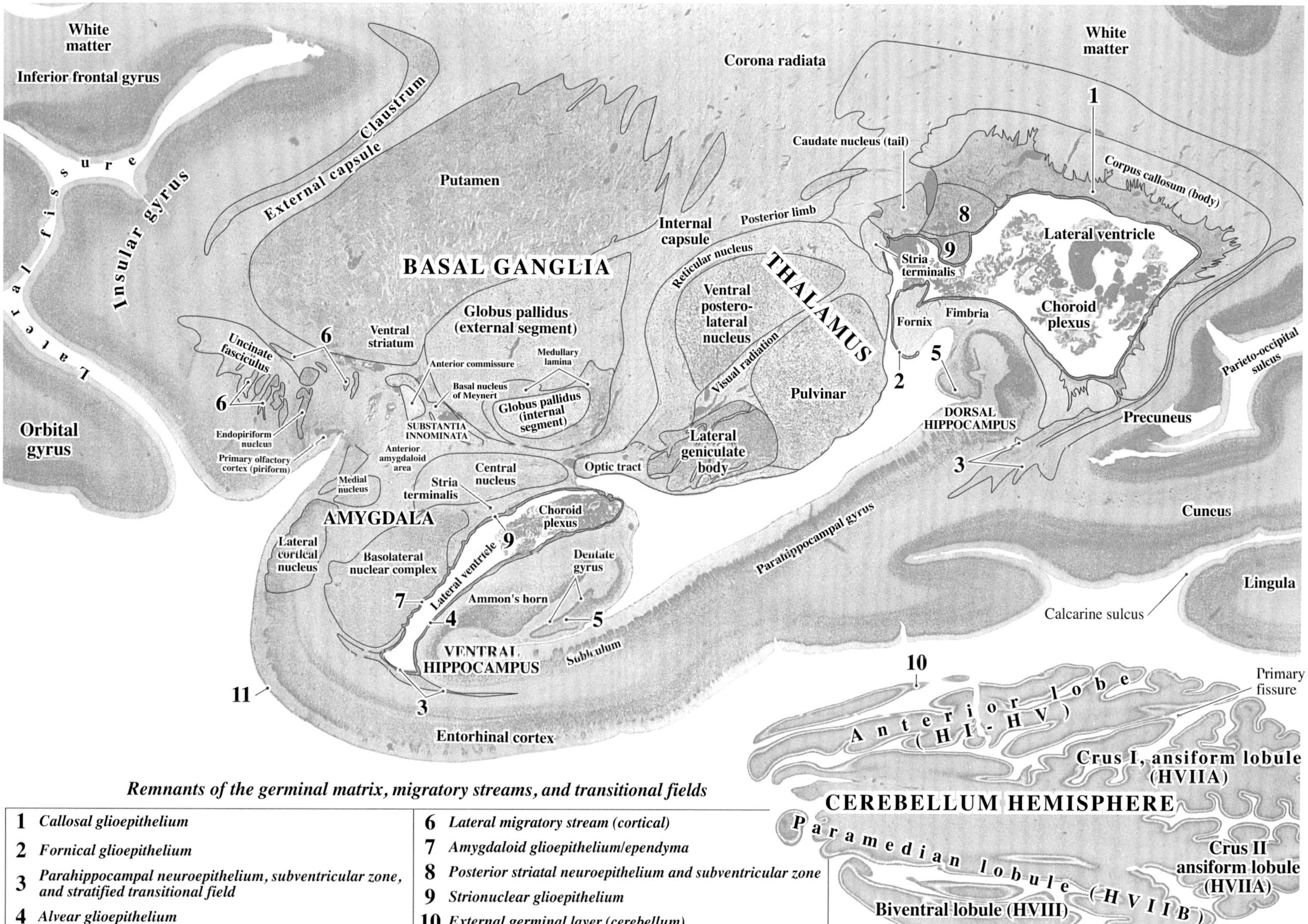

Remnants of the germinal matrix, migratory streams, and transitional fields

1	*Callosal glioepithelium*	**6**	*Lateral migratory stream (cortical)*
2	*Fornical glioepithelium*	**7**	*Amygdaloid glioepithelium/ependyma*
3	*Parahippocampal neuroepithelium, subventricular zone, and stratified transitional field*	**8**	*Posterior striatal neuroepithelium and subventricular zone*
4	*Alvear glioepithelium*	**9**	*Strionuclear glioepithelium*
5	*Subgranular zone (dentate)*	**10**	*External germinal layer (cerebellum)*
		11	*Subpial granular layer (cortical)*

PLATE 16A, CR 310 mm, GW 37, Y180-61, Sagittal, Section 501

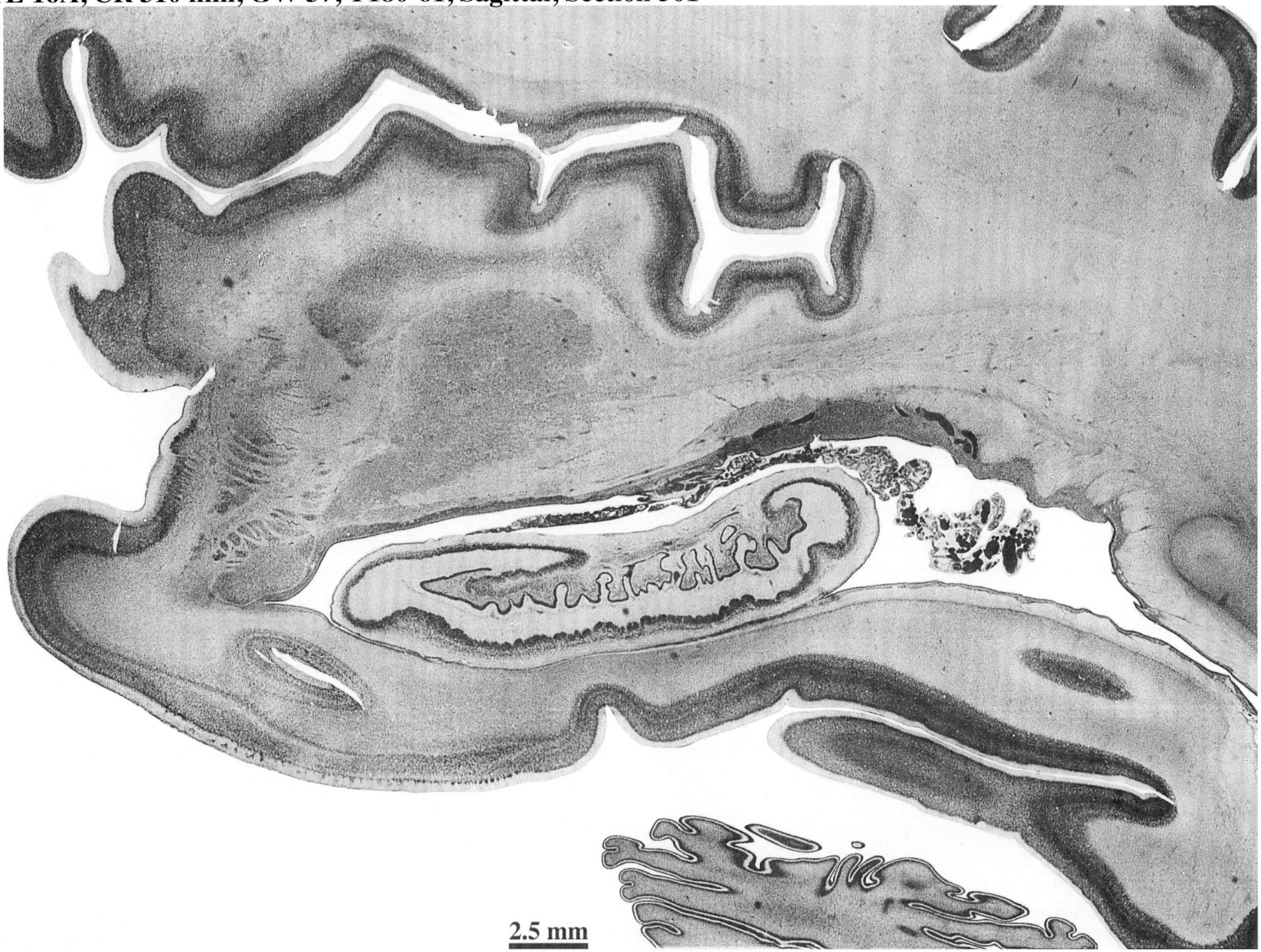

See the entire Section 501 in Plates 8A and B.

PLATE 16B

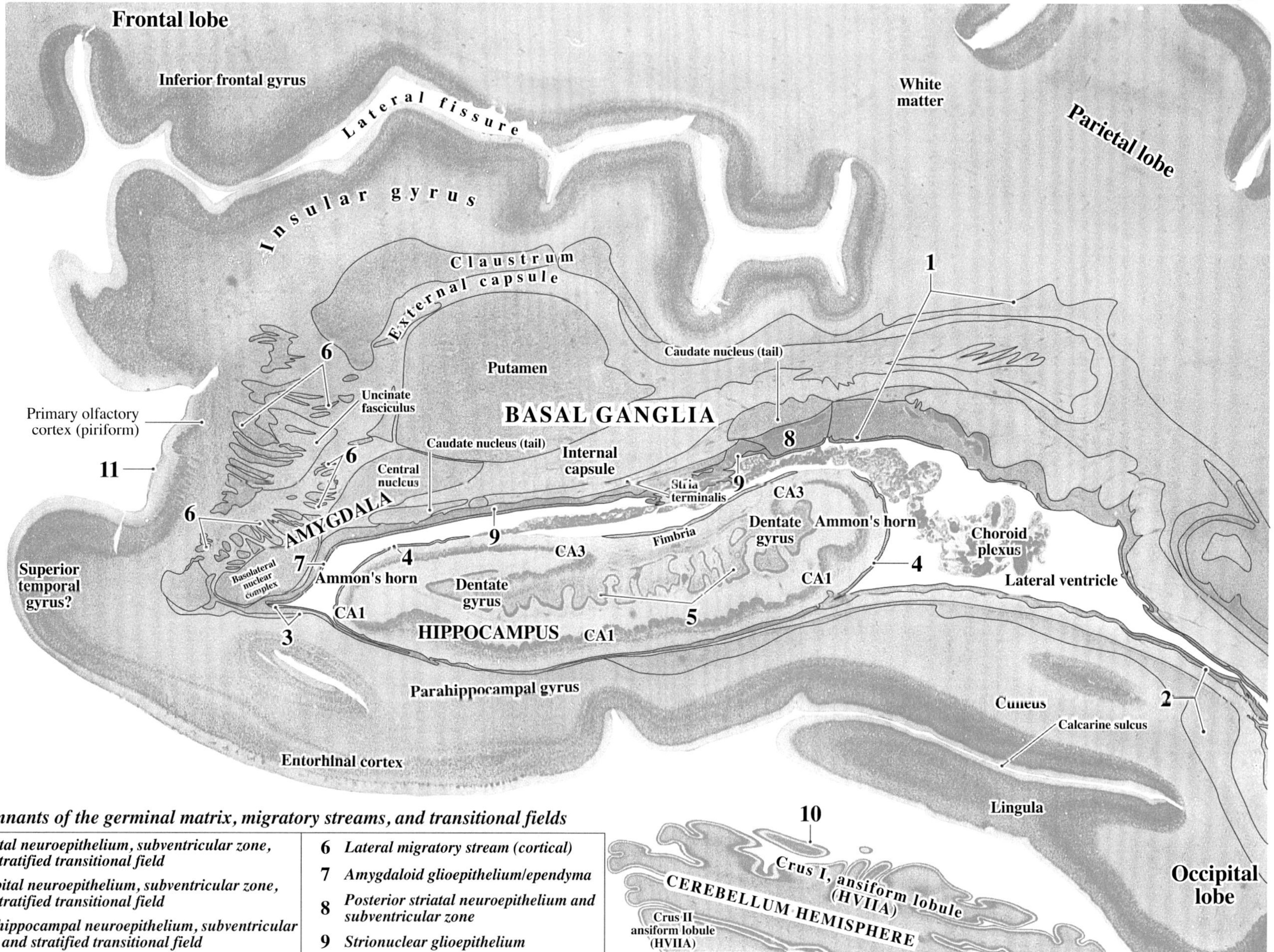

Remnants of the germinal matrix, migratory streams, and transitional fields

1	*Parietal neuroepithelium, subventricular zone, and stratified transitional field*	6	*Lateral migratory stream (cortical)*
2	*Occipital neuroepithelium, subventricular zone, and stratified transitional field*	7	*Amygdaloid glioepithelium/ependyma*
3	*Parahippocampal neuroepithelium, subventricular zone, and stratified transitional field*	8	*Posterior striatal neuroepithelium and subventricular zone*
		9	*Strionuclear glioepithelium*
4	*Alvear glioepithelium*	10	*External germinal layer (cerebellum)*
5	*Subgranular zone (dentate)*	11	*Subpial granular layer (cortical)*

PLATE 17, CR 310 mm
GW 37, Y180-61, Sagittal, Section 941

CEREBELLUM, VERMIS, INFERIOR LOBE, LOBULE X: NODULUS
(A relatively mature region of the cerebellar cortex)

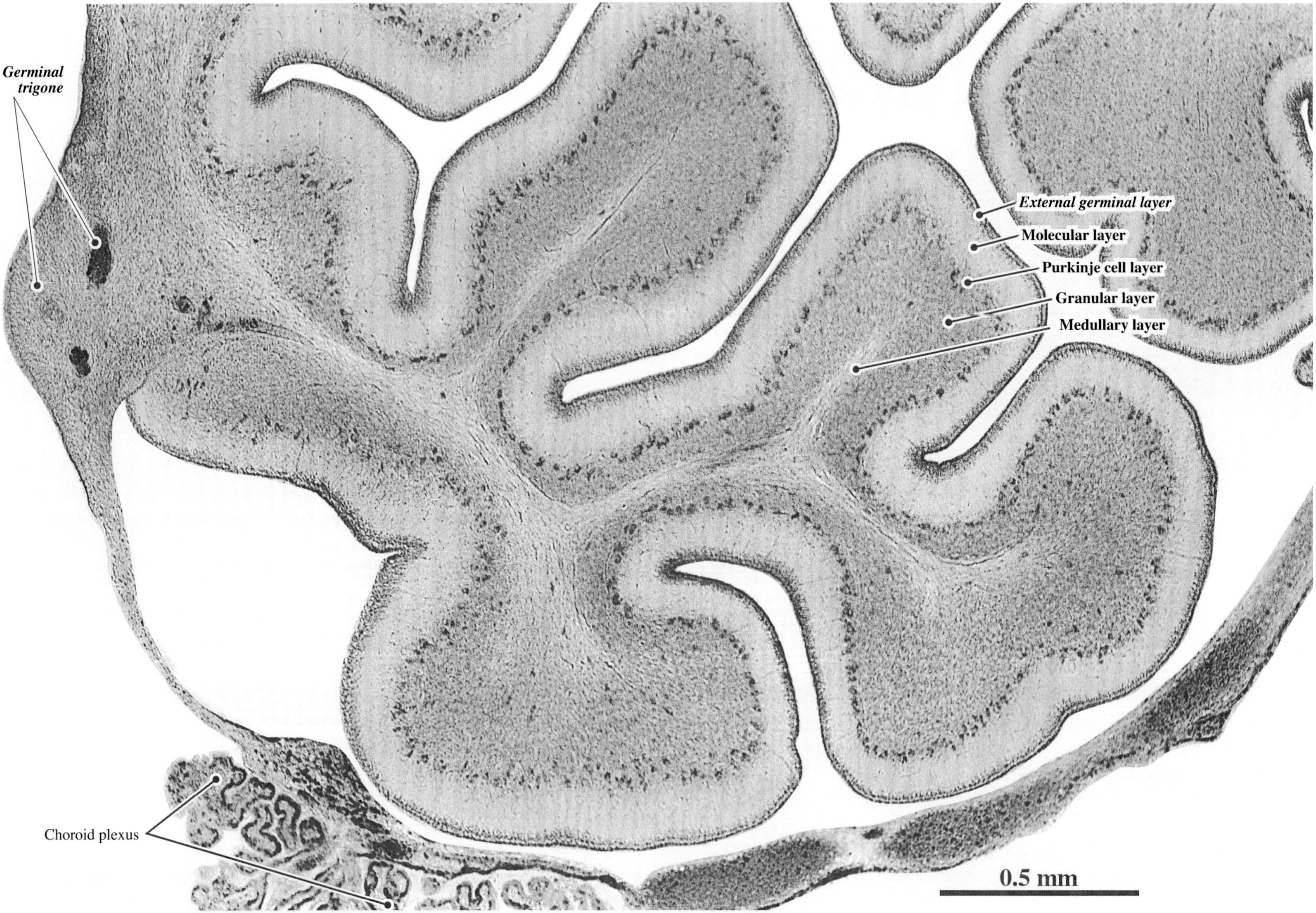

See the entire Section 941 in Plates 2A and B.
See larger view of the entire cerebellum in Plates 10A and B.

PLATE 18
CR 310 mm, GW 37, Y180-61

Sagittal, Section 941
CEREBELLUM, HEMISPHERE, PARAMEDIAN LOBULE
(A relatively immature region of the cerebellar cortex)

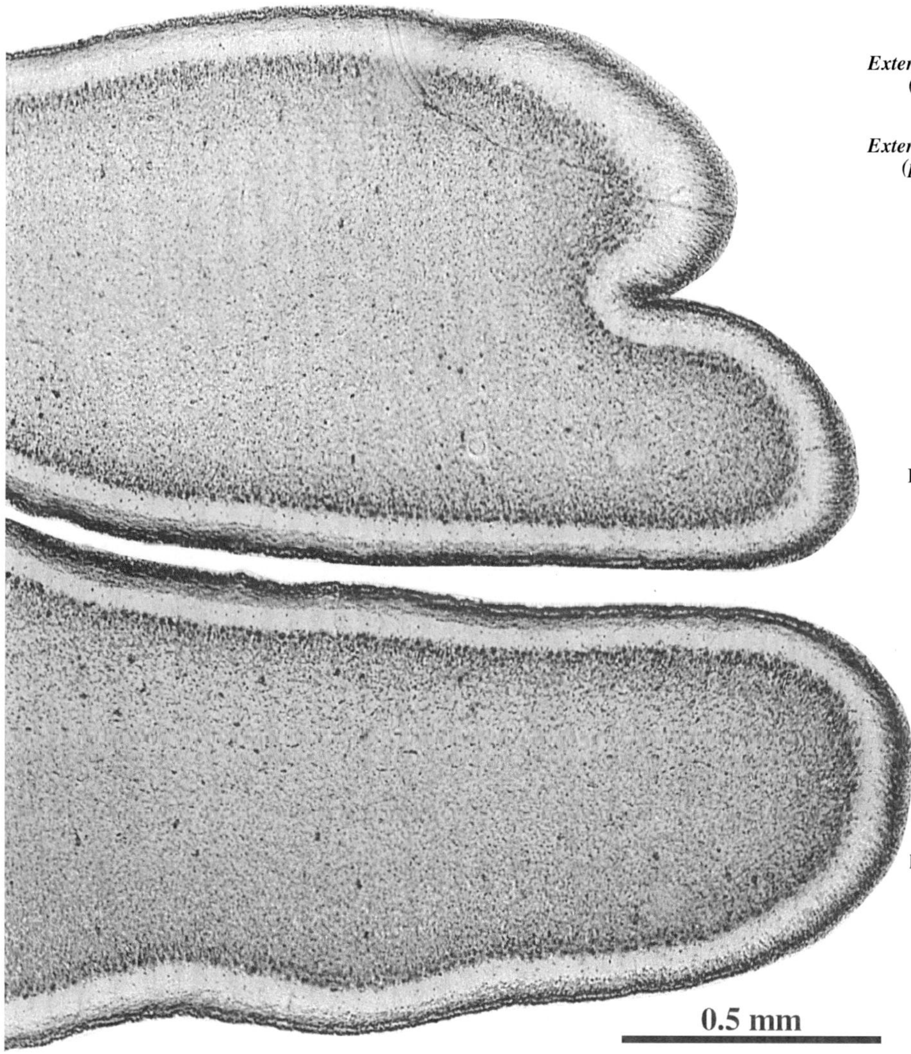

Sagittal, Section 781,CEREBELLUM, HEMISPHERE, CORTEX
(very high magnification)

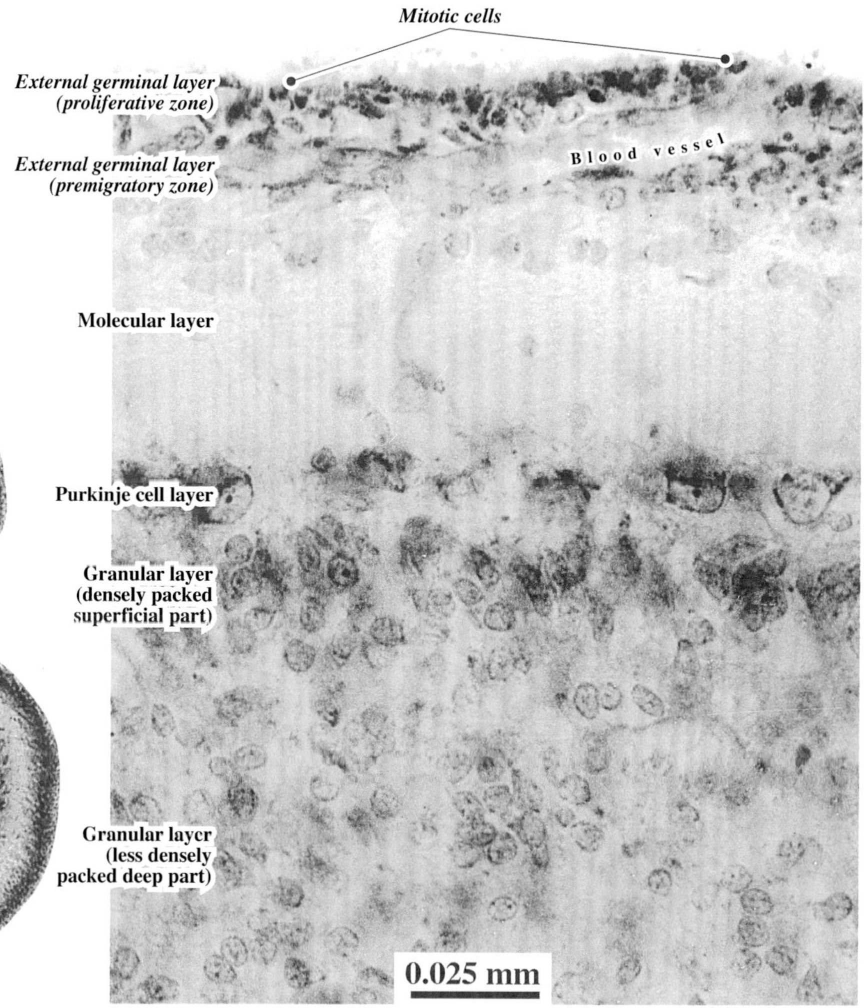

See the entire Section 941 in Plates 2A and B.
See larger view of the entire cerebellum in Plates 10A and B.

See the entire Section 781 in Plates 4A and B.
See larger view of the entire cerebellum in Plates 12A and B.

PART III: Y301-62 CR 320 mm (GW 37) Horizontal

This specimen is case number B-301-62 (Perinatal RPSL) in the Yakovlev Collection. A male infant was stillborn after a premature separation of the placenta. This brain is classified as a Normative Control in the Yakovlev Collection (Haleem, 1990). It was cut in the horizontal plane in 35-µm thick sections. Since there is no available photograph of this brain before it was embedded and cut, a photograph of the lateral view of another GW 37 brain that Larroche published in 1967 (**Figure 5**) is used to show gross anatomical features.

The approximate cutting plane of this brain is indicated in **Figure 6** (facing page) with lines superimposed on the GW 37 brain from the Larroche (1967) series. The anterior part of each section (on the left in all photographs) is slightly dorsal to the posterior part. In terms of the medial-lateral plane of sectioning, the part of each section near the bottom of the page shows structures cut slightly dorsal to those on the other side of the midline. As in all other specimens, the sections chosen for illustration are spaced closer together to show small structures in the diencephalon, midbrain, pons, and medulla. Illustrated sections are spaced farther apart when they contain only large brain structures, such as the cerebral cortex, basal ganglia, and cerebellum. Low-magnification photographs of 19 Nissl-stained sections are shown in **Plates 19–32**. **Plates 33–37** are below the cerebral cortex and show the brainstem and cerebellum at high magnification. **Plates 38–49** show high-magnification views of the brain core.

In the cortical regions of the telencephalon, remnants of the germinal matrices are present in all lobes of the cerebral cortex where the ***neuroepithelium/subventricular zone*** may still be generating neocortical interneurons. Remnants of migrating and sojourning neurons and/or glia are visible in the dwindling ***stratified transitional fields*** in all lobes of the cerebral cortex, more prominent in parietal, temporal, and occipital lobes. Because of its plane of sectioning, this specimen shows the most complete view of the occipital lobes at GW 37.

Many neurons, glia, and their mitotic precursor cells are still migrating through the olfactory peduncle toward the olfactory bulb (***rostral migratory stream***) from a presumed source area in the germinal matrix at the junction between the cerebral cortex, striatum, and nucleus accumbens. Within the lateral parts of the cerebral cortex, streams of neurons and glia are still in the ***lateral migratory stream***. That stream percolates through the claustrum, endopiriform nucleus, external capsule, and uncinate fasciculus,

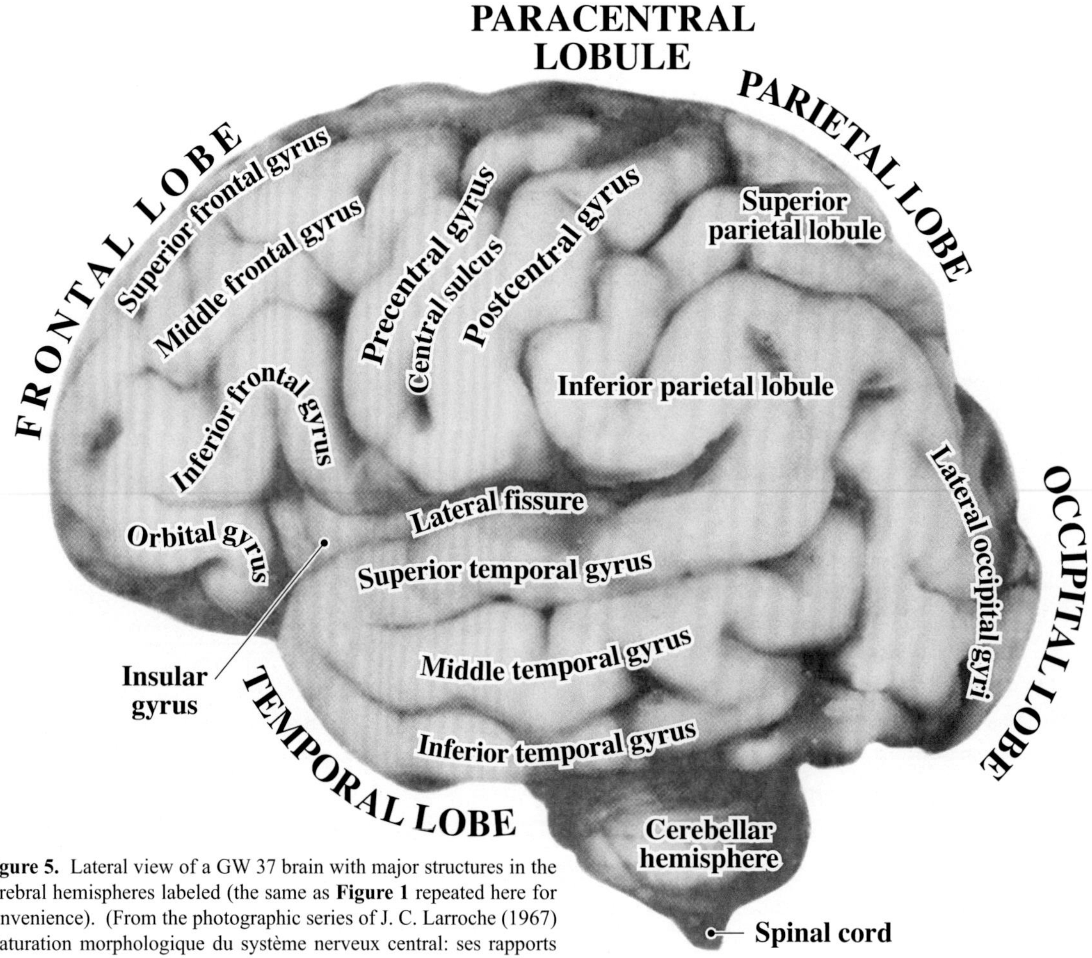

Figure 5. Lateral view of a GW 37 brain with major structures in the cerebral hemispheres labeled (the same as **Figure 1** repeated here for convenience). (From the photographic series of J. C. Larroche (1967) Maturation morphologique du système nerveux central: ses rapports avec le développement pondéral du foetus et son age gestationnel. In: *Regional Development of the Brain in Early Life*, A. Minkowski (ed.), London: Blackwell, page 248.)

GW 37 HORIZONTAL SECTION PLANES

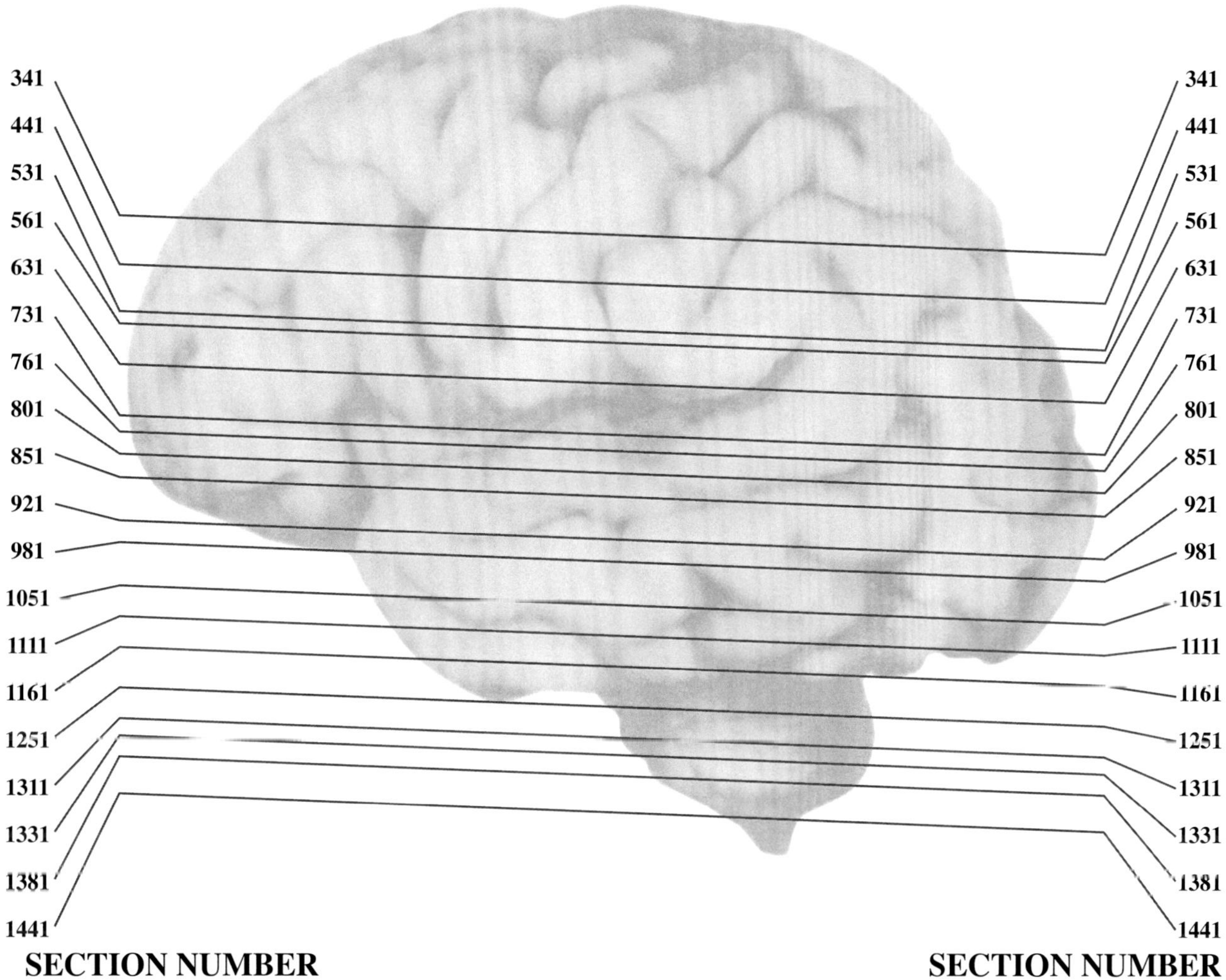

Figure 6. Lateral view of the same GW 37 brain shown in **Figure 4** with the approximate locations and cutting angle of the sections of Y301-62. (From the photographic series of J. C. Larroche (1967) Maturation morphologique du système nerveux central: ses rapports avec le développement pondéral du foetus et son age gestationnel. In: *Regional Development of the Brain in Early Life*, A. Minkowski (ed.), London: Blackwell, page 248.)

and cells appear to be heading toward the insular cortex, primary olfactory cortex, temporal cortex, and basolateral parts of the amygdaloid complex.

In the basal ganglia, there is a prominent ***neuroepithelium/subventricular zone*** overlying the striatum and nucleus accumbens where neurons are being generated. Another region of active neurogenesis in the telencephalon is the ***subgranular zone*** in the hilus of the dentate gyrus that is generating granule cells. Other structures in the telencephalon, such as the septum, fornix, and Ammon's horn have only a thin, darkly staining layer at the ventricle, and these are presumed to be generating glia, cells of the choroid plexus, and the ependymal lining of the ventricle.

Most of the structures in the diencephalon appear to be settled and are maturing, and the third ventricle is lined by a thin ***glioepithelium/ependyma***. In the midbrain and anterior pons, there is a slightly thicker and more convoluted ***glioepithelium/ependyma*** lining the posterior cerebral aqueduct and anterior fourth ventricle. The posterior pons and entire medulla have a thin ***glioepithelium/ependyma*** lining the rest of the fourth ventricle.

The ***external germinal layer*** is prominent over the entire surface of the cerebellar cortex and is still producing basket, stellate, and granule cells. The ***germinal trigone*** is still visible at the base of the nodulus and along the floccular peduncle; choroid plexus cells and glia may still be originating here.

PLATE 19A
CR 320 mm
GW 37
Y301-62
Horizontal
Section 341

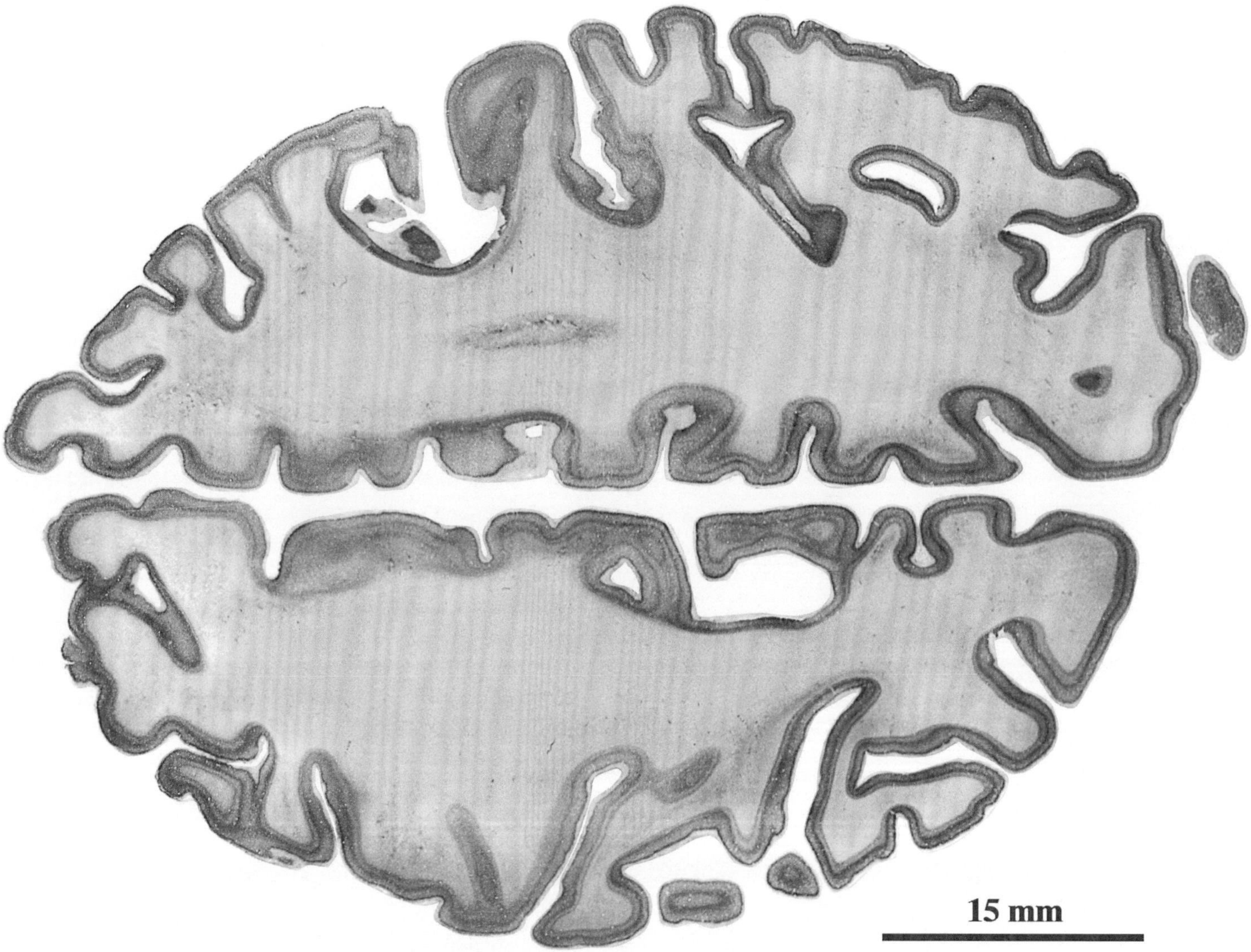

Remnants of the germinal matrix and transitional fields

1 ***Subpial granular layer***

PLATE 19B

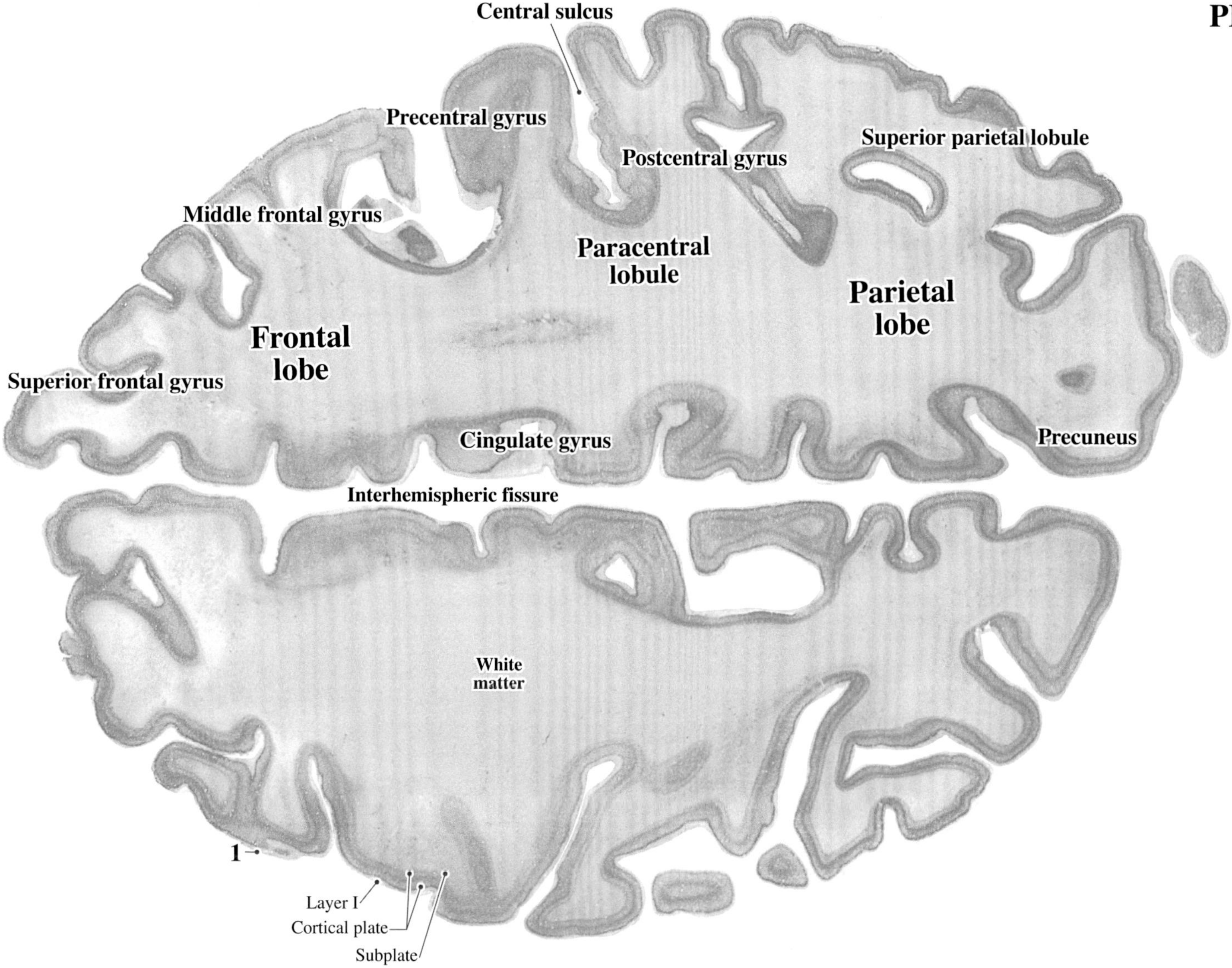
Central sulcus
Precentral gyrus
Postcentral gyrus
Superior parietal lobule
Middle frontal gyrus
Paracentral lobule
Parietal lobe
Frontal lobe
Superior frontal gyrus
Cingulate gyrus
Precuneus
Interhemispheric fissure
White matter
1
Layer I
Cortical plate
Subplate

PLATE 20A
CR 320 mm
GW 37
Y301-62
Horizontal
Section 441

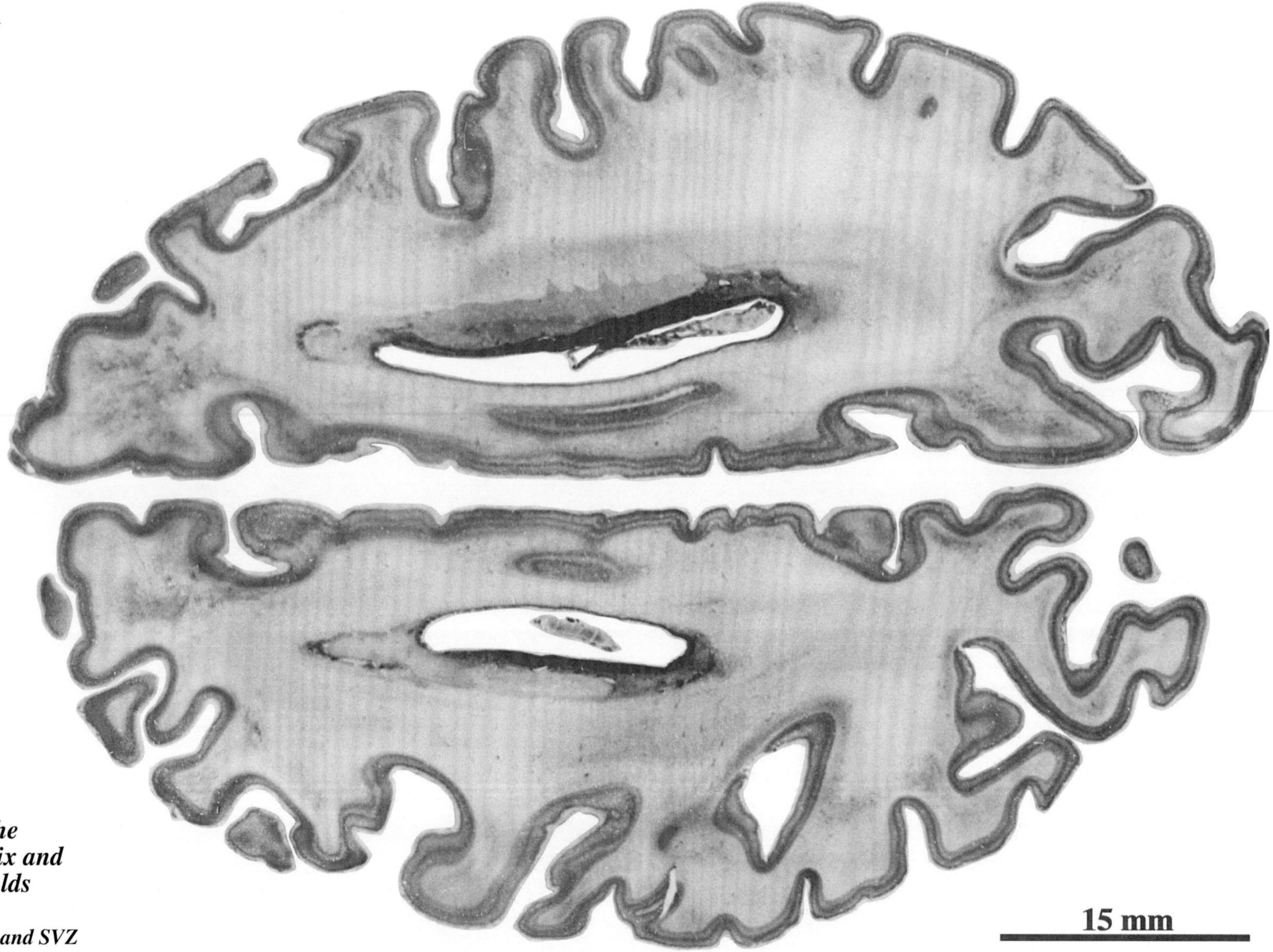

Remnants of the germinal matrix and transitional fields

1 *Frontal NEP and SVZ*
2 *Frontal STF*
3 *Callosal GEP*
4 *Parietal NEP and SVZ*
5 *Parietal STF*
6 *Paracentral STF*
7 *Anterolateral striatal NEP and SVZ*
8 *Posterior striatal NEP and SVZ*
9 *Strionuclear GEP*
10 *Subpial granular layer*

GEP - Glioepithelium
NEP - Neuroepithelium
STF - Stratified transitional field
SVZ - Subventricular zone

PLATE 20B

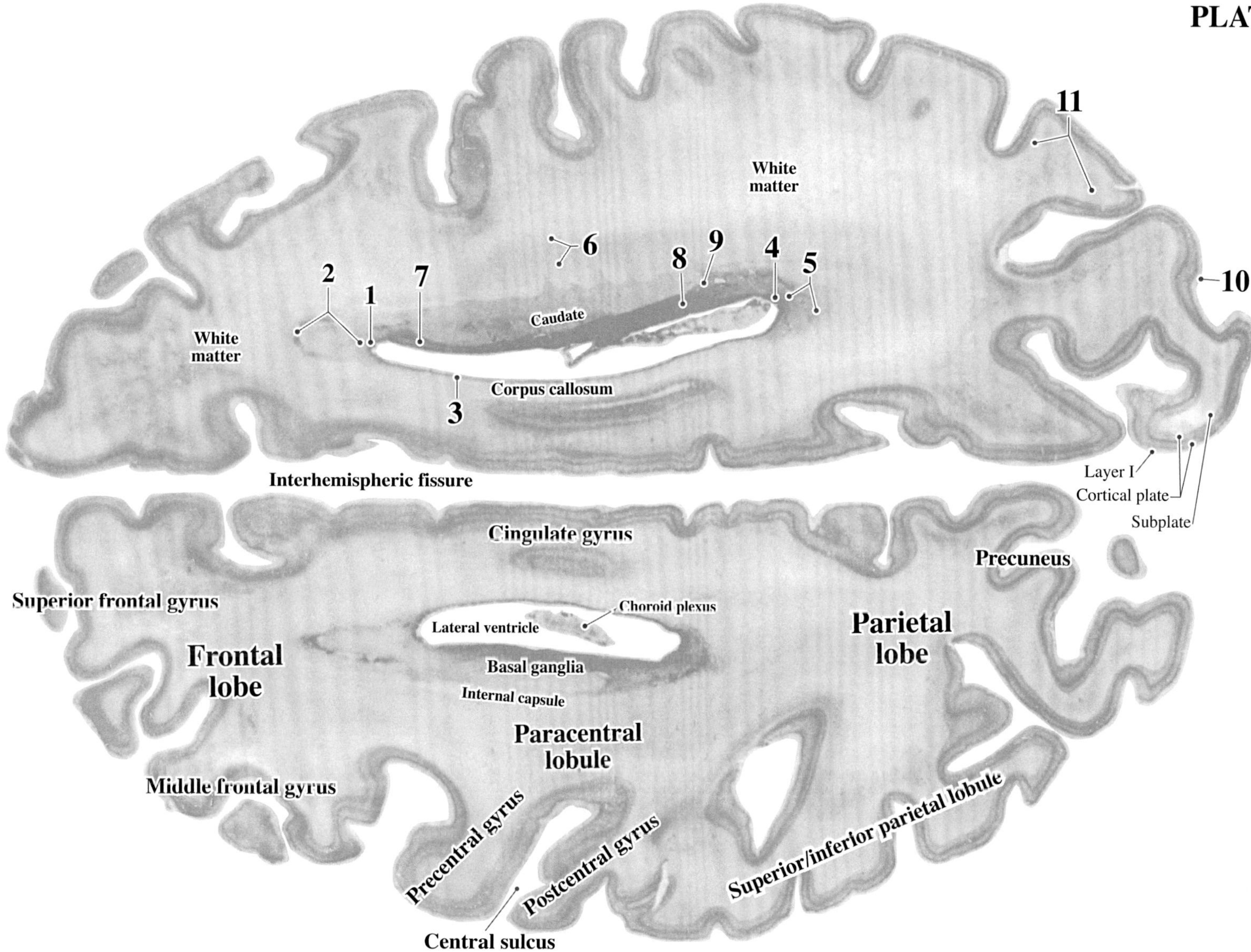
11
White matter
6
9
4
5
8
2
1
7
10
Caudate
White matter
Corpus callosum
3
Layer I
Cortical plate
Subplate
Interhemispheric fissure
Cingulate gyrus
Precuneus
Superior frontal gyrus
Choroid plexus
Lateral ventricle
Parietal lobe
Frontal lobe
Basal ganglia
Internal capsule
Paracentral lobule
Middle frontal gyrus
Precentral gyrus
Postcentral gyrus
Superior/inferior parietal lobule
Central sulcus

PLATE 21A
CR 320 mm
GW 37
Y301-62
Horizontal
Section 531

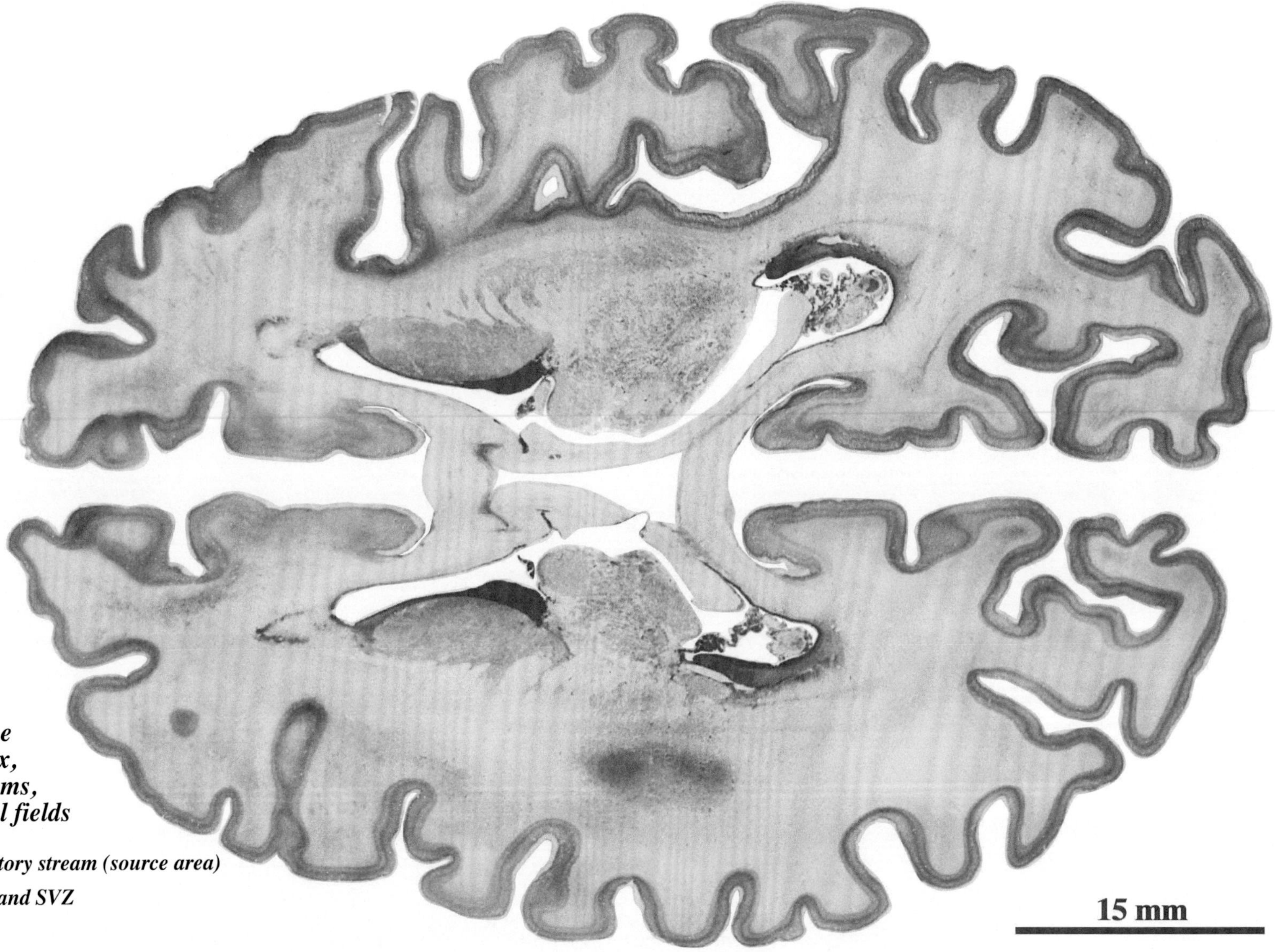

Remnants of the germinal matrix, migratory streams, and transitional fields

1 *Rostral migratory stream (source area)*
2 *Frontal NEP and SVZ*
3 *Frontal STF*
4 *Callosal GEP*
5 *Callosal sling*
6 *Fornical GEP*
7 *Parietal NEP and SVZ*
8 *Parietal STF*
9 *Posterior striatal NvEP and SVZ*
10 *Anterolateral striatal NEP and SVZ*
11 *Anteromedial striatal NEP and SVZ*
12 *Strionuclear GEP*

GEP - Glioepithelium
NEP - Neuroepithelium
STF - Stratified transitional field
SVZ - Subventricular zone

See detail of brain core in Plates 38A and B.

PLATE 21B

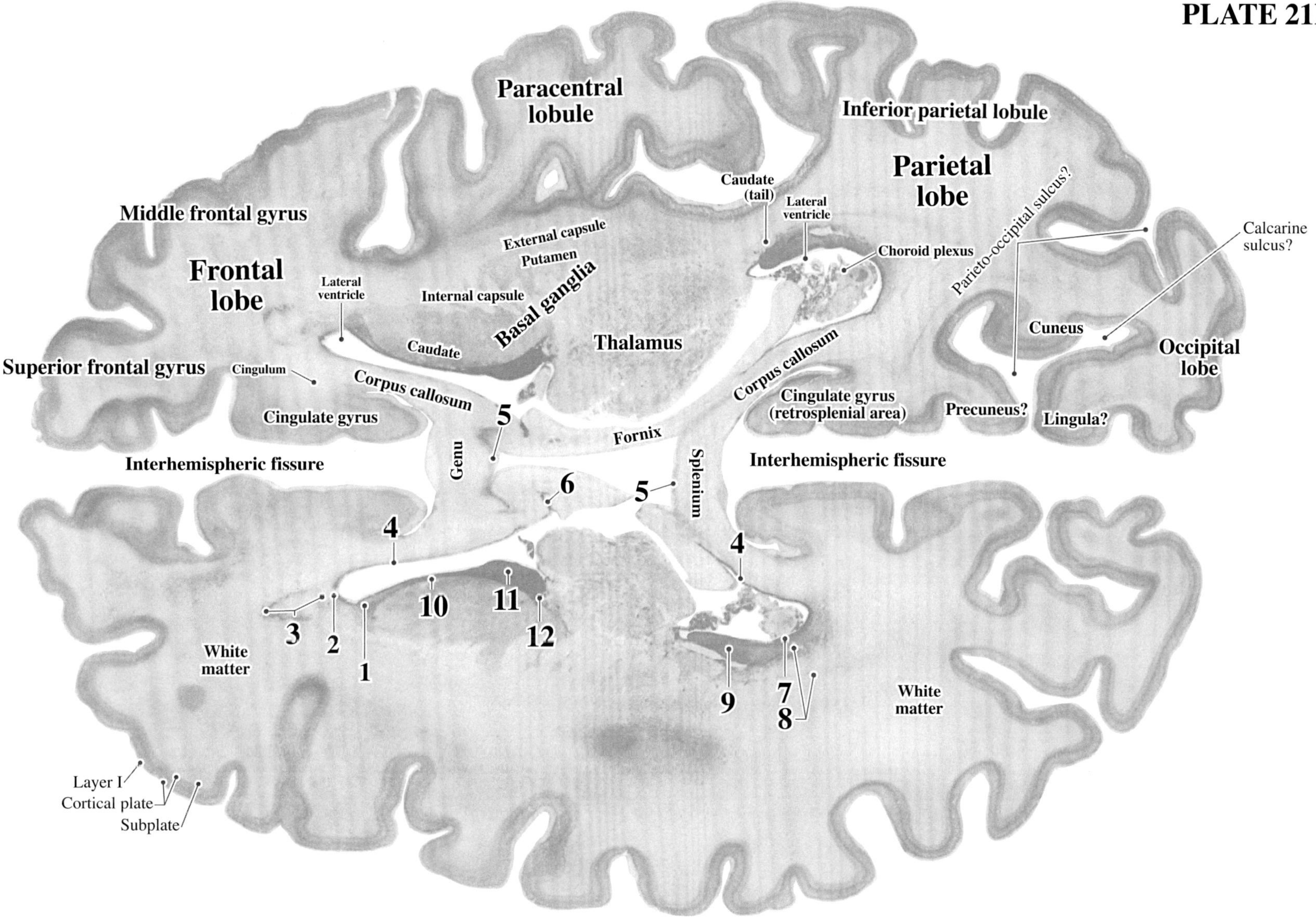
Paracentral lobule
Inferior parietal lobule
Parietal lobe
Parieto-occipital sulcus?
Caudate (tail)
Lateral ventricle
Choroid plexus
Calcarine sulcus?
Middle frontal gyrus
External capsule
Putamen
Frontal lobe
Lateral ventricle
Internal capsule
Basal ganglia
Thalamus
Caudate
Cuneus
Occipital lobe
Superior frontal gyrus
Cingulum
Corpus callosum
Corpus callosum
Cingulate gyrus
Cingulate gyrus (retrosplenial area)
Precuneus?
Lingula?
5
Fornix
Genu
Splenium
Interhemispheric fissure
Interhemispheric fissure
6
5
4
4
3
2
1
10
11
12
9
7
8
White matter
White matter
Layer I
Cortical plate
Subplate

PLATE 22A
CR 320 mm
GW 37
Y301-62
Horizontal
Section 561

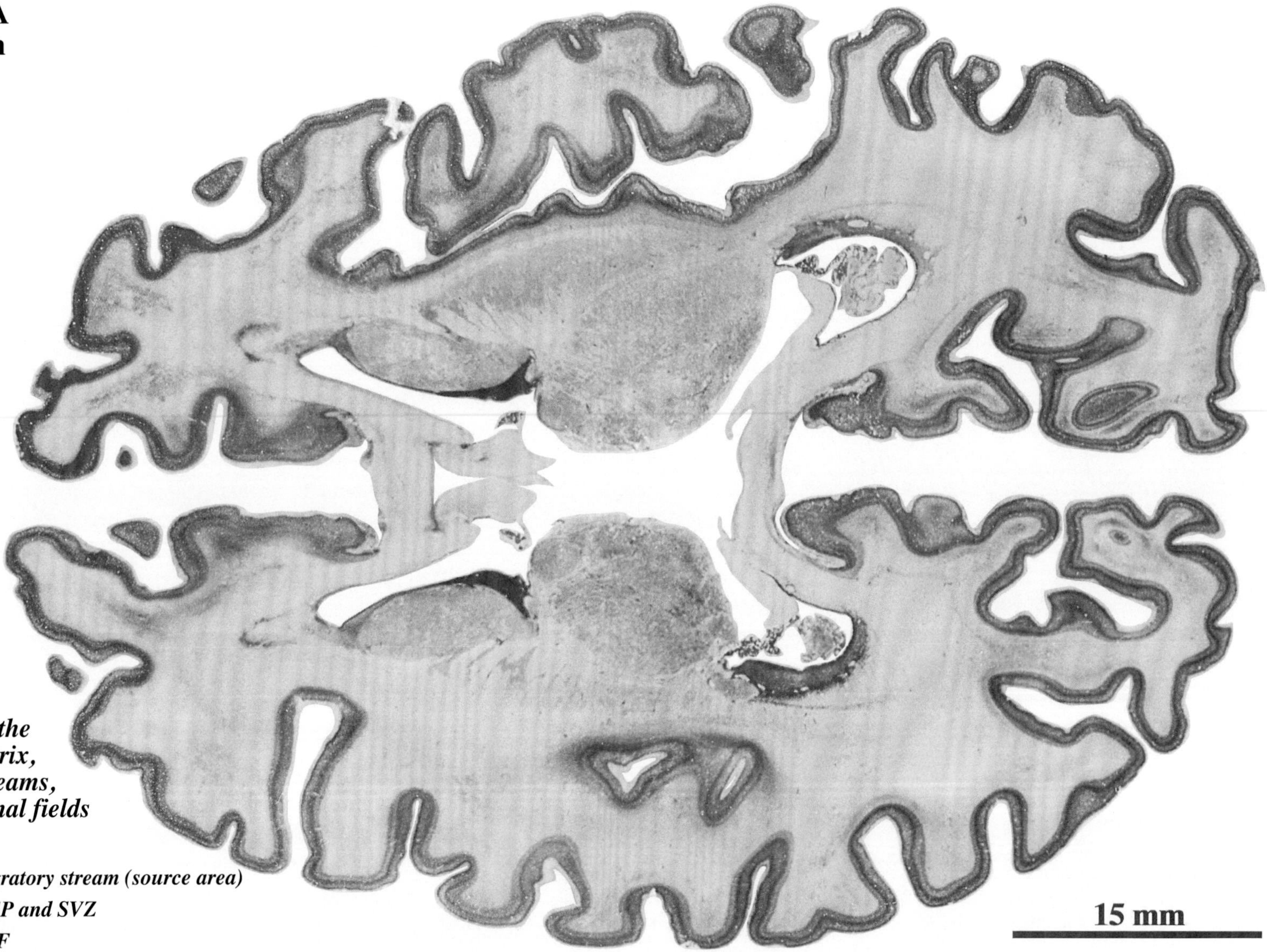

Remnants of the germinal matrix, migratory streams, and transitional fields

1 *Rostral migratory stream (source area)*
2 *Frontal NEP and SVZ*
3 *Frontal STF*
4 *Callosal GEP*
5 *Callosal sling*
6 *Fornical GEP*
7 *Parietal/occipital NEP and SVZ*
8 *Parietal/occipital STF*
9 *Posterior striatal NEP and SVZ*
10 *Anterolateral striatal NEP and SVZ*
11 *Anteromedial striatal NEP and SVZ*
12 *Strionuclear GEP*

GEP - Glioepithelium
NEP - Neuroepithelium
STF - Stratified transitional field
SVZ - Subventricular zone

See detail of brain core in Plates 39A and B.

PLATE 22B

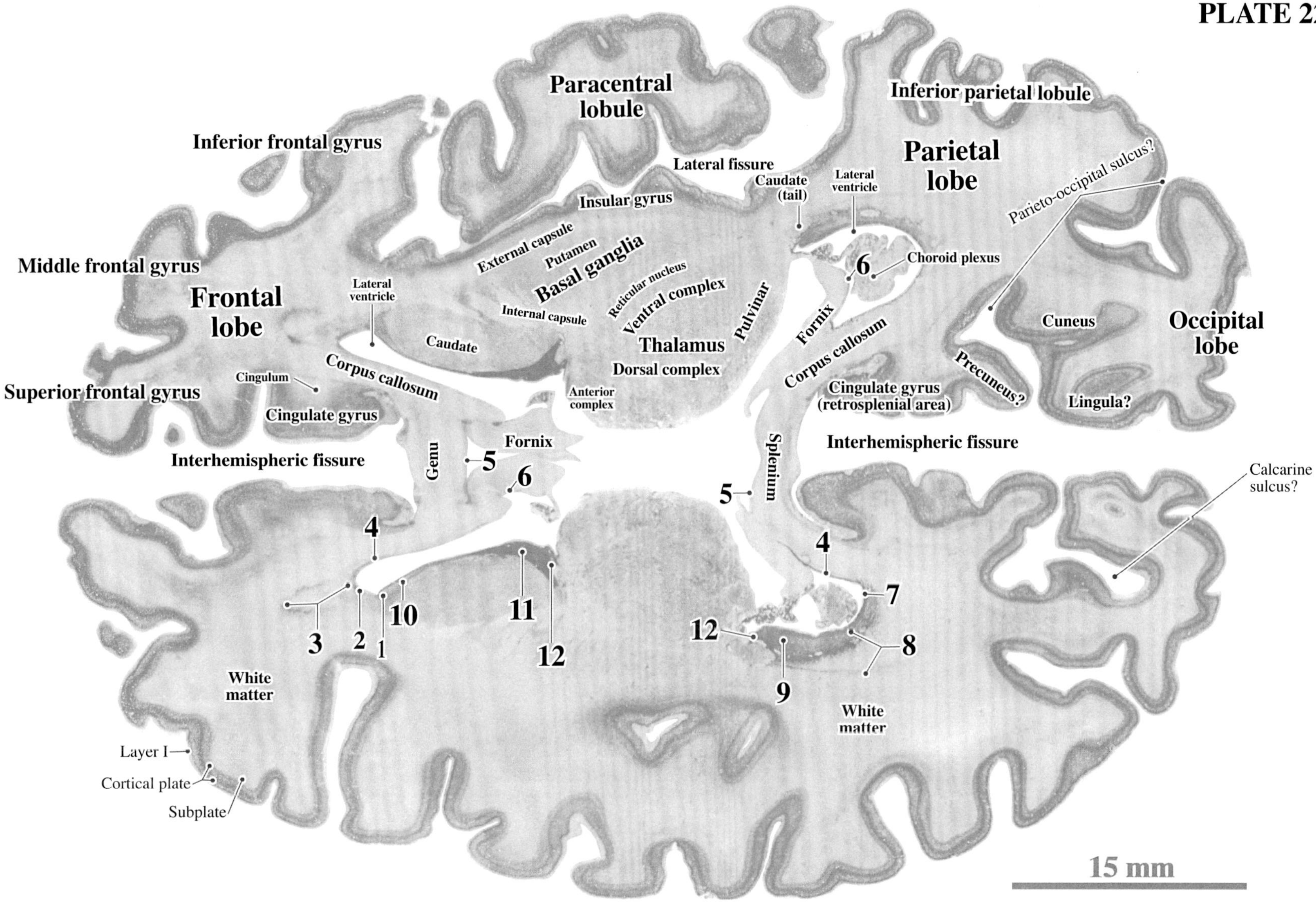

Paracentral lobule
Inferior parietal lobule
Inferior frontal gyrus
Lateral fissure
Parietal lobe
Caudate (tail)
Lateral ventricle
Insular gyrus
Parieto-occipital sulcus?
External capsule
Putamen
Basal ganglia
Choroid plexus
Middle frontal gyrus
Frontal lobe
Lateral ventricle
Reticular nucleus
Ventral complex
Internal capsule
Pulvinar
Fornix
Cuneus
Occipital lobe
Caudate
Thalamus
Corpus callosum
Dorsal complex
Corpus callosum
Precuneus?
Cingulum
Superior frontal gyrus
Cingulate gyrus
Anterior complex
Cingulate gyrus (retrosplenial area)
Lingula?
Fornix
Interhemispheric fissure
Interhemispheric fissure
Genu
Splenium
Calcarine sulcus?
White matter
White matter
Layer I
Cortical plate
Subplate
1
2
3
4
5
6
7
8
9
10
11
12
15 mm

PLATE 23A
CR 320 mm
GW 37
Y301-62
Horizontal
Section 631

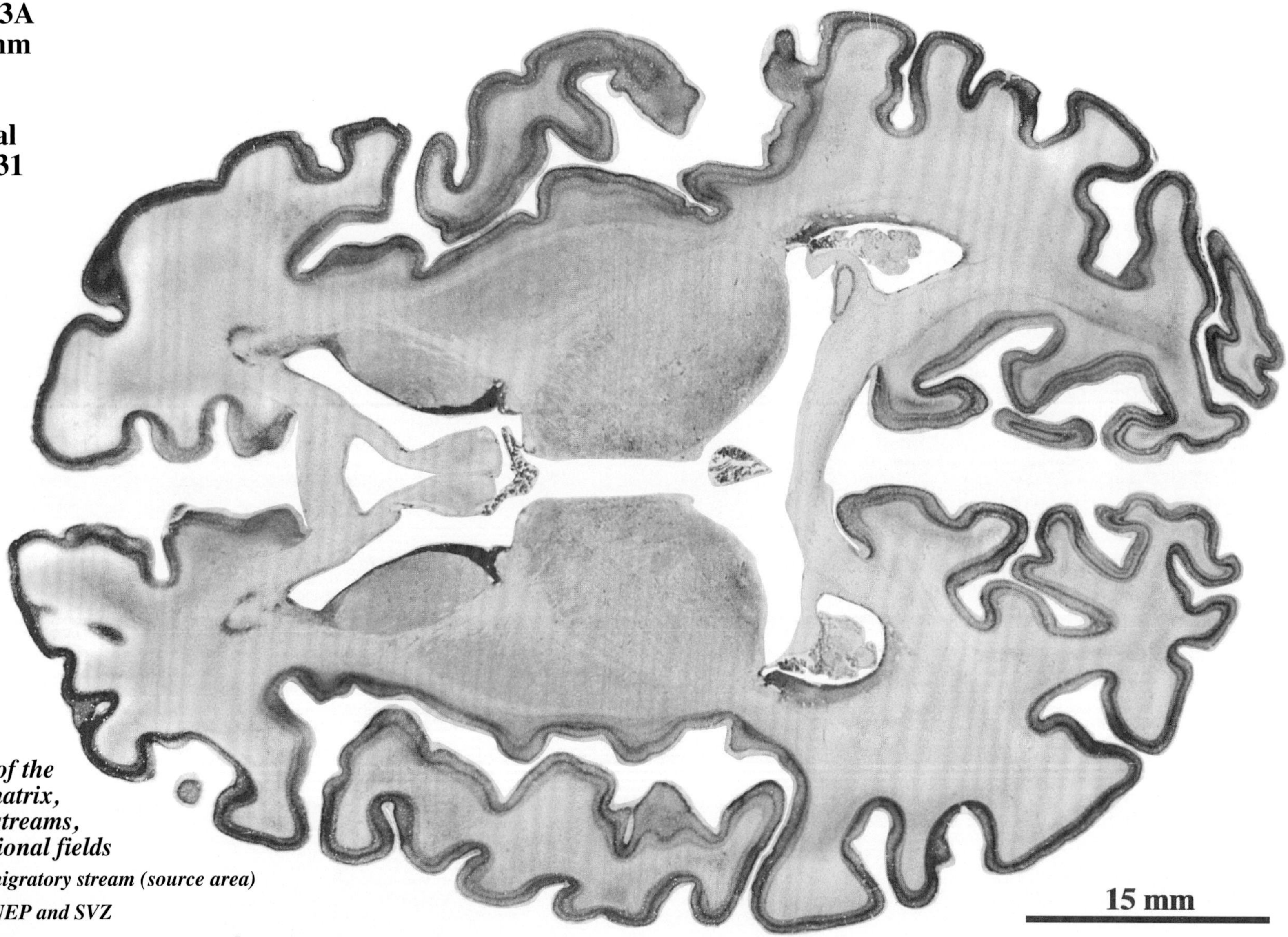

Remnants of the germinal matrix, migratory streams, and transitional fields

1 *Rostral migratory stream (source area)*
2 *Frontal NEP and SVZ*
3 *Frontal STF*
4 *Callosal GEP*
5 *Callosal sling*
6 *Fornical GEP*
7 *Parietal/occipital NEP and SVZ*
8 *Parietal/occipital STF*
9 *Posterior striatal NEP and SVZ*
10 *Anterolateral striatal NEP and SVZ*
11 *Anteromedial striatal NEP and SVZ*
12 *Strionuclear GEP*

GEP - Glioepithelium
NEP - Neuroepithelium
STF - Stratified transitional field
SVZ - Subventricular zone

See detail of brain core in Plates 40A and B.

PLATE 23B

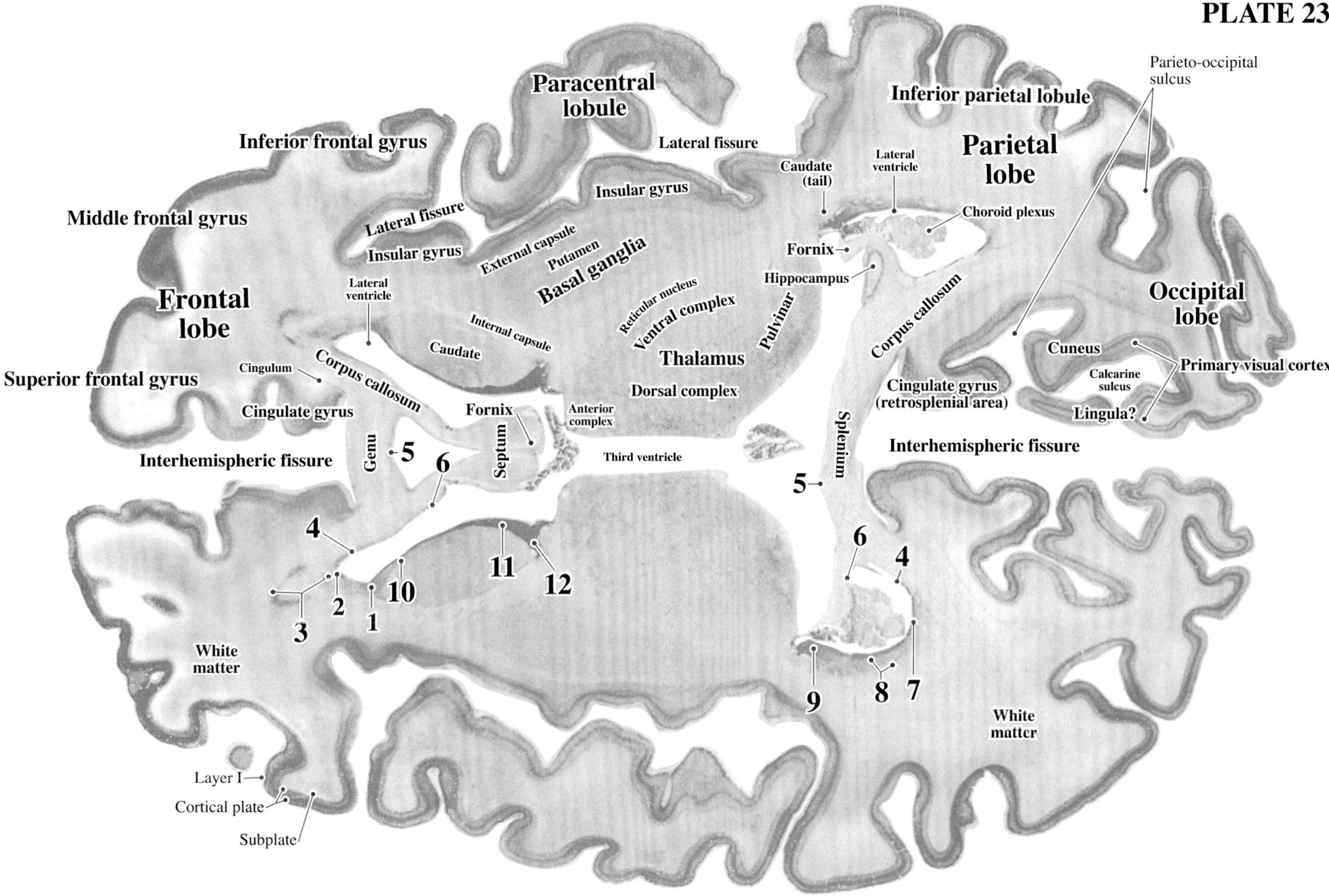
Parieto-occipital sulcus
Paracentral lobule
Inferior parietal lobule
Inferior frontal gyrus
Lateral fissure
Parietal lobe
Caudate (tail)
Lateral ventricle
Insular gyrus
Middle frontal gyrus
Lateral fissure
Choroid plexus
Insular gyrus
External capsule
Putamen
Fornix
Basal ganglia
Hippocampus
Frontal lobe
Lateral ventricle
Reticular nucleus
Ventral complex
Pulvinar
Corpus callosum
Occipital lobe
Internal capsule
Caudate
Thalamus
Cuneus
Cingulum
Corpus callosum
Primary visual cortex
Superior frontal gyrus
Calcarine sulcus
Cingulate gyrus (retrosplenial area)
Dorsal complex
Cingulate gyrus
Fornix
Anterior complex
Lingula?
Splenium
Interhemispheric fissure
Genu
5
6
Septum
Third ventricle
Interhemispheric fissure
5
4
6
4
11
12
10
2
1
3
White matter
9
8
7
White matter
Layer I
Cortical plate
Subplate

PLATE 24A
CR 320 mm
GW 37
Y301-62
Horizontal
Section 731

See detail of brain core in Plates 41A and B.

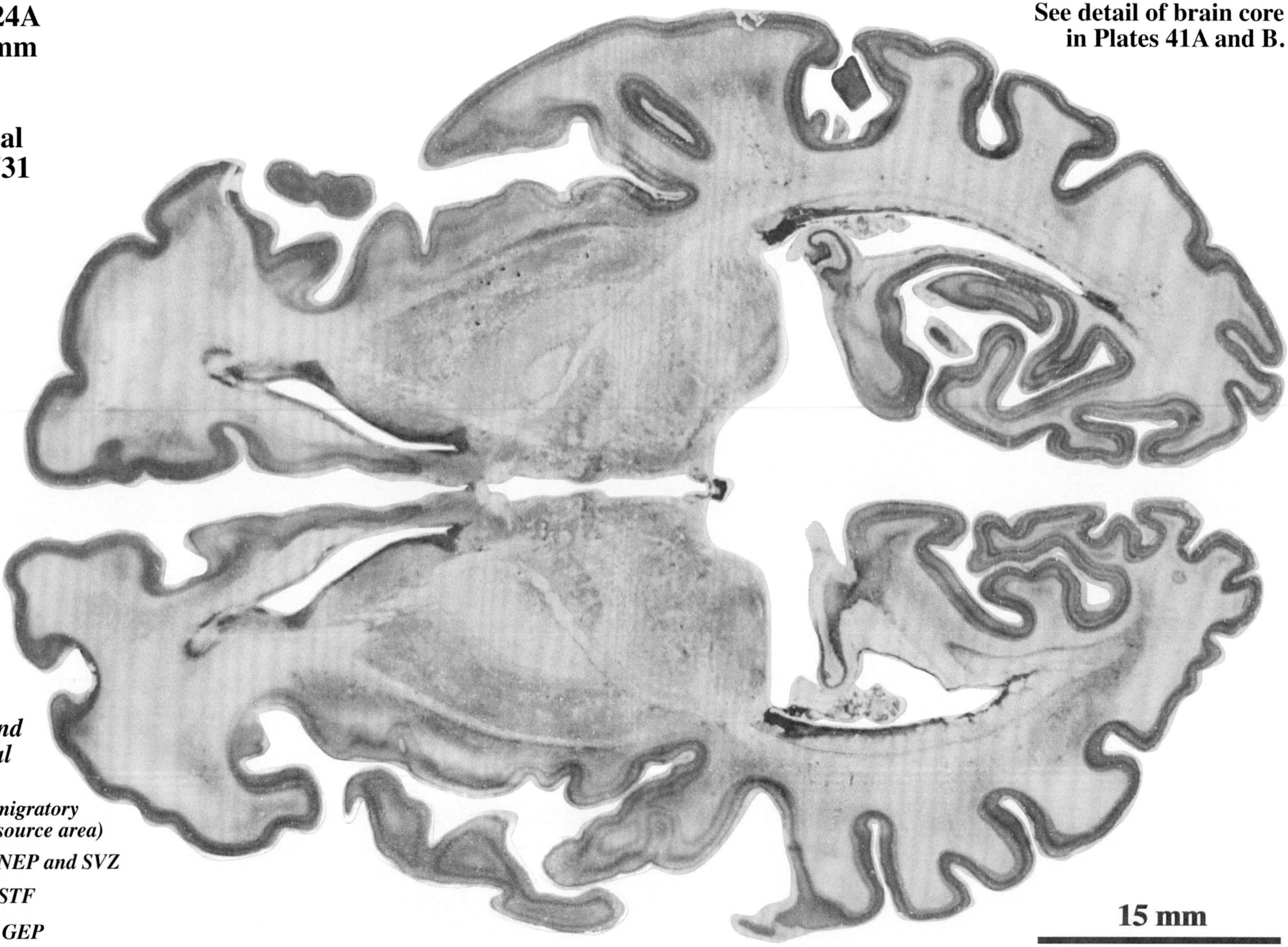

Remnants of the germinal matrix, migratory streams, and transitional fields

1 ***Rostral migratory stream (source area)***
2 ***Frontal NEP and SVZ***
3 ***Frontal STF***
4 ***Callosal GEP***
5 ***Fornical GEP***
6 ***Occipital NEP and SVZ***
7 ***Occipital STF***
8 ***Parietal/temporal NEP and SVZ***
9 ***Parietal/temporal STF***
10 ***Alvear GEP***
11 ***Posterior striatal NEP and SVZ***
12 ***Anterolateral striatal NEP and SVZ***
13 ***Anteromedial striatal NEP and SVZ***
14 ***Strionuclear GEP***
15 ***Septal G/EP***

15 mm

GEP - Glioepithelium
G/EP - Glioepithelium/ependyma
NEP - Neuroepithelium
STF - Stratified transitional field
SVZ - Subventricular zone

PLATE 24B

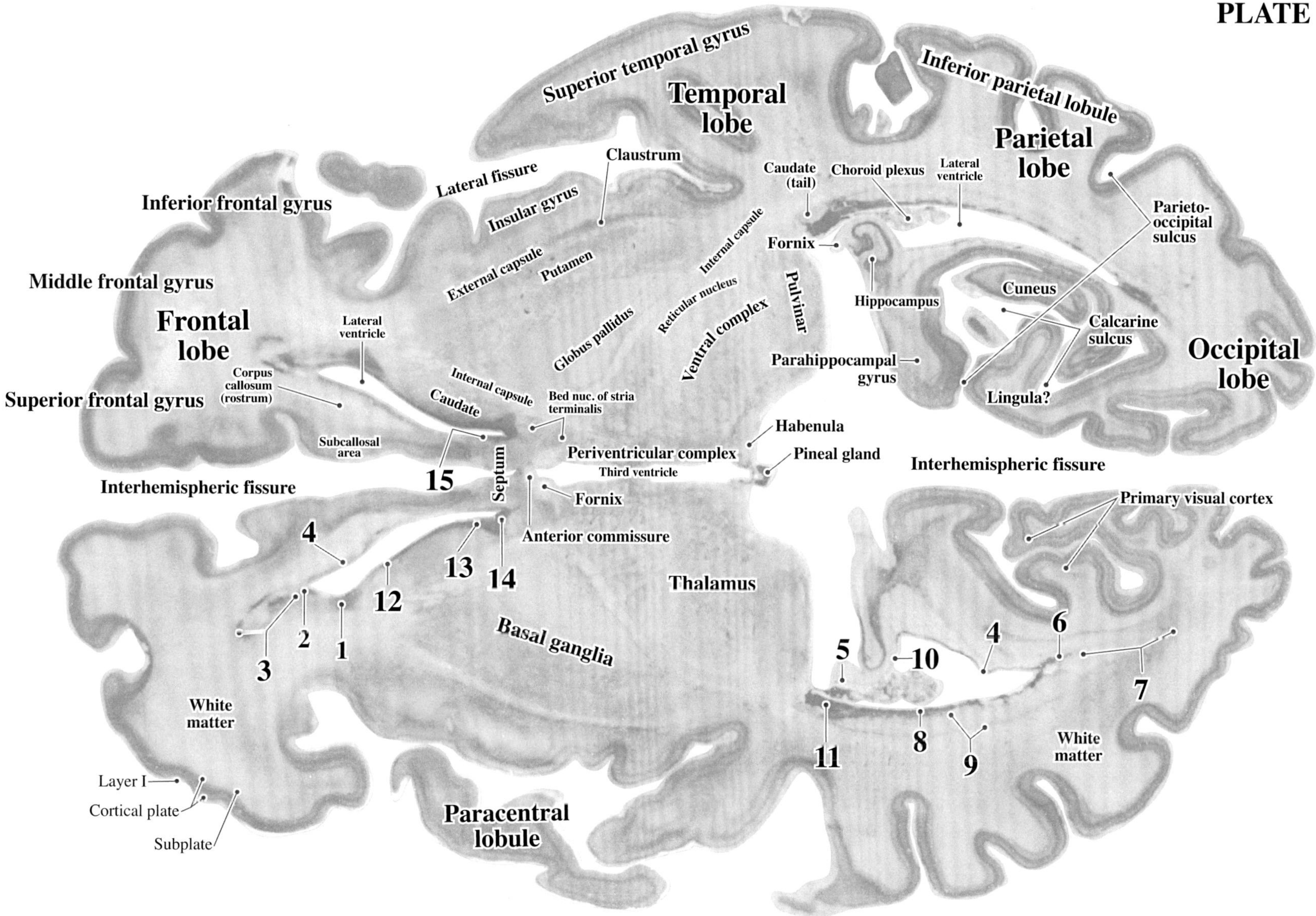

Superior temporal gyrus
Temporal lobe
Inferior parietal lobule
Parietal lobe
Claustrum
Lateral fissure
Insular gyrus
Inferior frontal gyrus
Caudate (tail)
Choroid plexus
Lateral ventricle
Parieto-occipital sulcus
Fornix
Internal capsule
External capsule
Putamen
Middle frontal gyrus
Reticular nucleus
Pulvinar
Hippocampus
Cuneus
Frontal lobe
Lateral ventricle
Ventral complex
Globus pallidus
Calcarine sulcus
Occipital lobe
Parahippocampal gyrus
Corpus callosum (rostrum)
Internal capsule
Superior frontal gyrus
Caudate
Bed nuc. of stria terminalis
Lingula?
Habenula
Subcallosal area
Periventricular complex
Pineal gland
Septum
Third ventricle
Interhemispheric fissure
15
Interhemispheric fissure
Fornix
Primary visual cortex
4
Anterior commissure
13
14
12
Thalamus
2
1
3
Basal ganglia
6
4
5
10
7
White matter
11
8
9
White matter
Layer I
Cortical plate
Subplate
Paracentral lobule

PLATE 25A
CR 320 mm
GW 37
Y301-62
Horizontal
Section 761

See detail of brain core in Plates 42A and B.

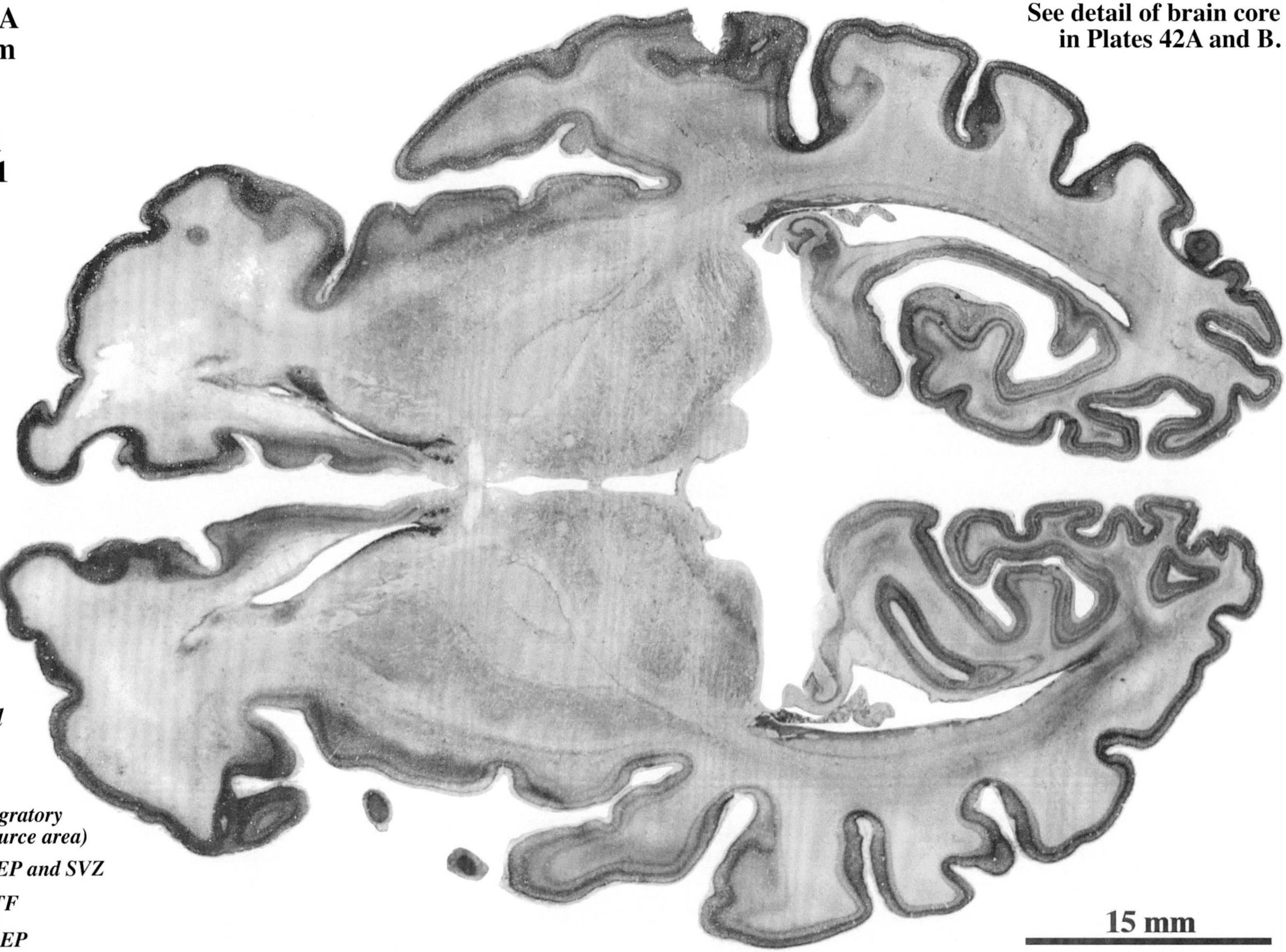

Remnants of the germinal matrix, migratory streams, and transitional fields

1 *Rostral migratory stream (source area)*
2 *Frontal NEP and SVZ*
3 *Frontal STF*
4 *Callosal GEP*
5 *Fornical GEP*
6 *Occipital NEP and SVZ*
7 *Occipital STF*
8 *Parietal/temporal NEP and SVZ*
9 *Parietal/temporal STF*
10 *Alvear GEP*
11 *Subgranular zone*
12 *Posterior striatal NEP and SVZ*
13 *Anterolateral striatal NEP and SVZ*
14 *Anteromedial striatal NEP and SVZ*
15 *Strionuclear GEP*

15 mm

GEP - Glioepithelium
NEP - Neuroepithelium
STF - Stratified transitional field
SVZ - Subventricular zone

PLATE 25B

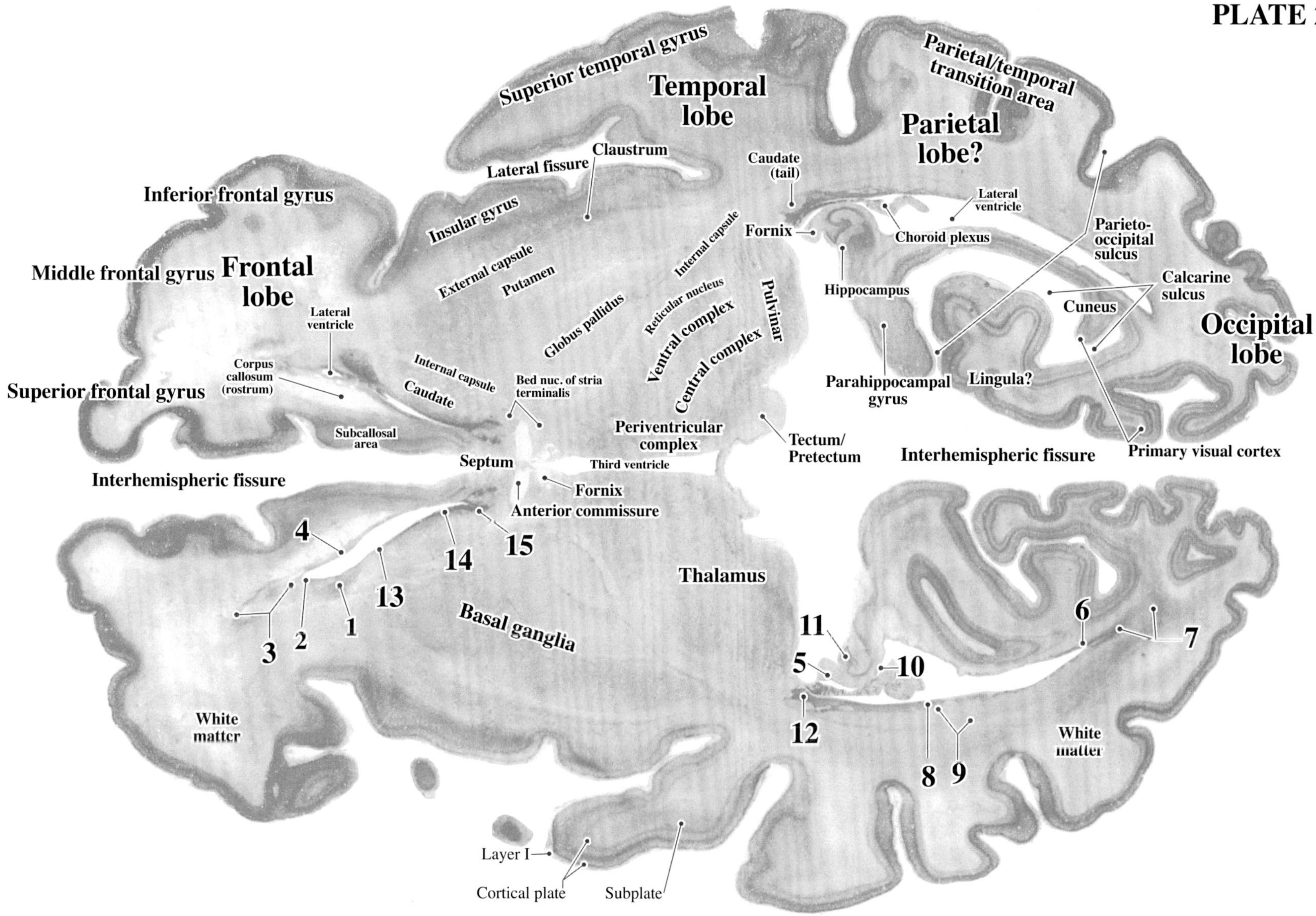
Superior temporal gyrus
Temporal lobe
Parietal/temporal transition area
Parietal lobe?
Claustrum
Lateral fissure
Caudate (tail)
Inferior frontal gyrus
Insular gyrus
Lateral ventricle
Fornix
Choroid plexus
Parieto-occipital sulcus
Internal capsule
Middle frontal gyrus
Frontal lobe
External capsule
Putamen
Reticular nucleus
Pulvinar
Hippocampus
Calcarine sulcus
Cuneus
Lateral ventricle
Globus pallidus
Ventral complex
Central complex
Occipital lobe
Corpus callosum (rostrum)
Internal capsule
Bed nuc. of stria terminalis
Parahippocampal gyrus
Lingula?
Superior frontal gyrus
Caudate
Subcallosal area
Periventricular complex
Tectum/ Pretectum
Interhemispheric fissure
Primary visual cortex
Septum
Third ventricle
Interhemispheric fissure
Fornix
Anterior commissure
4
15
14
13
Thalamus
3
2
1
Basal ganglia
11
6
7
5
10
12
8
9
White matter
White matter
Layer I
Cortical plate
Subplate

PLATE 26A
CR 320 mm
GW 37
Y301-62
Horizontal
Section 801

See detail of brain core in Plates 43A and B.

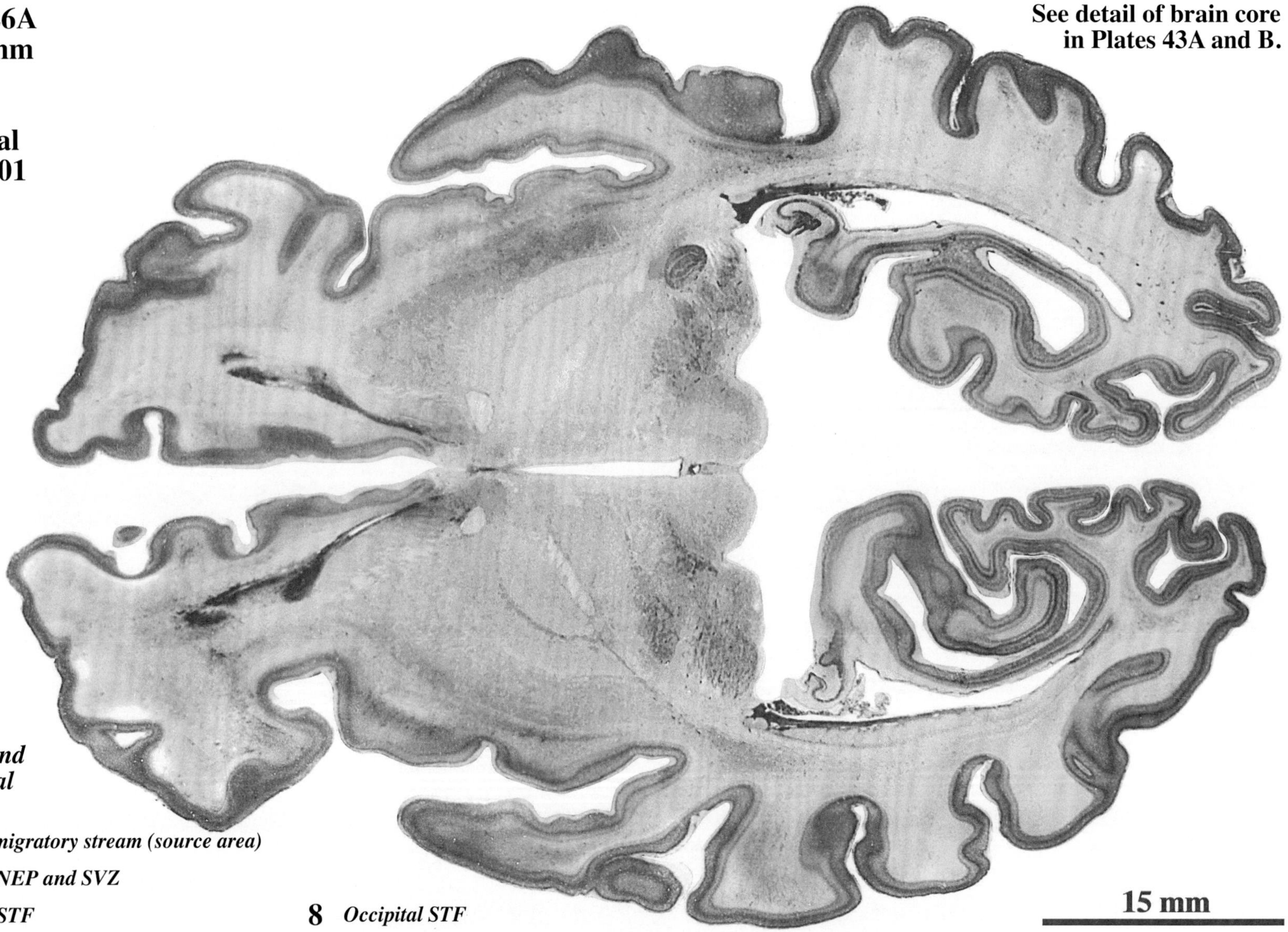

Remnants of the germinal matrix, migratory streams, and transitional fields

1 *Rostral migratory stream (source area)*
2 *Frontal NEP and SVZ*
3 *Frontal STF*
4 *Callosal GEP*
5 *Fornical GEP*
6 *Parahippocampal NEP, SVZ, and STF*
7 *Occipital NEP and SVZ*
8 *Occipital STF*
9 *Temporal NEP and SVZ*
10 *Temporal STF*
11 *Alvear GEP*
12 *Subgranular zone (dentate)*
13 *Lateral migratory stream (cortical)*
14 *Posterior striatal NEP and SVZ*
15 *Accumbent NEP and SVZ (infiltrated by the rostral migratory stream)*

15 mm

GEP - Glioepithelium
NEP - Neuroepithelium
STF - Stratified transitional field
SVZ - Subventricular zone

PLATE 26B

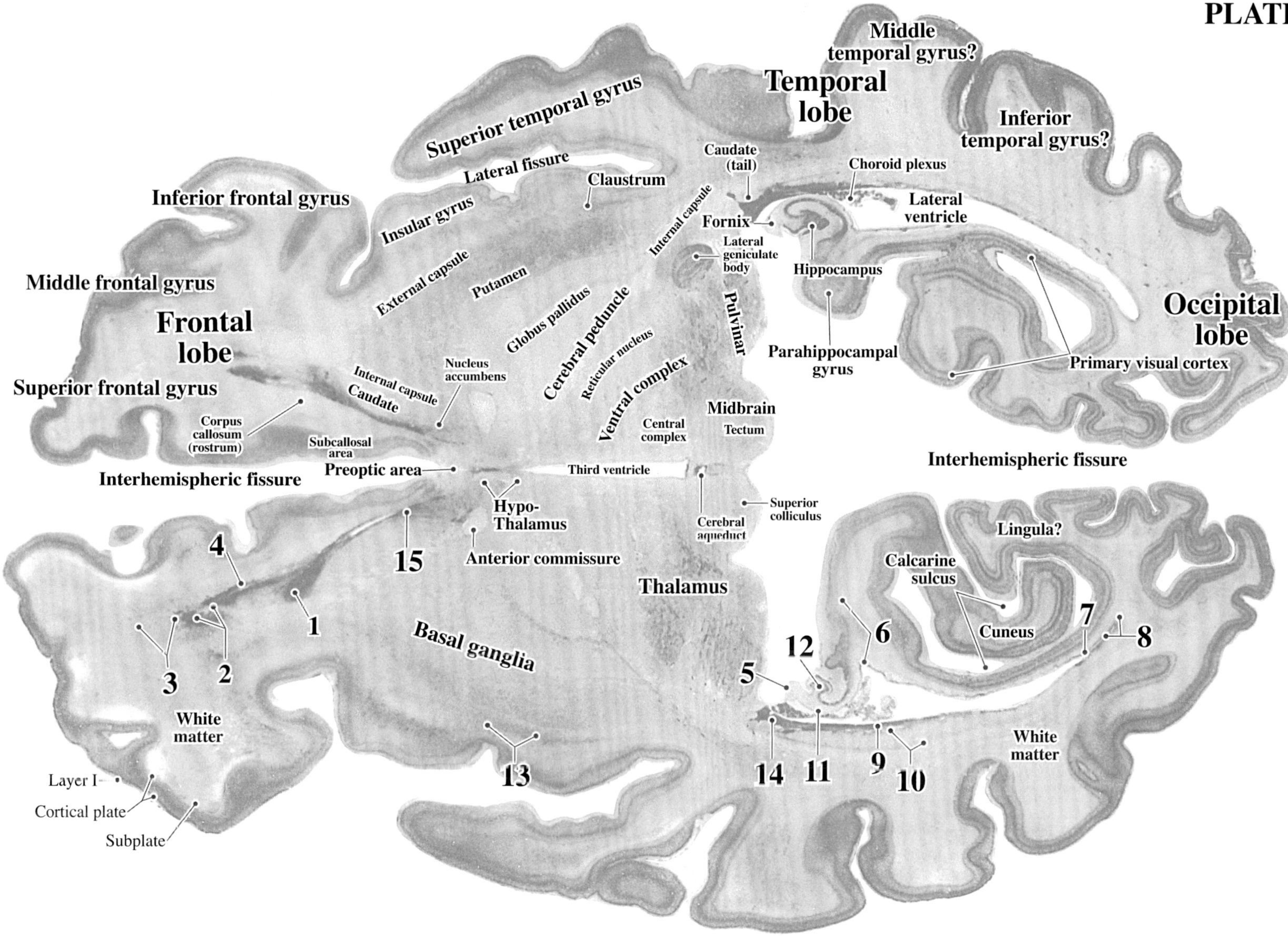
Middle temporal gyrus?
Temporal lobe
Superior temporal gyrus
Inferior temporal gyrus?
Caudate (tail)
Choroid plexus
Lateral fissure
Claustrum
Inferior frontal gyrus
Insular gyrus
Internal capsule
Fornix
Lateral ventricle
Lateral geniculate body
Hippocampus
External capsule
Putamen
Middle frontal gyrus
Globus pallidus
Cerebral peduncle
Pulvinar
Parahippocampal gyrus
Occipital lobe
Frontal lobe
Reticular nucleus
Nucleus accumbens
Internal capsule
Primary visual cortex
Superior frontal gyrus
Caudate
Ventral complex
Midbrain
Corpus callosum (rostrum)
Central complex
Tectum
Subcallosal area
Interhemispheric fissure
Preoptic area
Third ventricle
Interhemispheric fissure
Hypo-Thalamus
Superior colliculus
Cerebral aqueduct
Lingula?
4
15
Anterior commissure
Calcarine sulcus
Thalamus
1
Cuneus
7
8
6
Basal ganglia
12
3
2
5
White matter
14
11
9
10
13
White matter
Layer I
Cortical plate
Subplate

PLATE 27A
CR 320 mm
GW 37
Y301-62
Horizontal
Section 851

See detail of brain core in Plates 44A and B.

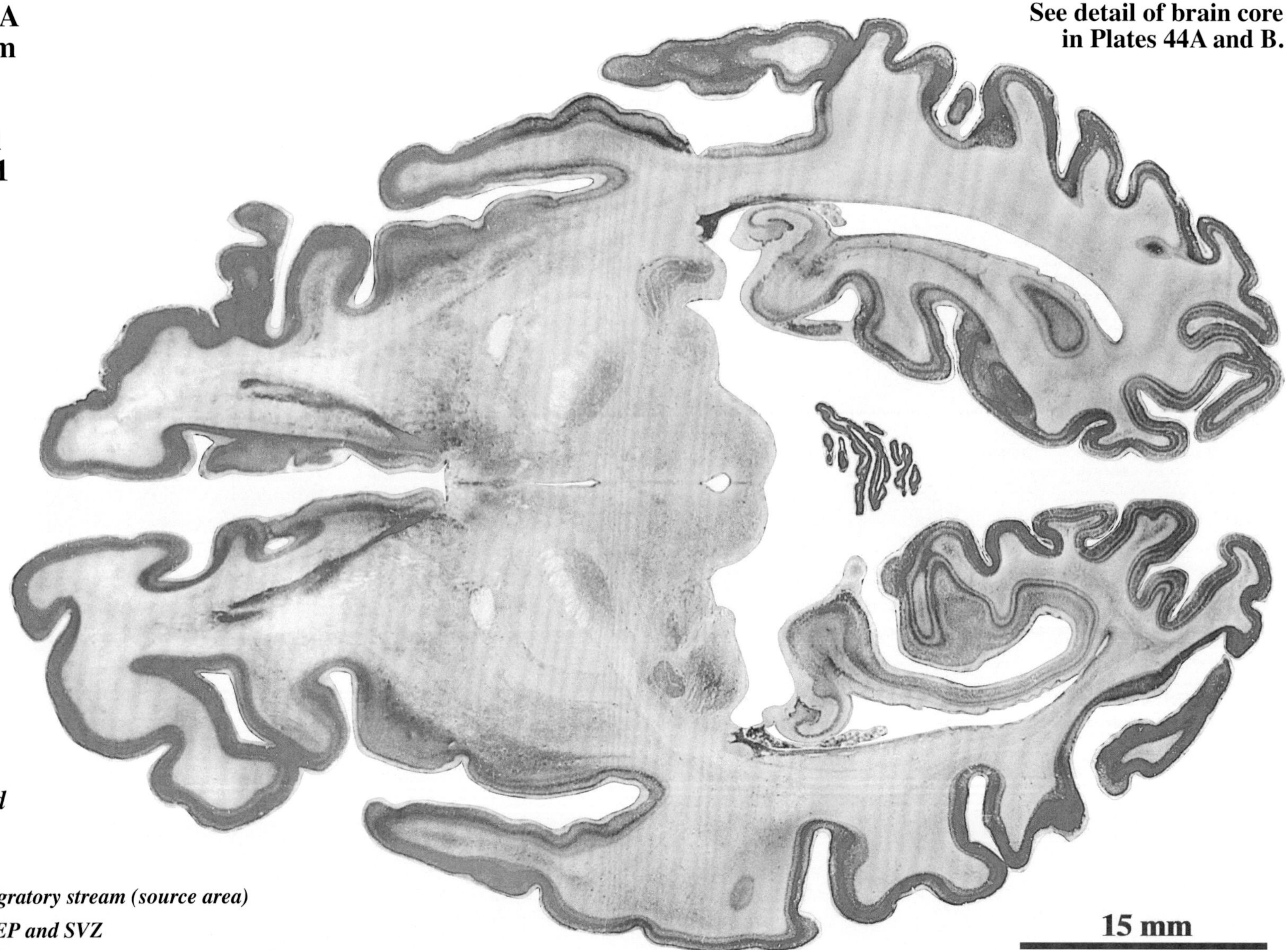

Remnants of the germinal matrix, migratory streams, and transitional fields

1 *Rostral migratory stream (source area)*
2 *Frontal NEP and SVZ*
3 *Frontal STF*
4 *Fornical GEP*
5 *Parahippocampal NEP, SVZ, and STF*
6 *Occipital NEP and SVZ*
7 *Occipital STF*
8 *Temporal NEP and SVZ*
9 *Temporal STF*
10 *Alvear GEP*
11 *Subgranular zone (dentate)*
12 *Lateral migratory stream (cortical)*
13 *Posterior striatal NEP and SVZ*
14 *Accumbent NEP and SVZ (infiltrated by the rostral migratory stream)*
15 *External germinal layer (cerebellum)*

GEP - Glioepithelium
NEP - Neuroepithelium
STF - Stratified transitional field
SVZ - Subventricular zone

PLATE 27B

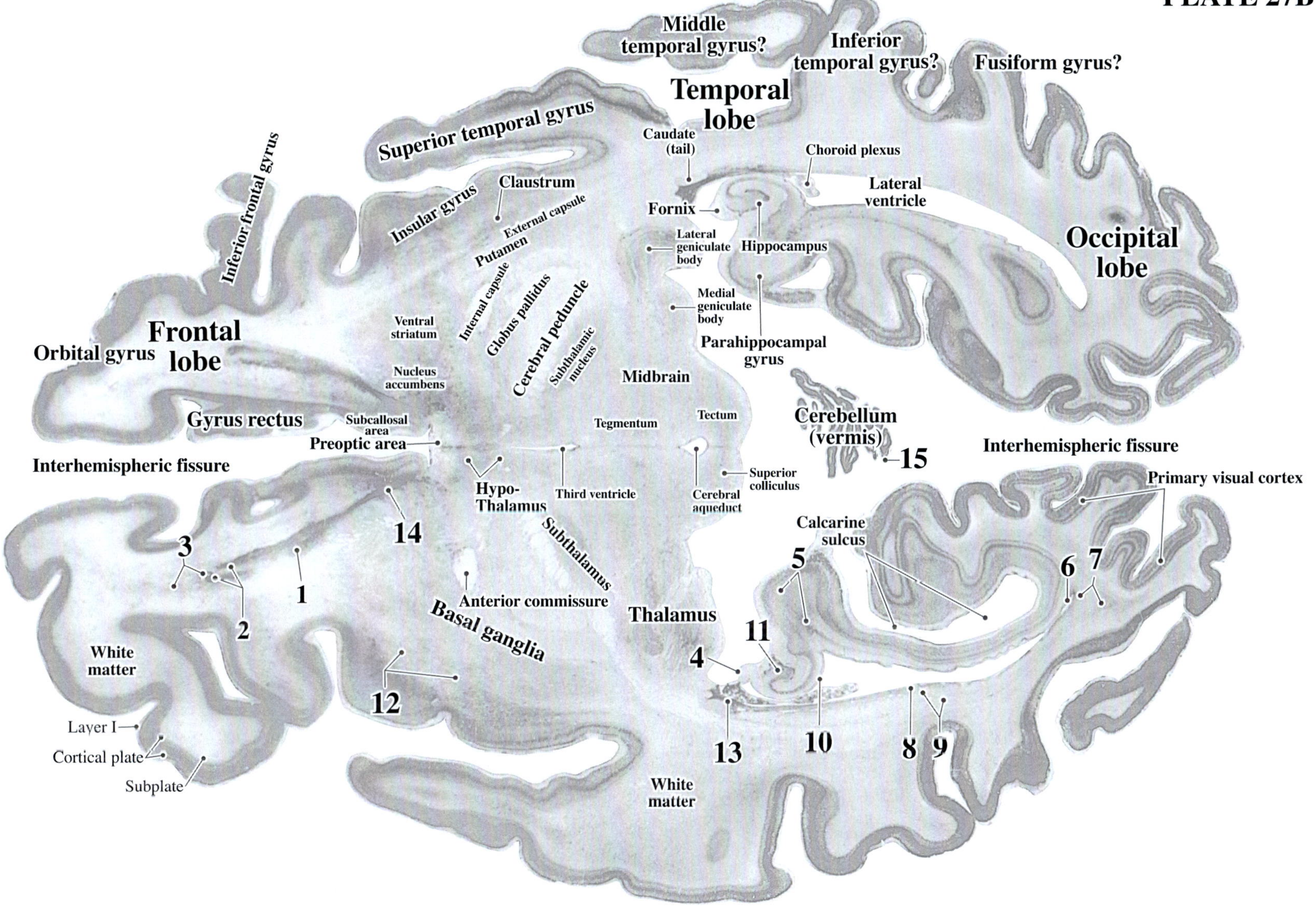
Middle
temporal gyrus?
Inferior
temporal gyrus?
Fusiform gyrus?
Temporal
lobe
Superior temporal gyrus
Inferior frontal gyrus
Caudate
(tail)
Choroid plexus
Claustrum
Lateral
ventricle
Fornix
Insular gyrus
External capsule
Lateral
geniculate
body
Hippocampus
Occipital
lobe
Putamen
Internal capsule
Globus pallidus
Cerebral peduncle
Medial
geniculate
body
Ventral
striatum
Frontal
lobe
Parahippocampal
gyrus
Orbital gyrus
Subthalamic
nucleus
Nucleus
accumbens
Midbrain
Gyrus rectus
Subcallosal
area
Tegmentum
Tectum
Cerebellum
(vermis)
Preoptic area
Interhemispheric fissure
Interhemispheric fissure
15
Superior
colliculus
Primary visual cortex
Hypo-
Thalamus
Third ventricle
Cerebral
aqueduct
14
Subthalamus
Calcarine
sulcus
3
5
6
7
1
Anterior commissure
2
Basal ganglia
Thalamus
11
White
matter
4
12
Layer I
Cortical plate
Subplate
13
10
8
9
White
matter

PLATE 28A
CR 320 mm
GW 37
Y301-62
Horizontal
Section 921

See detail of brain core in Plates 45A and B.

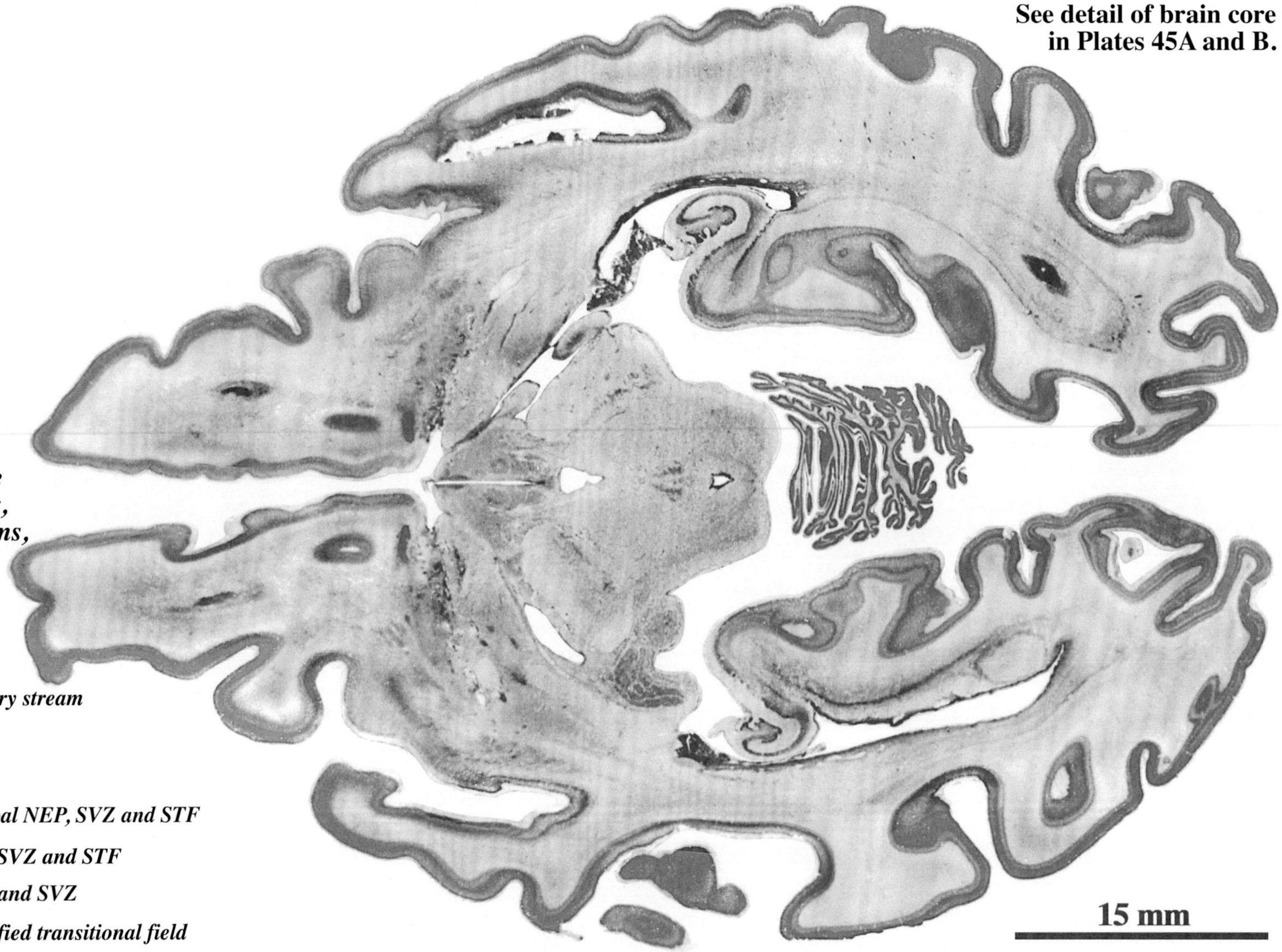

Remnants of the germinal matrix, migratory streams, and transitional fields

1 *Rostral migratory stream*
2 *Frontal STF*
3 *Fornical GEP*
4 *Parahippocampal NEP, SVZ and STF*
5 *Occipital NEP, SVZ and STF*
6 *Temporal NEP and SVZ*
7 *Temporal stratified transitional field*
8 *Alvear glioepithelium*
9 *Subgranular zone (dentate)*
10 *Lateral migratory stream (cortical)*
11 *Posterior striatal NEP and SVZ*
12 *External germinal layer (cerebellum)*

GEP - Glioepithelium
NEP - Neuroepithelium
STF - Stratified transitional field
SVZ - Subventricular zone

PLATE 28B

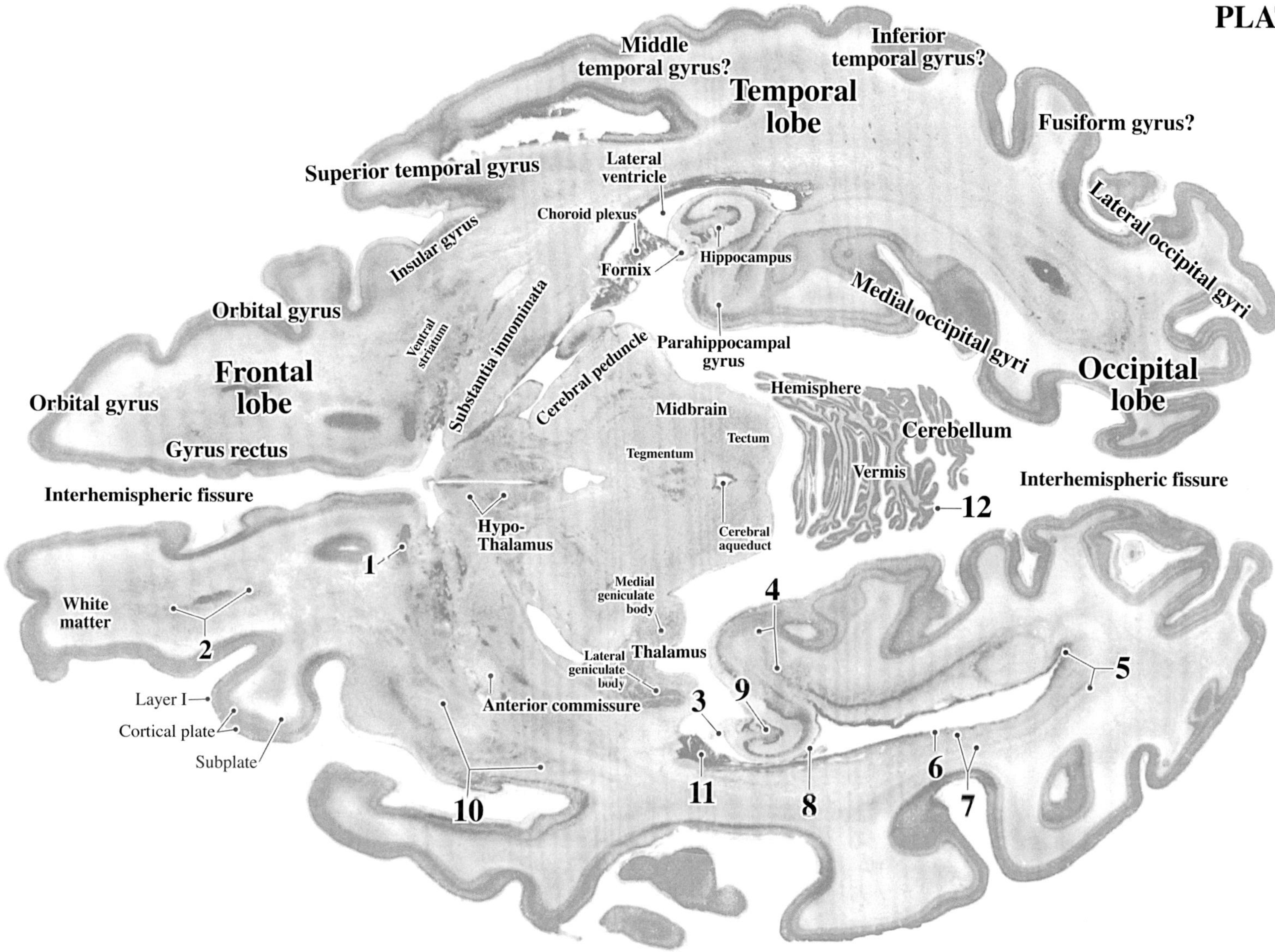
Middle temporal gyrus?
Inferior temporal gyrus?
Temporal lobe
Fusiform gyrus?
Superior temporal gyrus
Lateral ventricle
Choroid plexus
Hippocampus
Fornix
Insular gyrus
Lateral occipital gyri
Medial occipital gyri
Orbital gyrus
Ventral striatum
Substantia innominata
Cerebral peduncle
Parahippocampal gyrus
Frontal lobe
Occipital lobe
Orbital gyrus
Hemisphere
Midbrain
Cerebellum
Gyrus rectus
Tectum
Tegmentum
Vermis
Interhemispheric fissure
Interhemispheric fissure
Hypo-Thalamus
Cerebral aqueduct
12
1
White matter
Medial geniculate body
4
2
Thalamus
Lateral geniculate body
5
Layer I
Cortical plate
Subplate
Anterior commissure
3
9
6
7
10
11
8

PLATE 29A
CR 320 mm
GW 37
Y301-62
Horizontal
Section 981

See detail of brain core in Plates 46A and B.

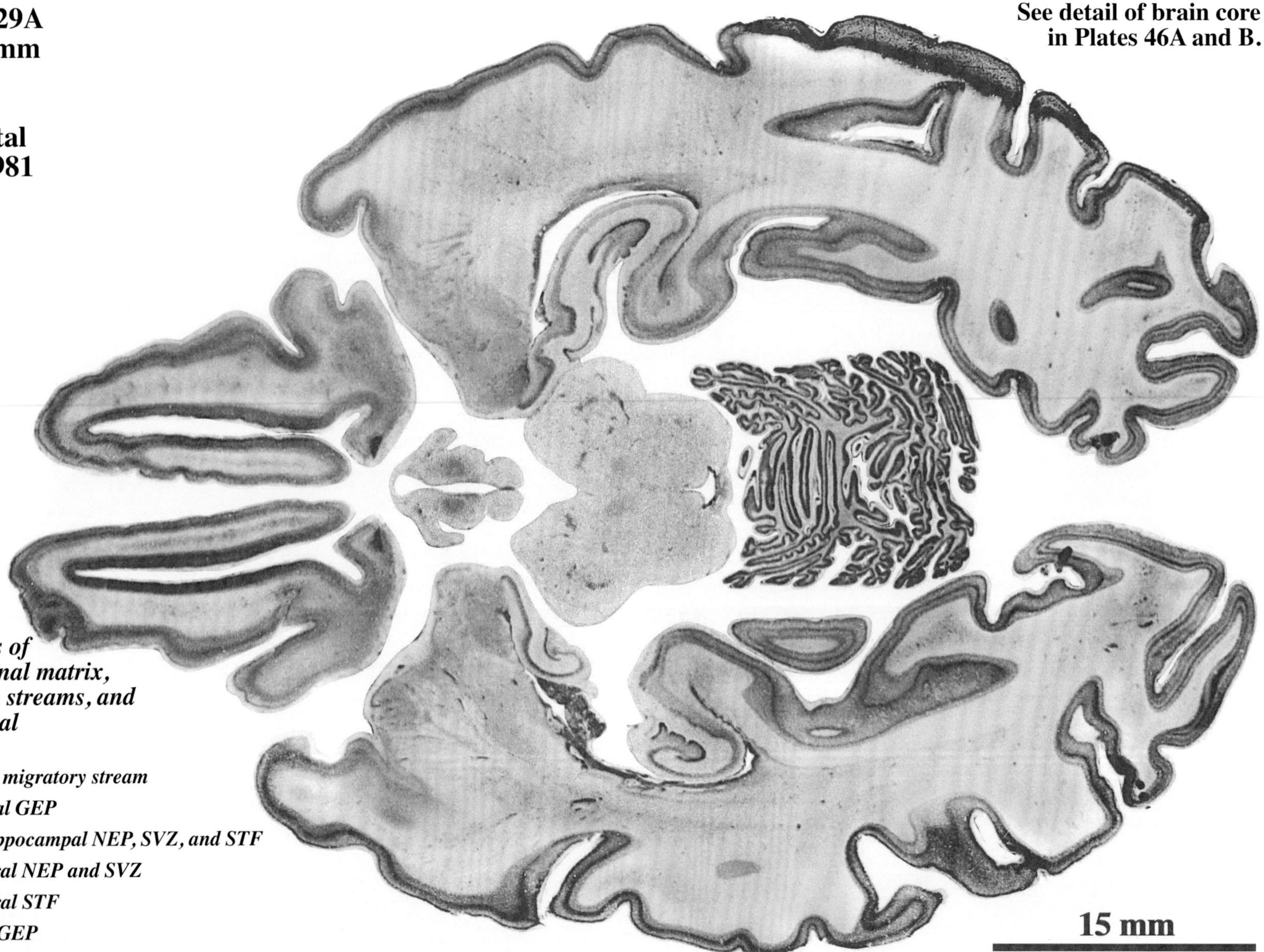

15 mm

Remnants of the germinal matrix, migratory streams, and transitional fields

1 *Rostral migratory stream*
2 *Fornical GEP*
3 *Parahippocampal NEP, SVZ, and STF*
4 *Temporal NEP and SVZ*
5 *Temporal STF*
6 *Alvear GEP*
7 *Subgranular zone (dentate)*
8 *Lateral migratory stream (cortical)*
9 *Amygdaloid G/EP*
10 *External germinal layer (cerebellum)*

GEP - Glioepithelium
G/EP - Glioepithelium/ependyma
NEP - Neuroepithelium
STF - Stratified transitional field
SVZ - Subventricular zone

PLATE 29B

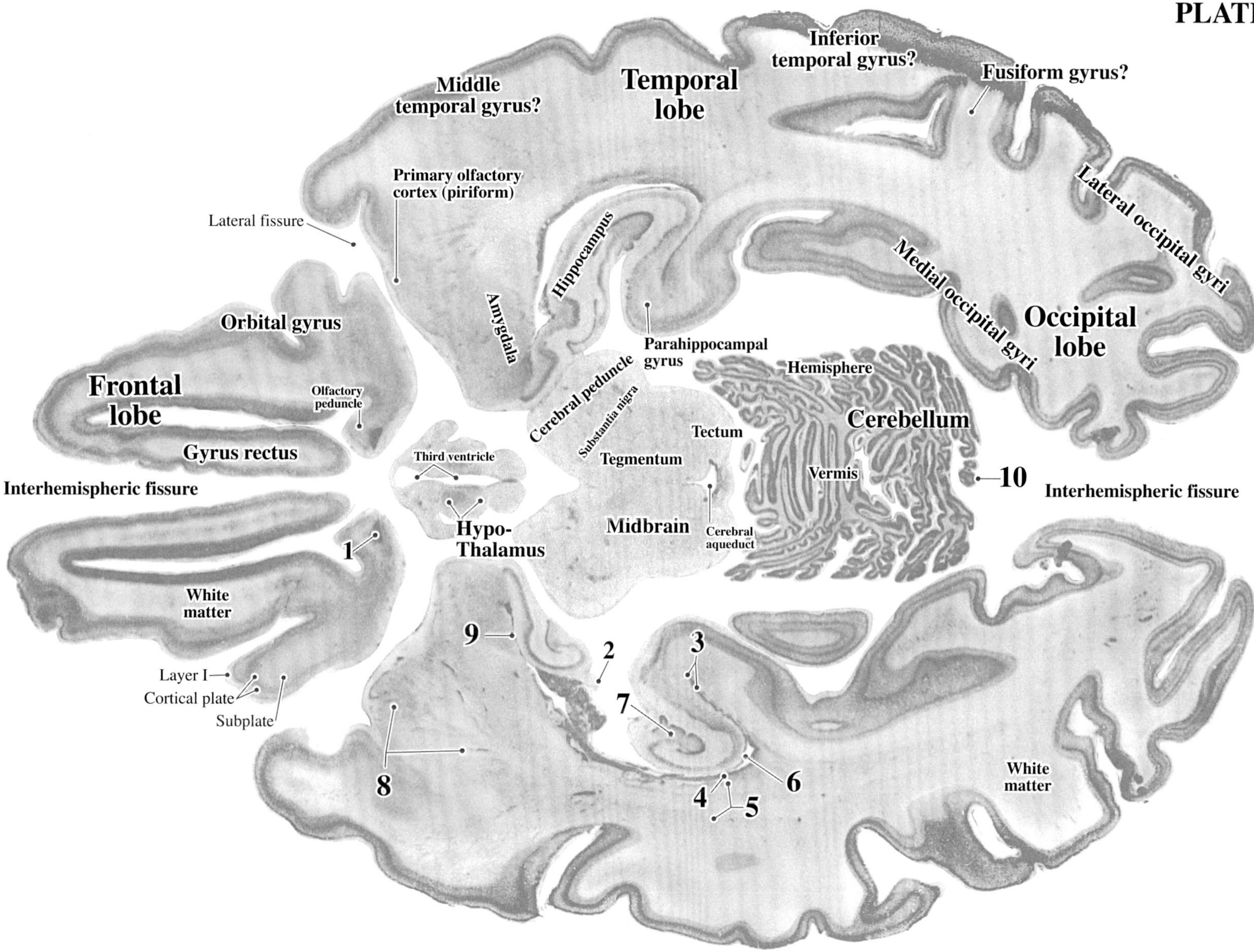
Inferior
temporal gyrus?
Fusiform gyrus?
Middle
temporal gyrus?
Temporal
lobe
Primary olfactory
cortex (piriform)
Lateral fissure
Lateral occipital gyri
Medial occipital gyri
Hippocampus
Amygdala
Occipital
lobe
Orbital gyrus
Parahippocampal
gyrus
Hemisphere
Cerebral peduncle
Frontal
lobe
Olfactory
peduncle
Substantia nigra
Cerebellum
Tectum
Third ventricle
Gyrus rectus
Tegmentum
Vermis
10
Interhemispheric fissure
Interhemispheric fissure
Hypo-
Thalamus
Midbrain
Cerebral
aqueduct
1
White
matter
9
2
3
Layer I
Cortical plate
Subplate
7
6
8
4
5
White
matter

PLATE 30A
CR 320 mm
GW 37
Y301-62
Horizontal
Section 1051

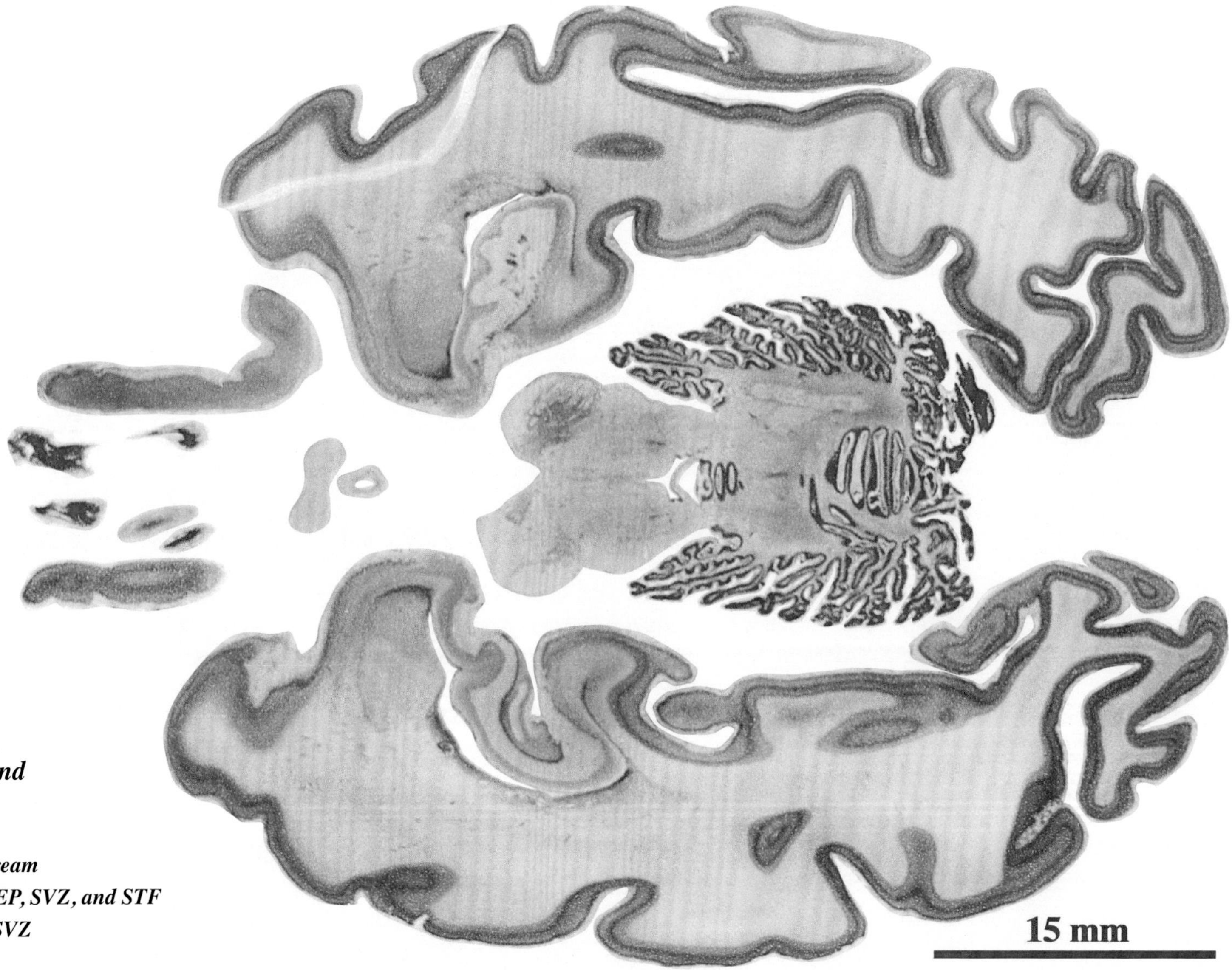

Remnants of the germinal matrix, migratory streams, and transitional fields

1 *Rostral migratory stream*
2 *Parahippocampal NEP, SVZ, and STF*
3 *Temporal NEP and SVZ*
4 *Temporal STF*
5 *Alvear GEP*
6 *Subgranular zone (dentate)*
7 *Lateral migratory stream (cortical)*
8 *Amygdaloid G/EP*
9 *External germinal layer (cerebellum)*

GEP - Glioepithelium
G/EP - Glioepithelium/ependyma
NEP - Neuroepithelium
STF - Stratified transitional field
SVZ - Subventricular zone

See detail of brain core and cerebellum in Plates 47A and B.

PLATE 30B

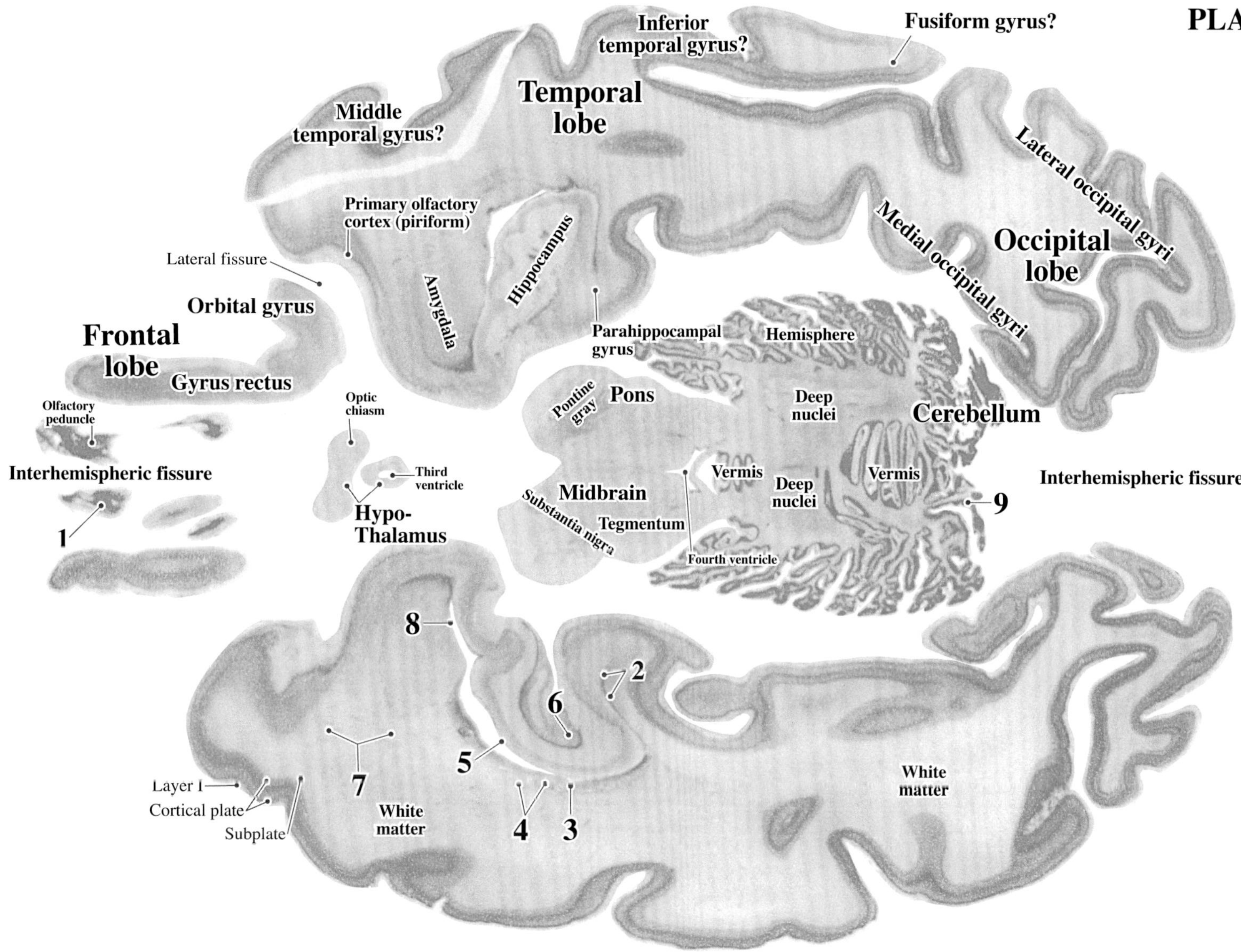

Inferior temporal gyrus?
Fusiform gyrus?
Temporal lobe
Middle temporal gyrus?
Lateral occipital gyri
Primary olfactory cortex (piriform)
Medial occipital gyri
Hippocampus
Occipital lobe
Lateral fissure
Amygdala
Orbital gyrus
Parahippocampal gyrus
Hemisphere
Frontal lobe
Gyrus rectus
Pontine gray
Pons
Deep nuclei
Cerebellum
Olfactory peduncle
Optic chiasm
Third ventricle
Vermis
Vermis
Interhemispheric fissure
Deep nuclei
Interhemispheric fissure
Midbrain
9
Hypo-Thalamus
Substantia nigra
Tegmentum
1
Fourth ventricle
8
2
6
7
5
White matter
Layer I
Cortical plate
White matter
Subplate
4
3

PLATE 31A
CR 320 mm
GW 37
Y301-62
Horizontal
Section 1111

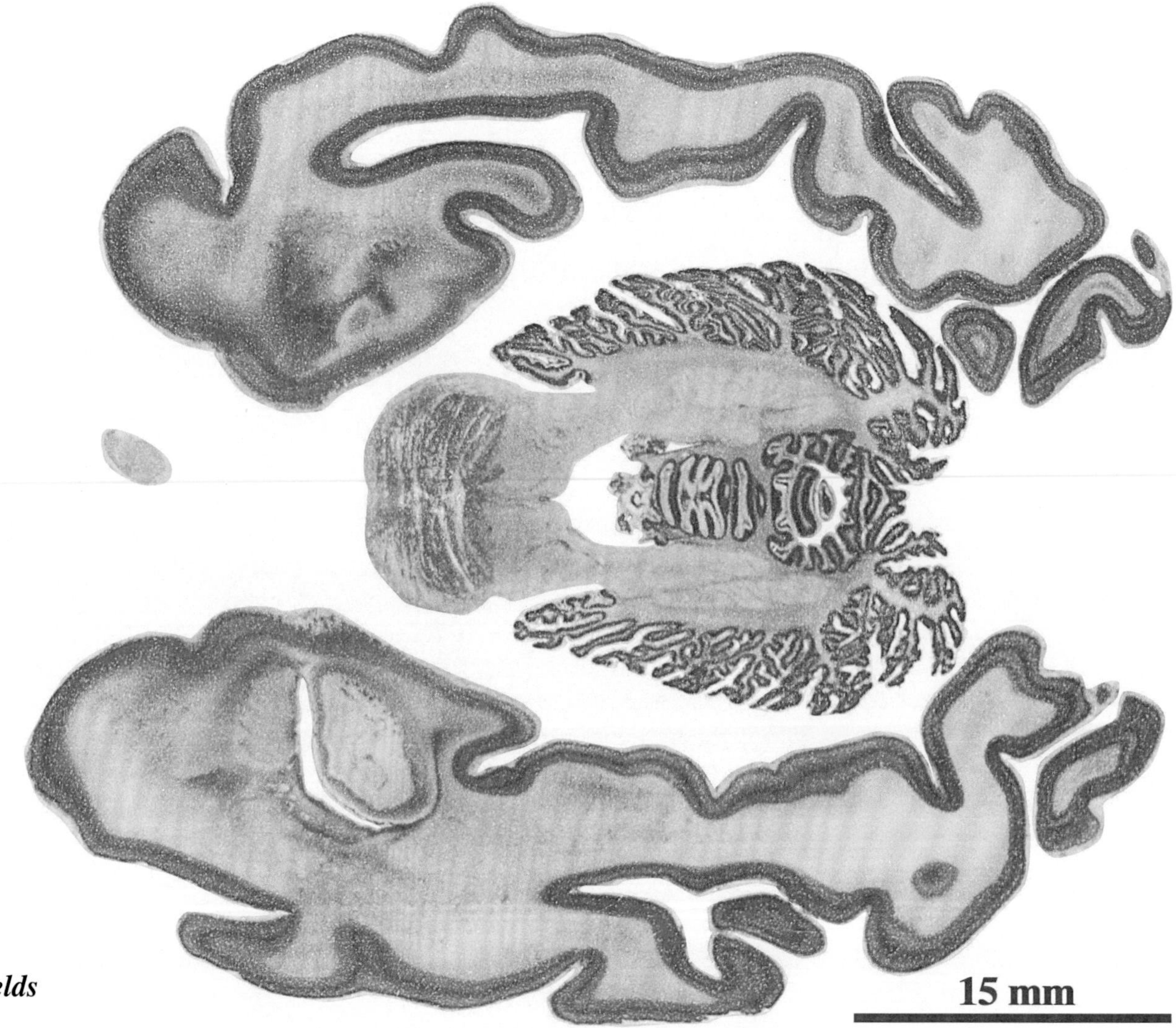

Remnants of the germinal matrix, migratory streams, and transitional fields

1 *Parahippocampal NEP, SVZ, and STF*
2 *Temporal NEP and SVZ*
3 *Temporal STF*
4 *Alvear GEP*
5 *Lateral migratory stream (cortical)*
6 *Amygdaloid G/EP*
7 *External germinal layer (cerebellum)*

GEP - Glioepithelium
G/EP - Glioepithelium/ependyma
NEP - Neuroepithelium
STF - Stratified transitional field
SVZ - Subventricular zone

See detail of brain core and cerebellum in Plates 48A and B.

PLATE 31B

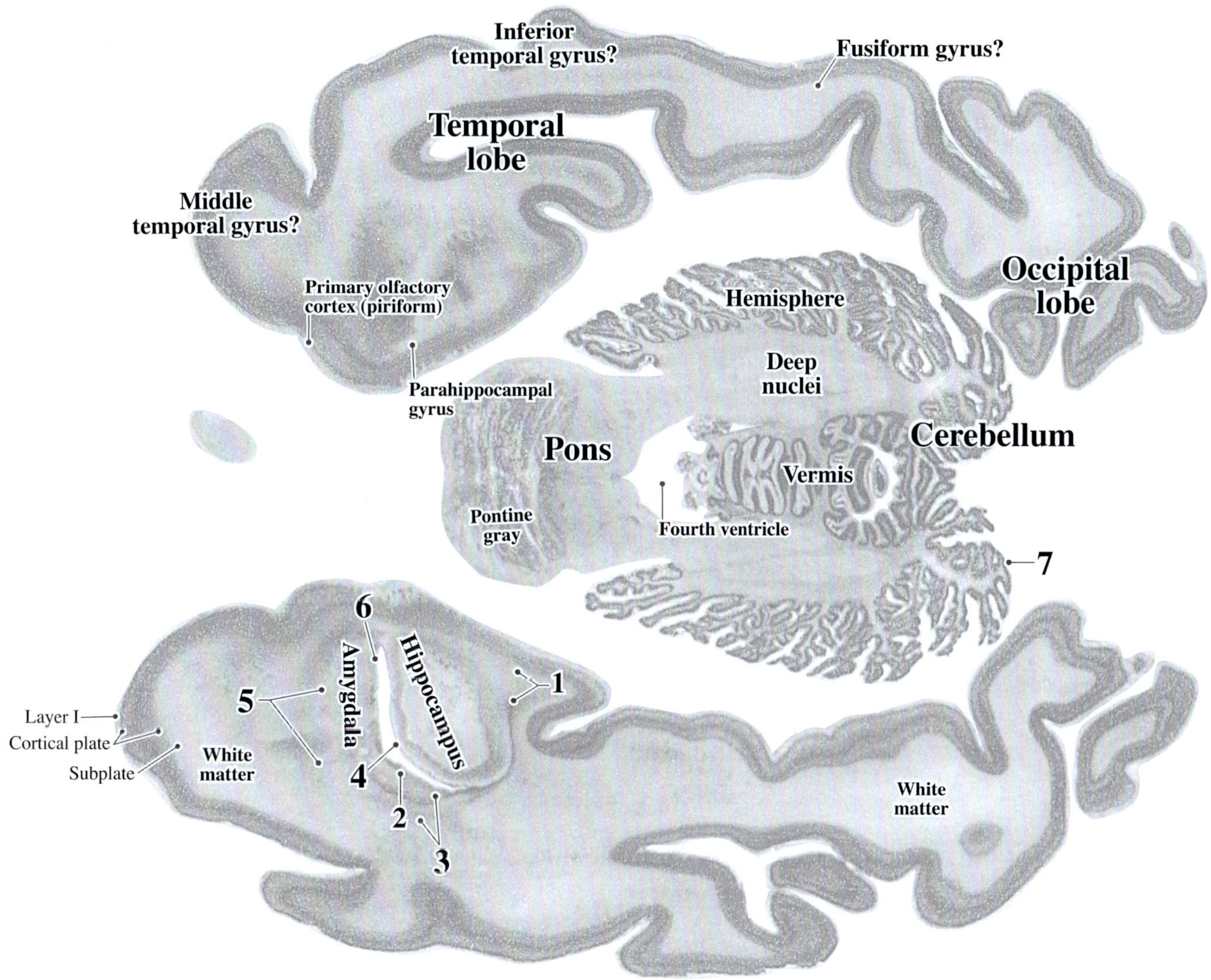

Inferior temporal gyrus?
Fusiform gyrus?
Temporal lobe
Middle temporal gyrus?
Primary olfactory cortex (piriform)
Parahippocampal gyrus
Hemisphere
Deep nuclei
Occipital lobe
Pons
Cerebellum
Vermis
Pontine gray
Fourth ventricle
7
6
Amygdala
Hippocampus
1
5
Layer I
Cortical plate
Subplate
White matter
4
2
3
White matter

PLATE 32A
CR 320 mm
GW 37
Y301-62
Horizontal
Section 1161

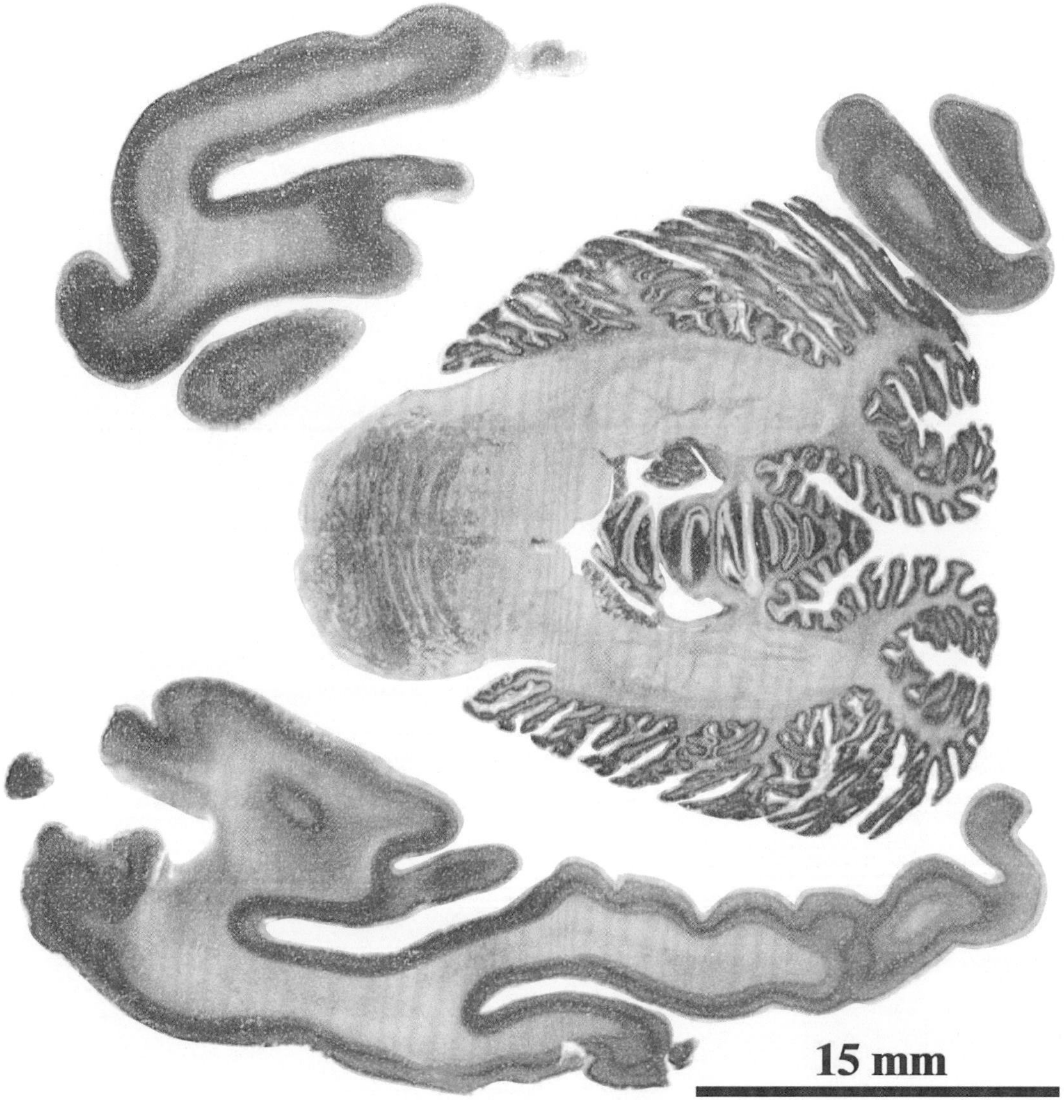

See detail of brain core and cerebellum in Plates 49A and B.

PLATE 32B

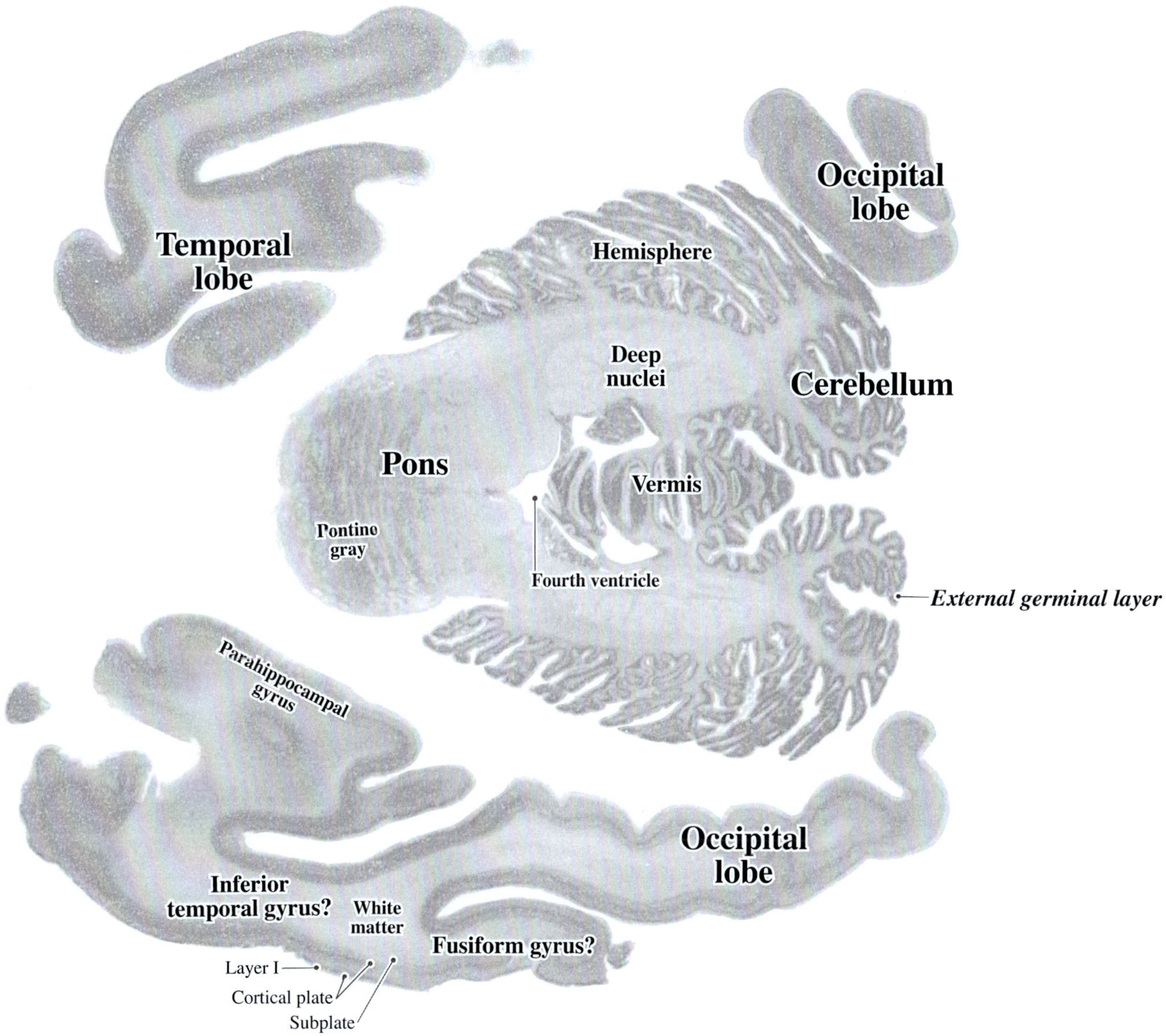

Temporal lobe
Occipital lobe
Hemisphere
Deep nuclei
Cerebellum
Pons
Vermis
Pontine gray
Fourth ventricle
External germinal layer
Parahippocampal gyrus
Occipital lobe
Inferior temporal gyrus?
White matter
Fusiform gyrus?
Layer I
Cortical plate
Subplate

PLATE 33A
CR 320 mm
GW 37
Y301-62
Horizontal
Section 1251

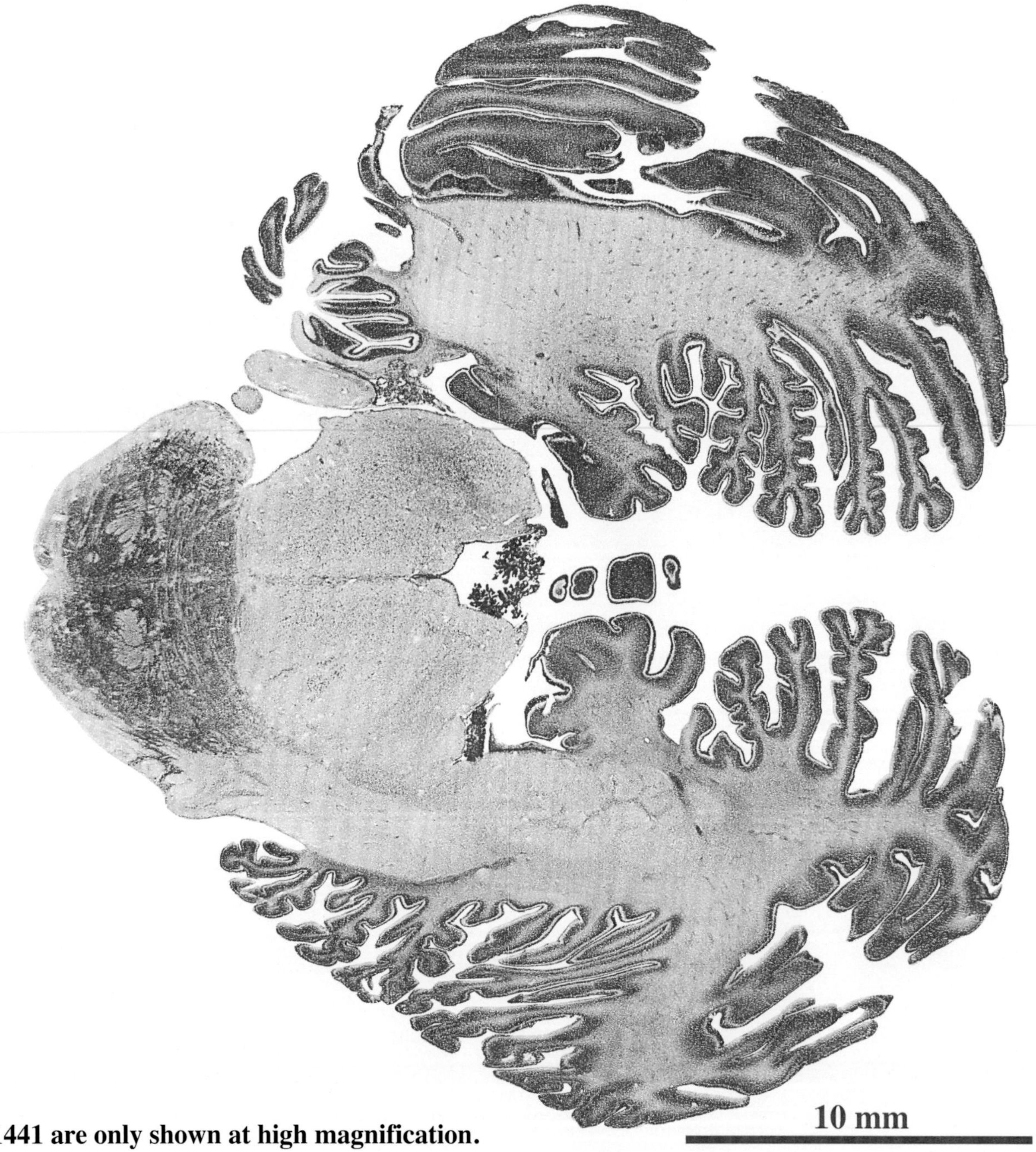

Sections 1251 to 1441 are only shown at high magnification.

PLATE 33B

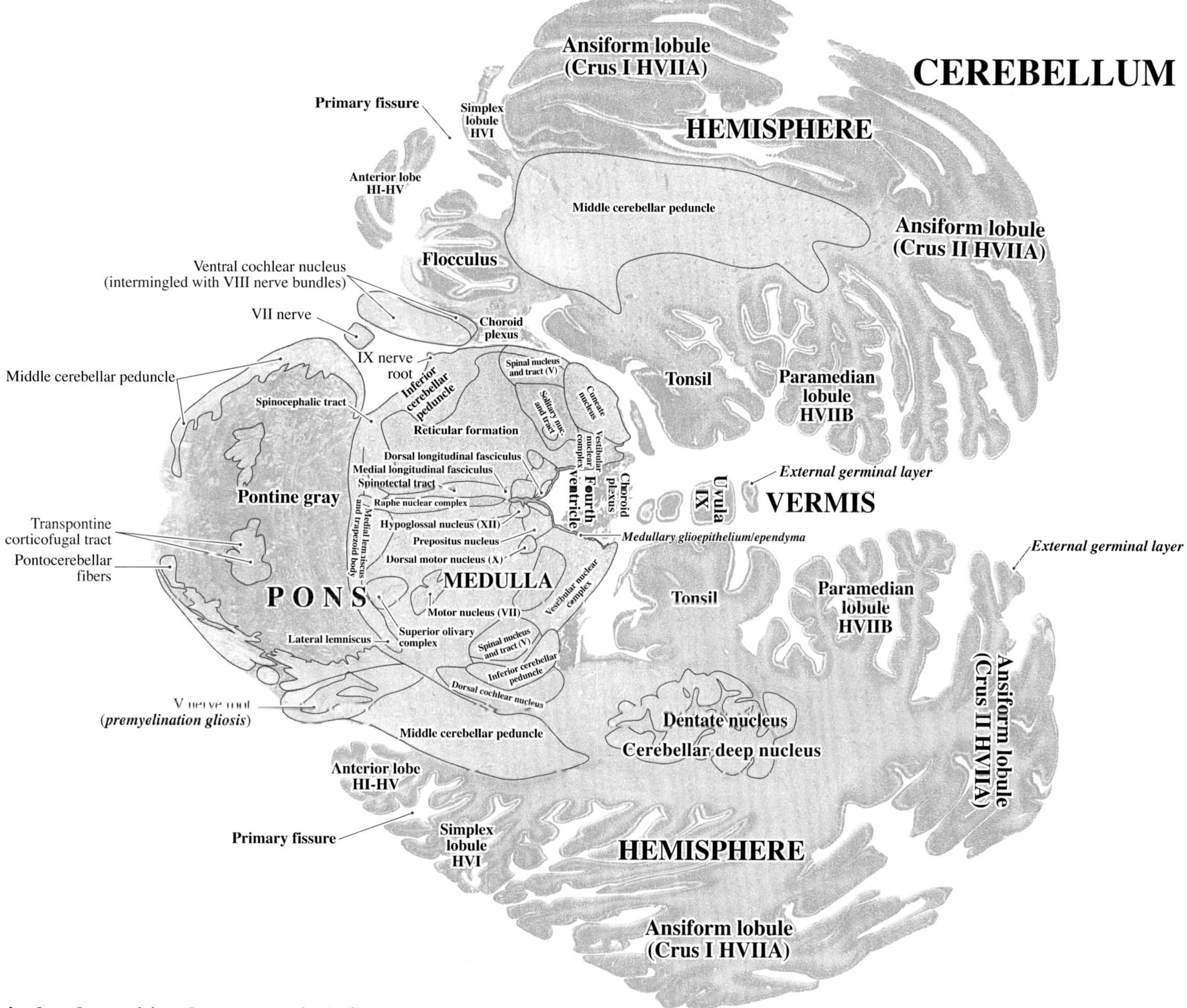

Germinal and transitional structures in *italics*

PLATE 34A
CR 320 mm
GW 37
Y301-62
Horizontal
Section 1311

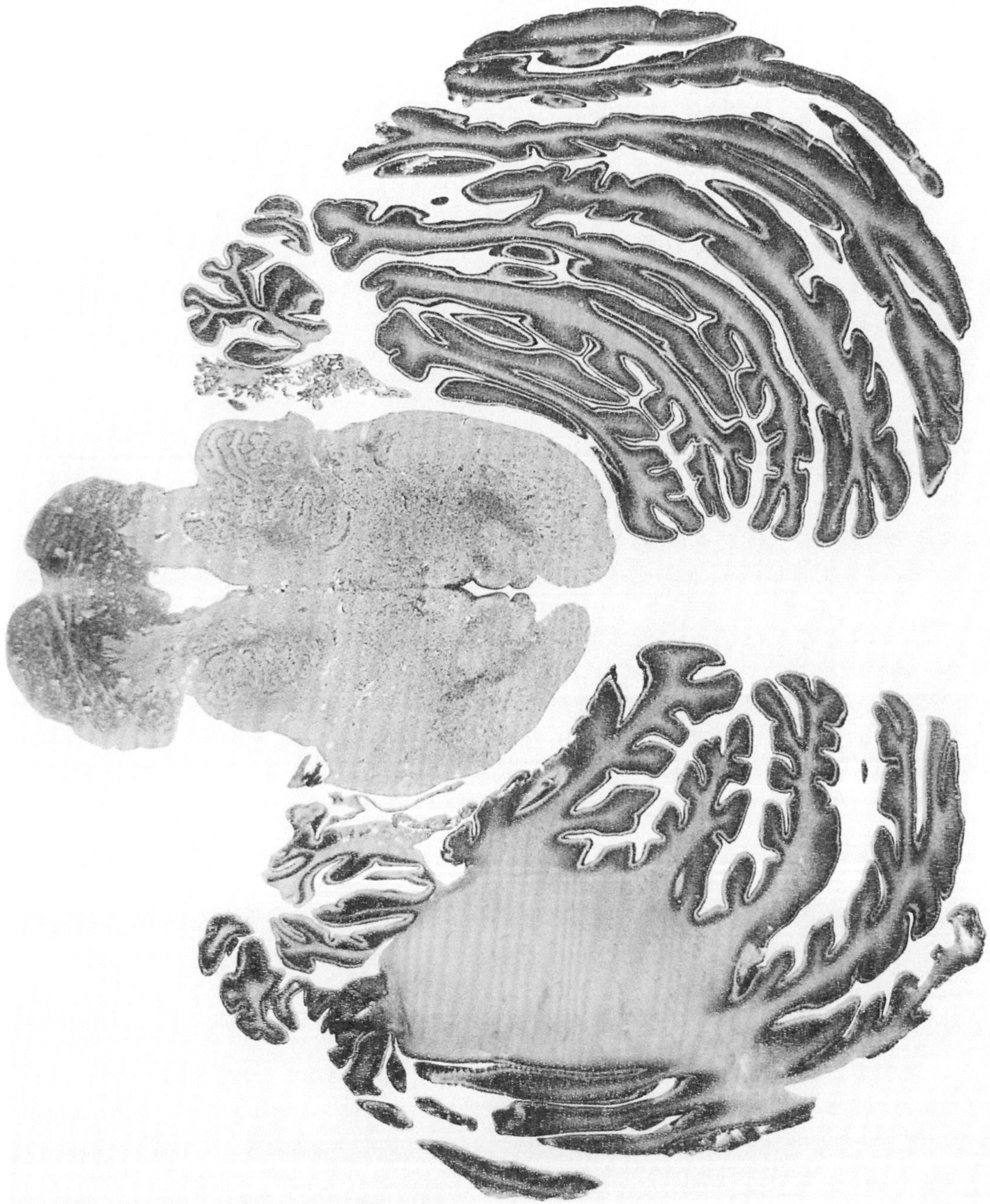

Sections 1251 to 1441 are only shown at high magnification.

PLATE 34B

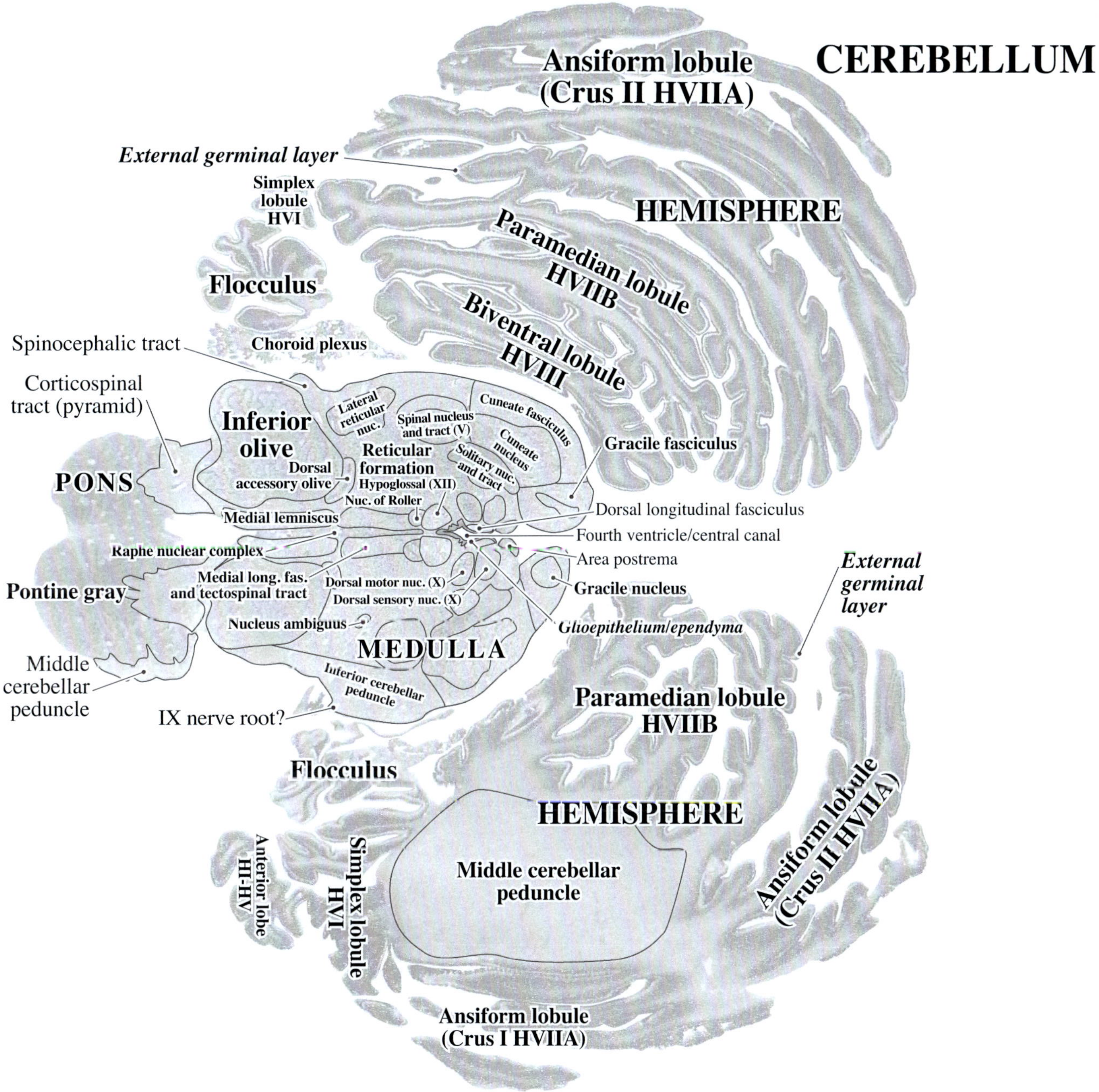

Germinal and transitional structures in *italics*

PLATE 35A
CR 320 mm
GW 37
Y301-62
Horizontal
Section 1331

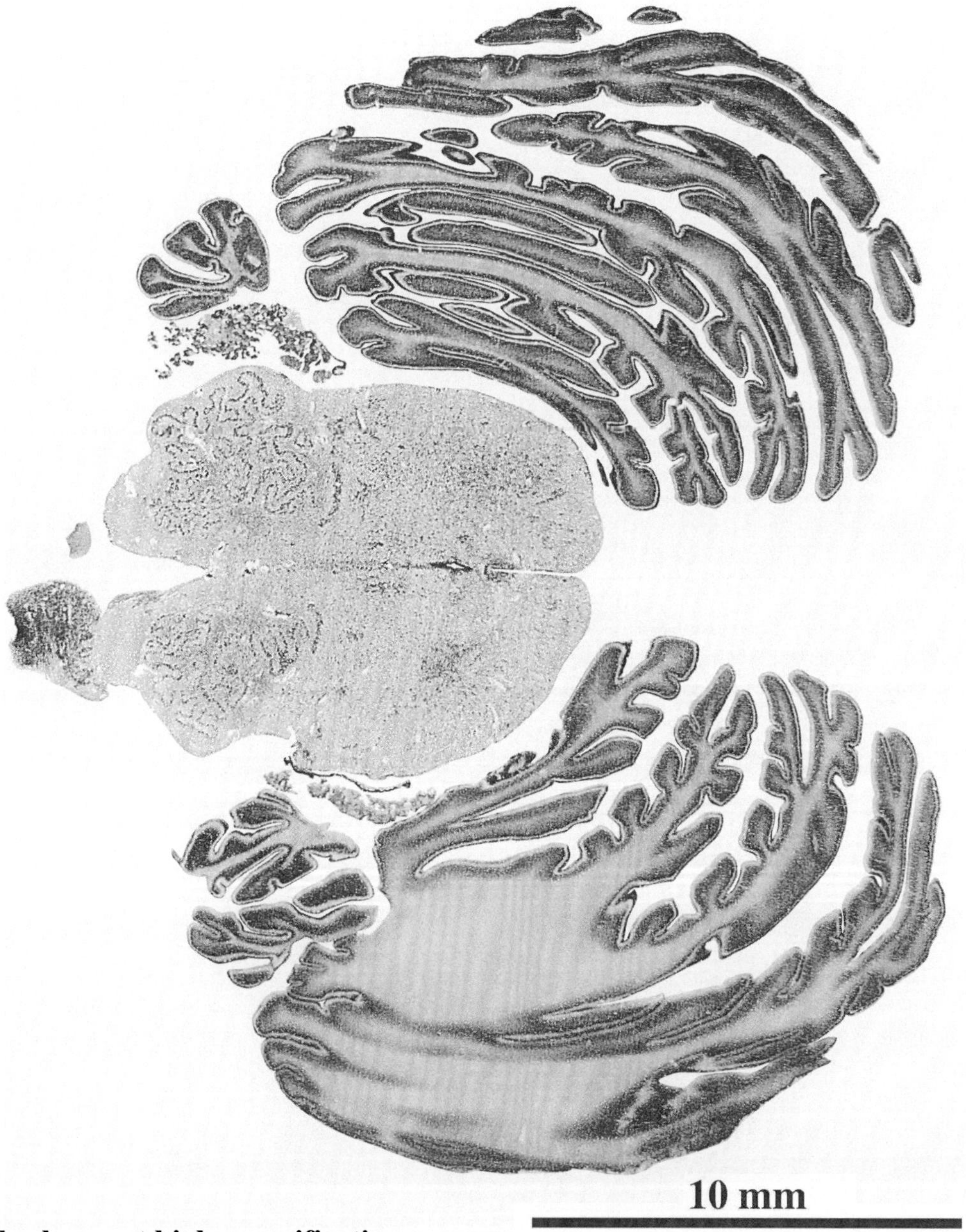

Sections 1251 to 1441 are only shown at high magnification.

PLATE 35B

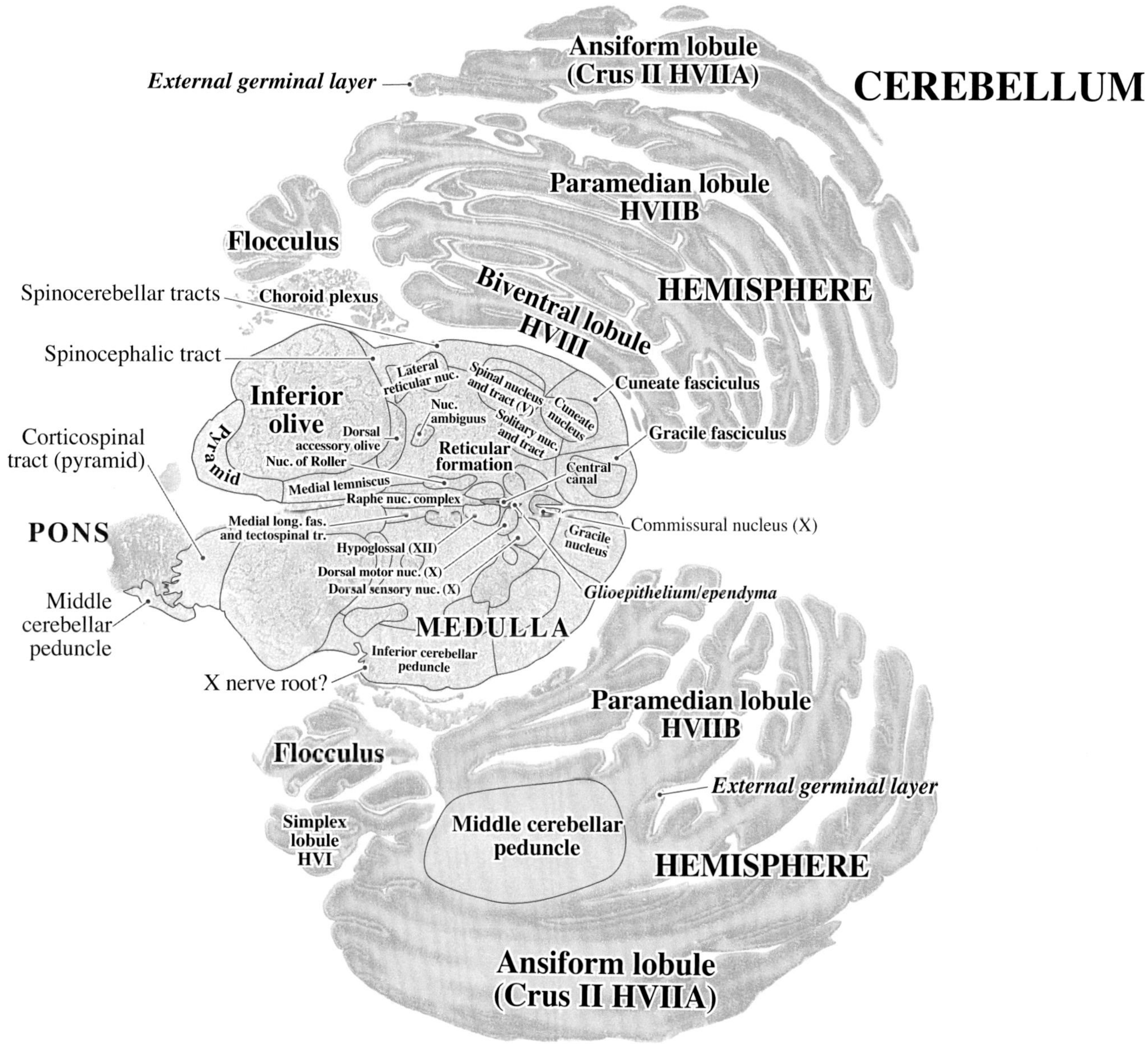

Germinal and transitional structures in *italics*

PLATE 36A
CR 320 mm
GW 37
Y301-62
Horizontal
Section 1381

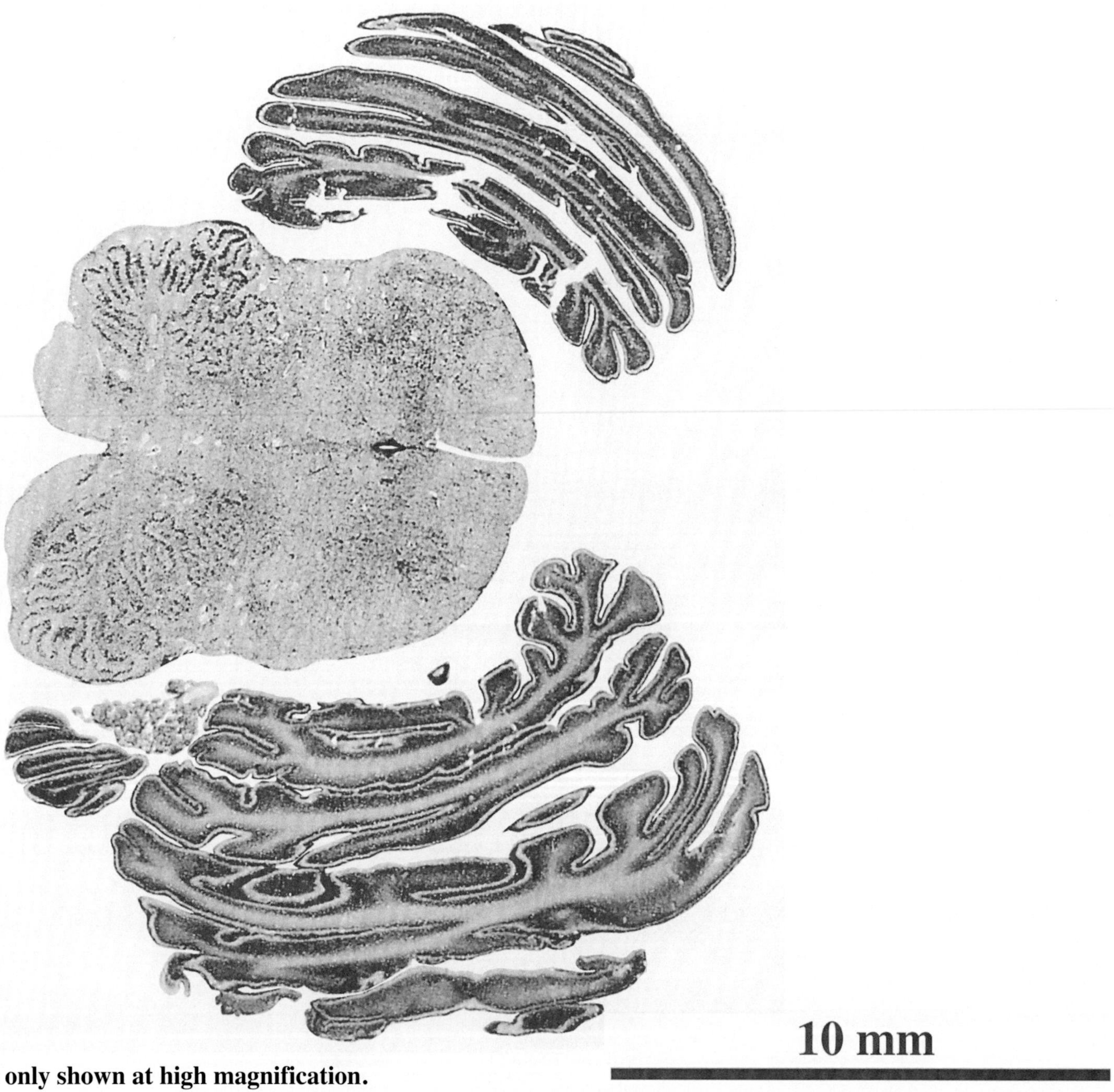

Sections 1251 to 1441 are only shown at high magnification.

PLATE 36B

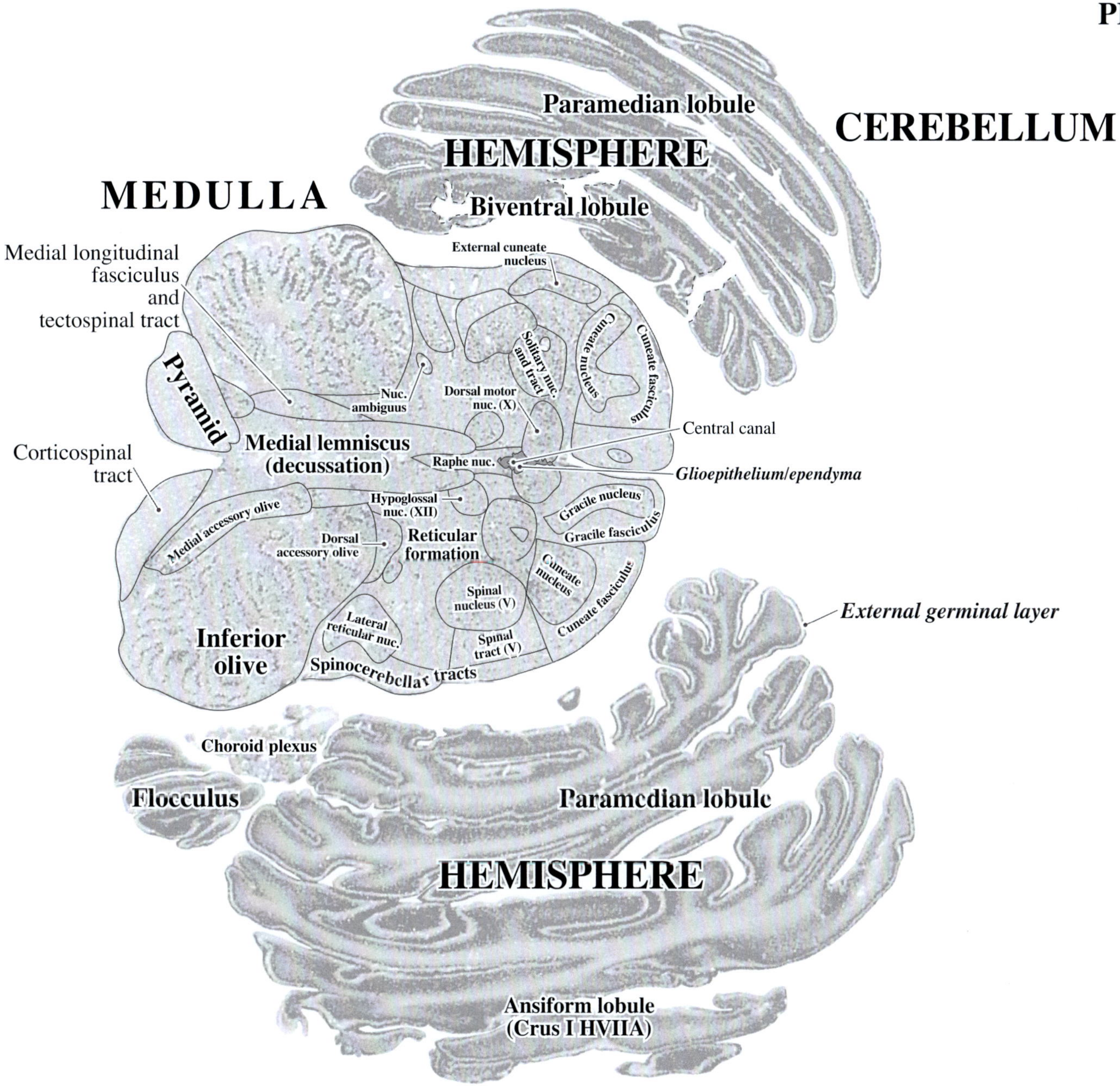

Germinal and transitional structures in *italics*

PLATE 37A
CR 320 mm
GW 37
Y301-62
Horizontal
Section 1441

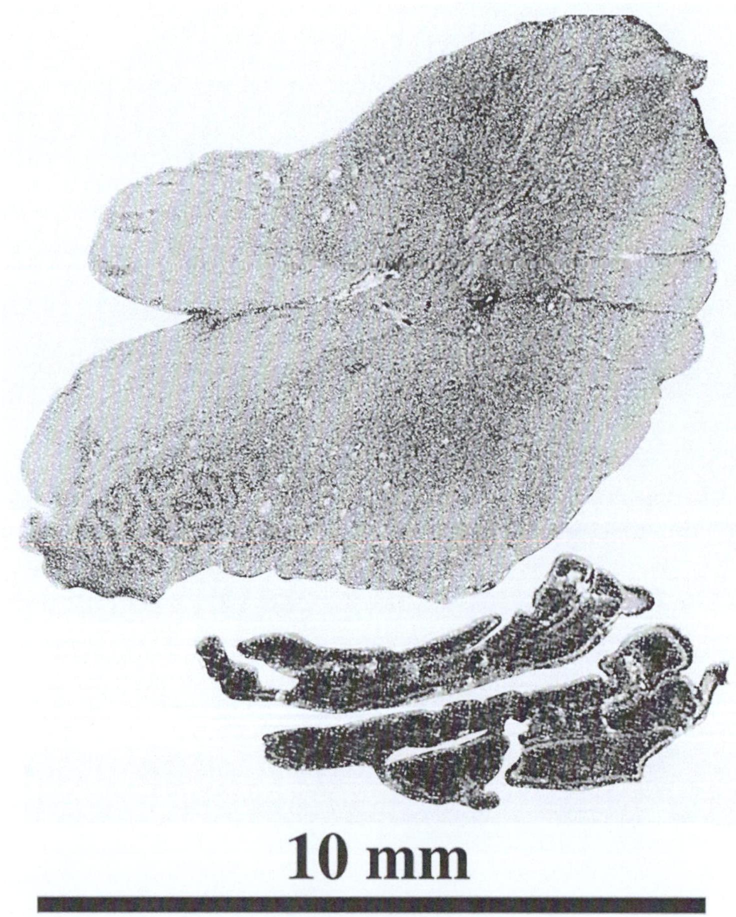

Sections 1251 to 1441 are only shown at high magnification.

PLATE 37B

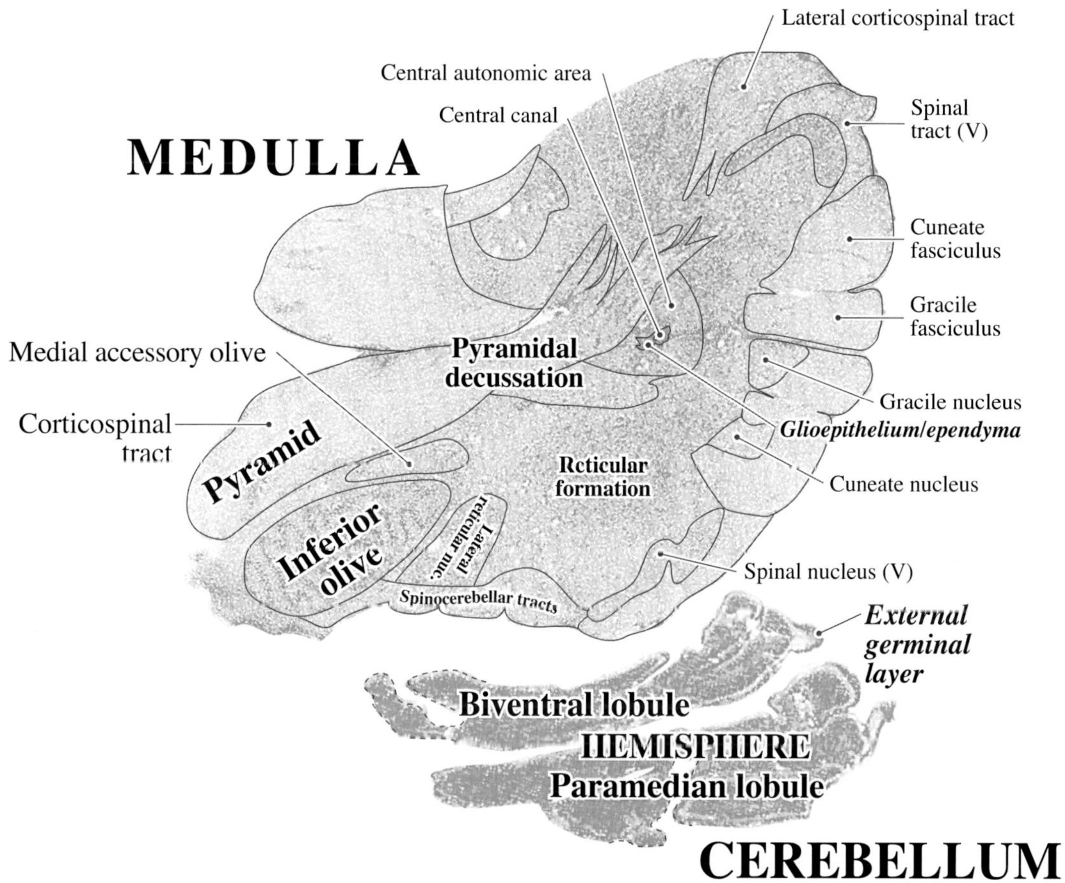

Germinal and transitional structures in *italics*

PLATE 38A
CR 320 mm
GW 37
Y301-62
Horizontal
Section 531

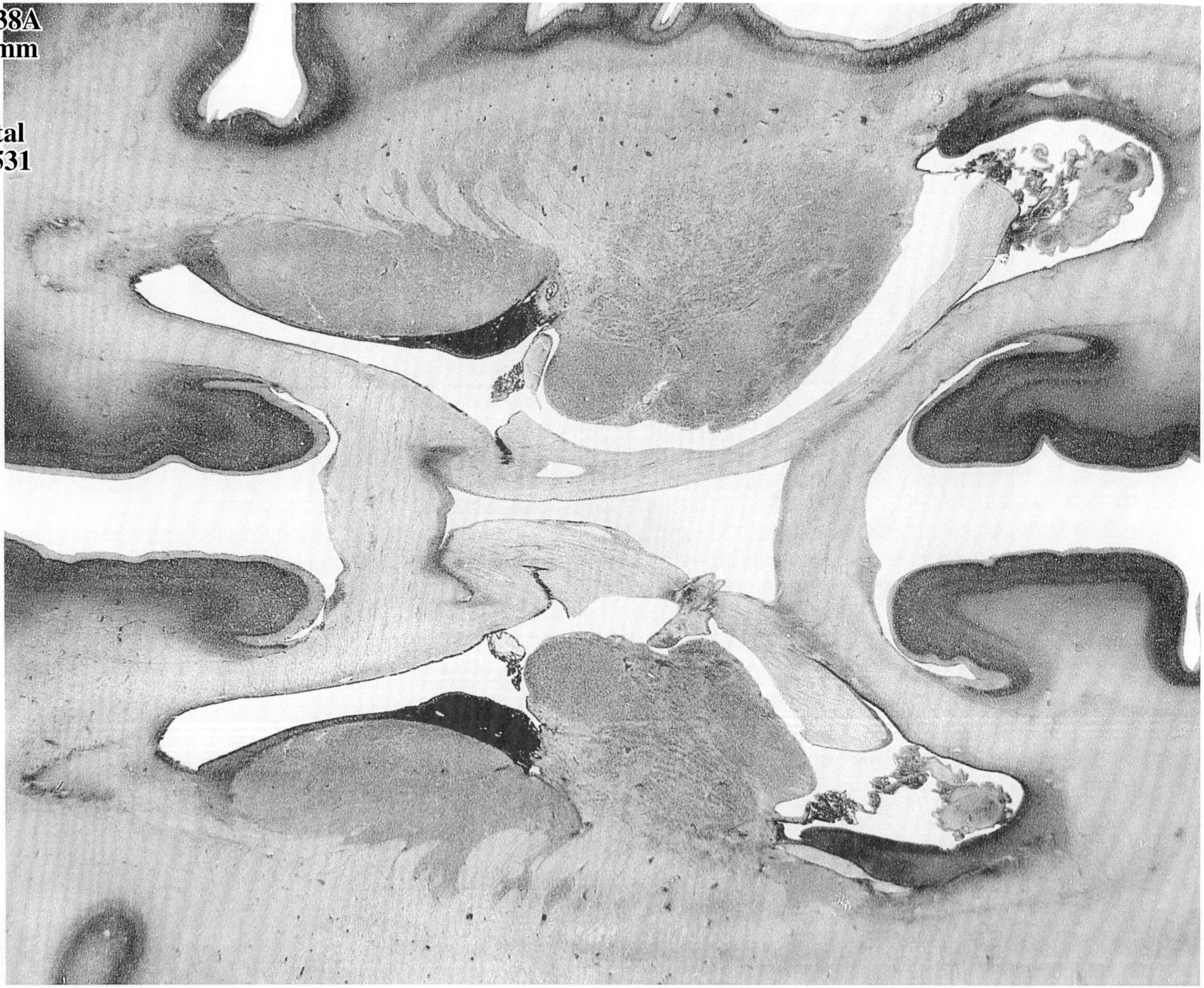

See the entire Section 531 in Plate 21.

10 mm

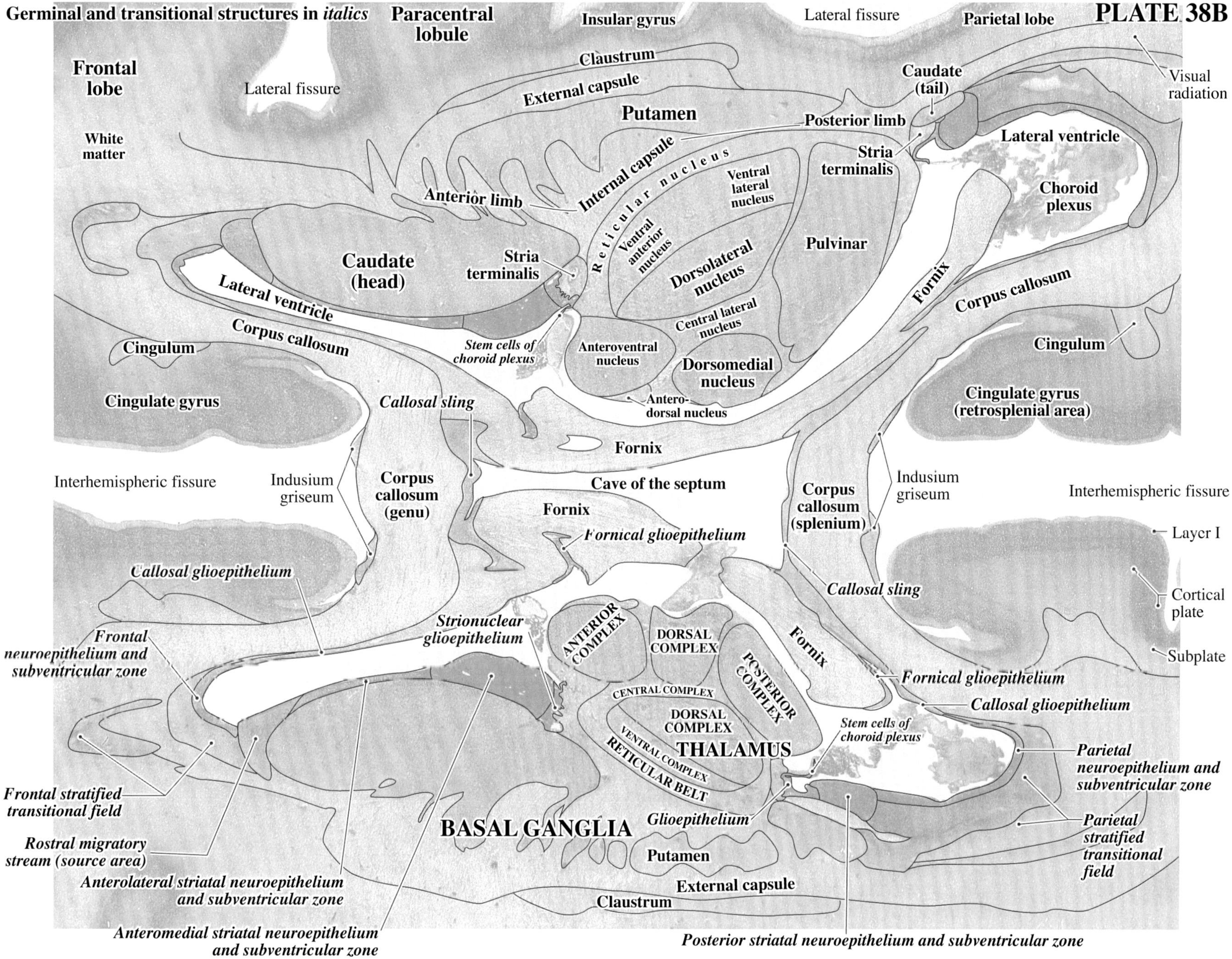

Germinal and transitional structures in *italics*
PLATE 38B
Paracentral lobule
Insular gyrus
Lateral fissure
Parietal lobe
Frontal lobe
Lateral fissure
White matter
Claustrum
External capsule
Putamen
Caudate (tail)
Visual radiation
Posterior limb
Lateral ventricle
Stria terminalis
Internal capsule
Reticular nucleus
Ventral lateral nucleus
Choroid plexus
Anterior limb
Ventral anterior nucleus
Caudate (head)
Stria terminalis
Dorsolateral nucleus
Pulvinar
Lateral ventricle
Fornix
Corpus callosum
Corpus callosum
Central lateral nucleus
Cingulum
Stem cells of choroid plexus
Anteroventral nucleus
Dorsomedial nucleus
Cingulum
Cingulate gyrus
Antero-dorsal nucleus
Cingulate gyrus (retrosplenial area)
Callosal sling
Fornix
Interhemispheric fissure
Indusium griseum
Corpus callosum (genu)
Cave of the septum
Corpus callosum (splenium)
Indusium griseum
Interhemispheric fissure
Fornix
Fornical glioepithelium
Layer I
Callosal glioepithelium
Callosal sling
Cortical plate
ANTERIOR COMPLEX
DORSAL COMPLEX
Fornix
Strionuclear glioepithelium
Frontal neuroepithelium and subventricular zone
POSTERIOR COMPLEX
Subplate
Fornical glioepithelium
CENTRAL COMPLEX
Callosal glioepithelium
DORSAL COMPLEX
Stem cells of choroid plexus
THALAMUS
VENTRAL COMPLEX
Parietal neuroepithelium and subventricular zone
RETICULAR BELT
Frontal stratified transitional field
Glioepithelium
Parietal stratified transitional field
BASAL GANGLIA
Rostral migratory stream (source area)
Putamen
Anterolateral striatal neuroepithelium and subventricular zone
External capsule
Claustrum
Anteromedial striatal neuroepithelium and subventricular zone
Posterior striatal neuroepithelium and subventricular zone

PLATE 39A
CR 320 mm
GW 37
Y301-62
Horizontal
Section 561

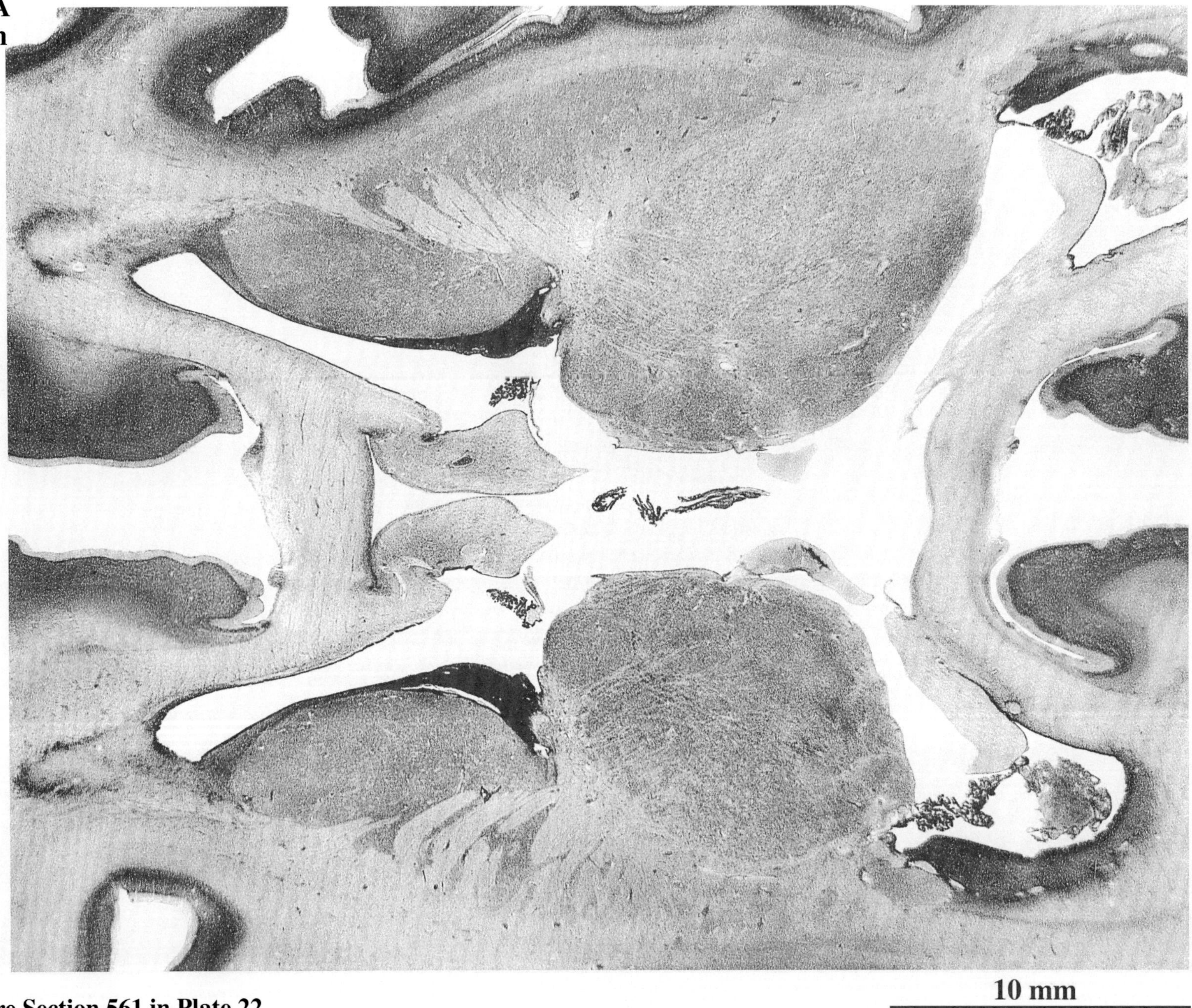

See the entire Section 561 in Plate 22.

PLATE 39B

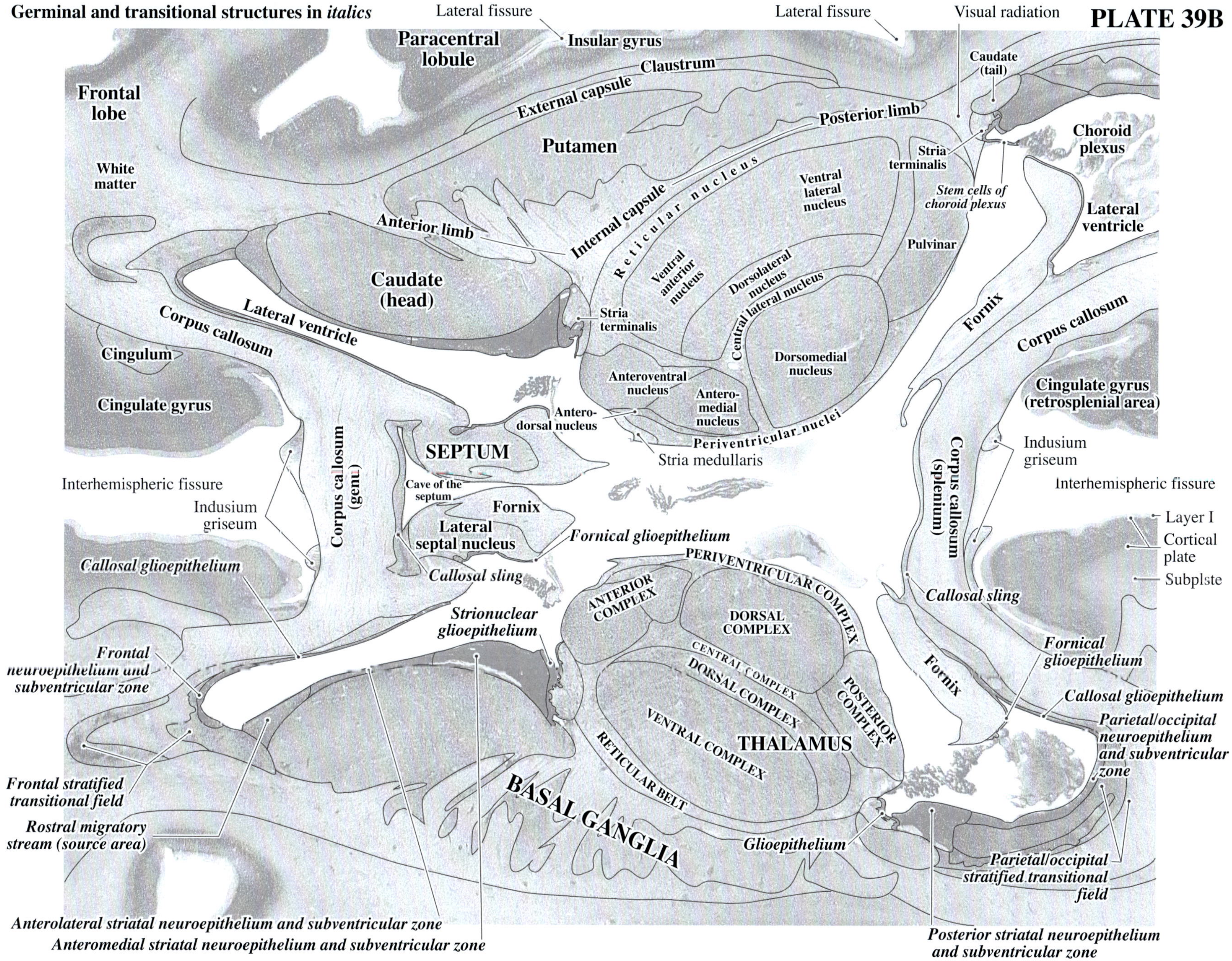
Germinal and transitional structures in *italics*
Lateral fissure
Lateral fissure
Visual radiation
Paracentral lobule
Insular gyrus
Claustrum
Frontal lobe
External capsule
Caudate (tail)
Posterior limb
Choroid plexus
Putamen
White matter
Stria terminalis
Stem cells of choroid plexus
Ventral lateral nucleus
Reticular nucleus
Lateral ventricle
Internal capsule
Anterior limb
Pulvinar
Ventral anterior nucleus
Dorsolateral nucleus
Caudate (head)
Central lateral nucleus
Fornix
Stria terminalis
Lateral ventricle
Corpus callosum
Corpus callosum
Cingulum
Dorsomedial nucleus
Anteroventral nucleus
Antero-medial nucleus
Cingulate gyrus
Cingulate gyrus (retrosplenial area)
Antero-dorsal nucleus
Periventricular nuclei
Corpus callosum (genu)
SEPTUM
Stria medullaris
Corpus callosum (splenium)
Indusium griseum
Interhemispheric fissure
Cave of the septum
Interhemispheric fissure
Indusium griseum
Fornix
Layer I
Lateral septal nucleus
Fornical glioepithelium
Cortical plate
Callosal glioepithelium
PERIVENTRICULAR COMPLEX
Subplste
Callosal sling
Callosal sling
ANTERIOR COMPLEX
DORSAL COMPLEX
Strionuclear glioepithelium
Fornical glioepithelium
Frontal neuroepithelium and subventricular zone
CENTRAL COMPLEX
Fornix
DORSAL COMPLEX
POSTERIOR COMPLEX
Callosal glioepithelium
VENTRAL COMPLEX
THALAMUS
Parietal/occipital neuroepithelium and subventricular zone
Frontal stratified transitional field
RETICULAR BELT
BASAL GANGLIA
Rostral migratory stream (source area)
Glioepithelium
Parietal/occipital stratified transitional field
Anterolateral striatal neuroepithelium and subventricular zone
Anteromedial striatal neuroepithelium and subventricular zone
Posterior striatal neuroepithelium and subventricular zone

PLATE 40A
CR 320 mm
GW 37
Y301-62
Horizontal
Section 631

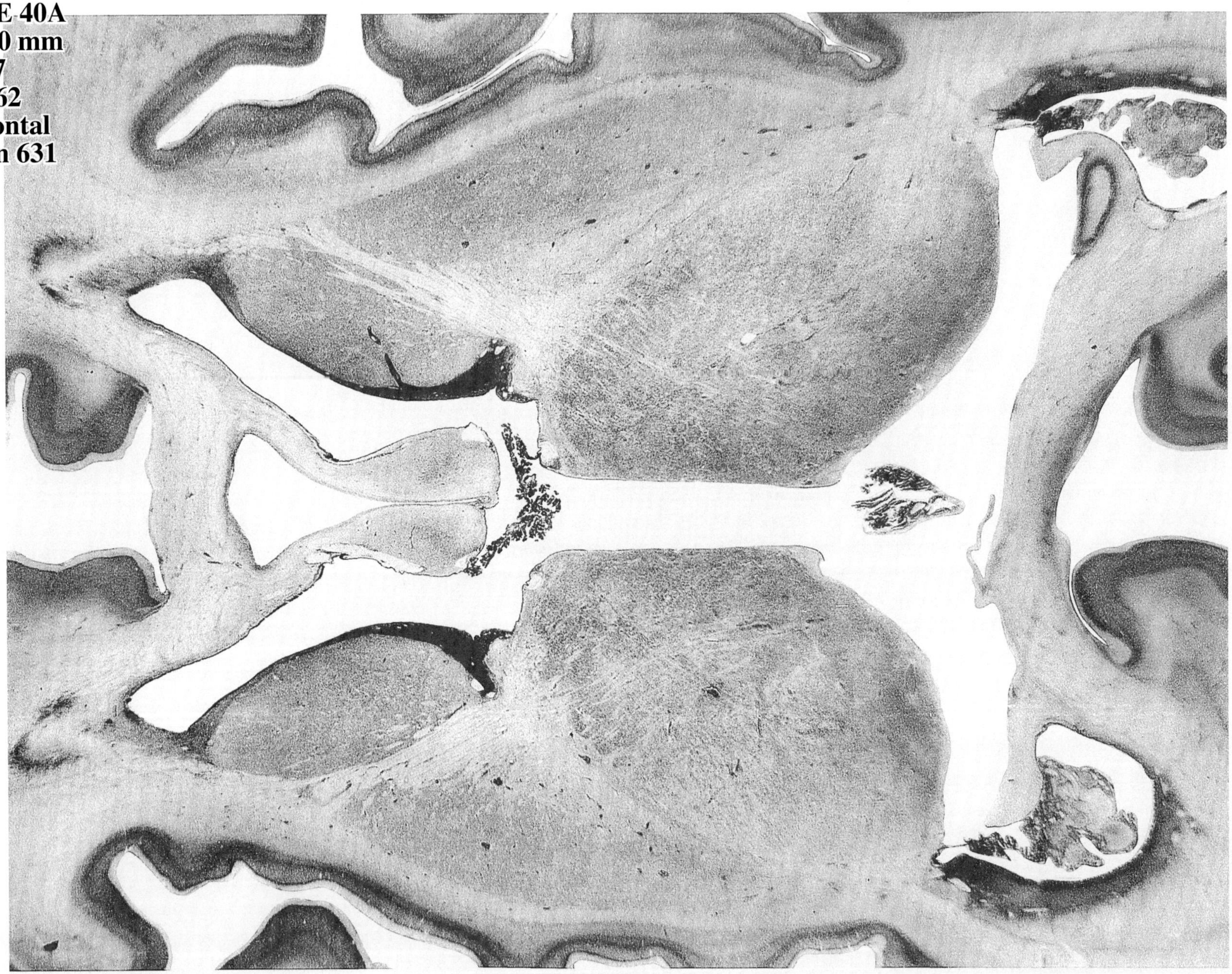

See the entire Section 631 in Plate 23.

10 mm

PLATE 40B

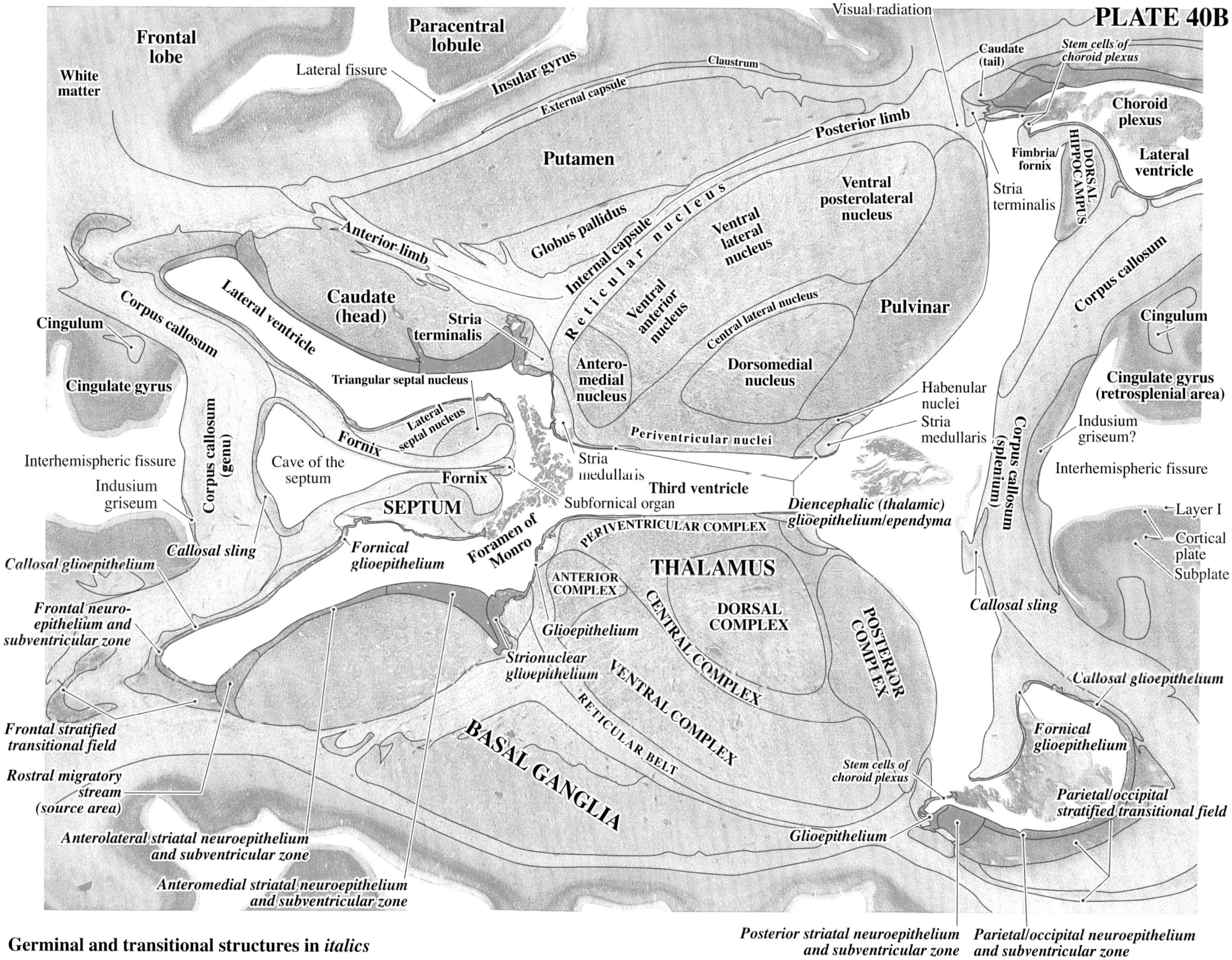

Germinal and transitional structures in *italics*

PLATE 41A
CR 320 mm
GW 37
Y301-62
Horizontal
Section 731

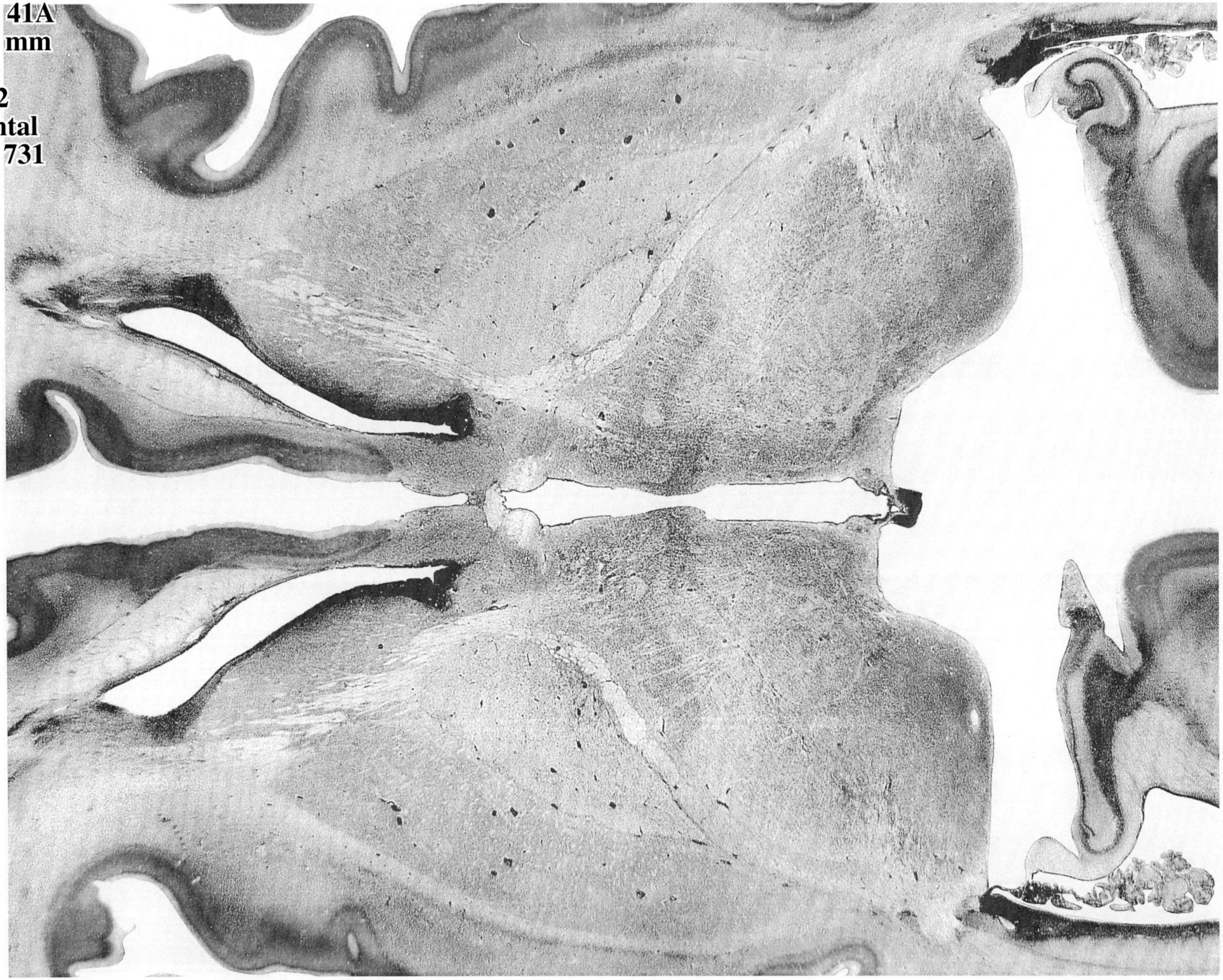

10 mm

See the entire Section 731 in Plate 24.

PLATE 41B

Germinal and transitional structures in *italics*

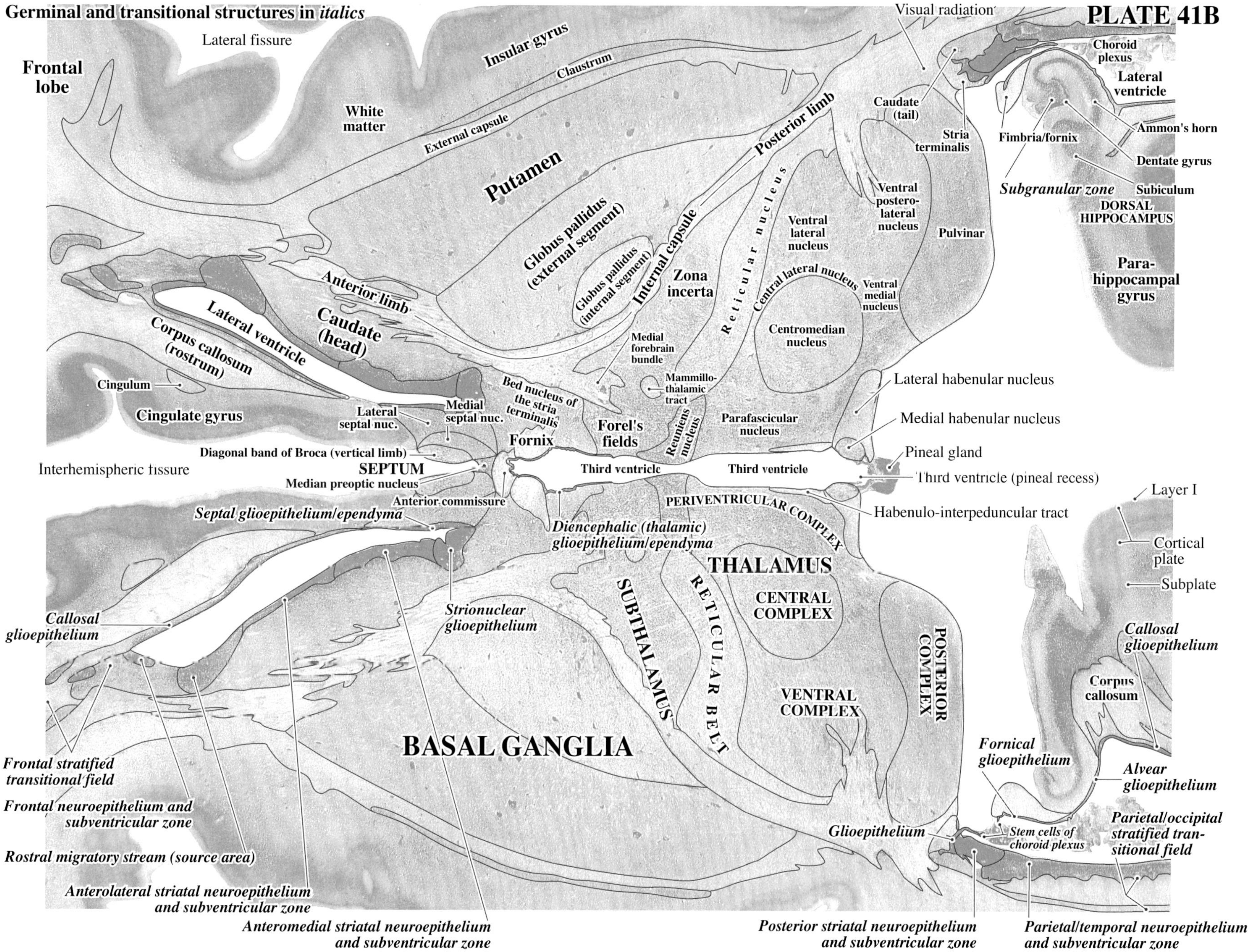

PLATE 42A
CR 320 mm
GW 37
Y301-62
Horizontal
Section 761

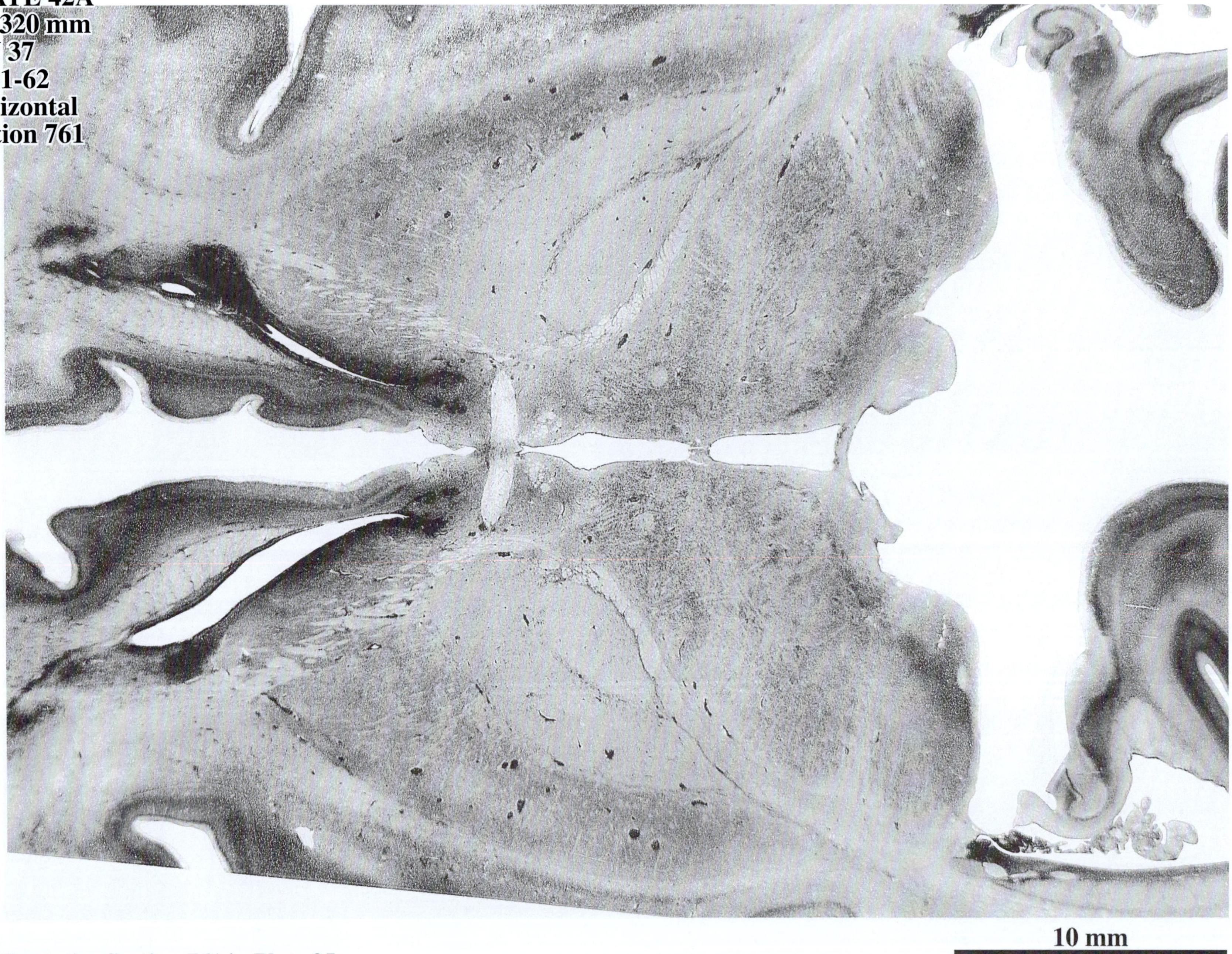

See the entire Section 761 in Plate 25.

PLATE 42B

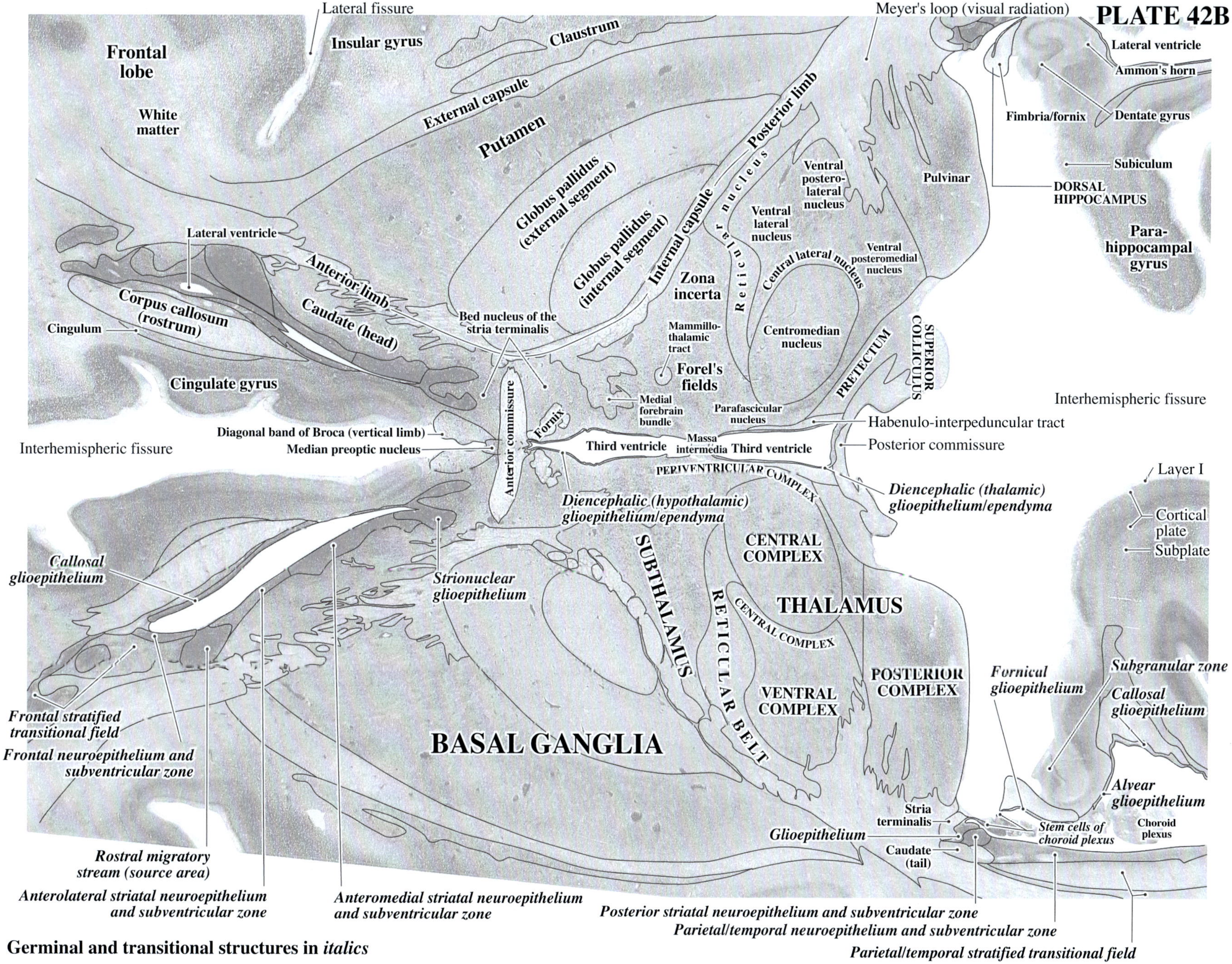

Germinal and transitional structures in *italics*

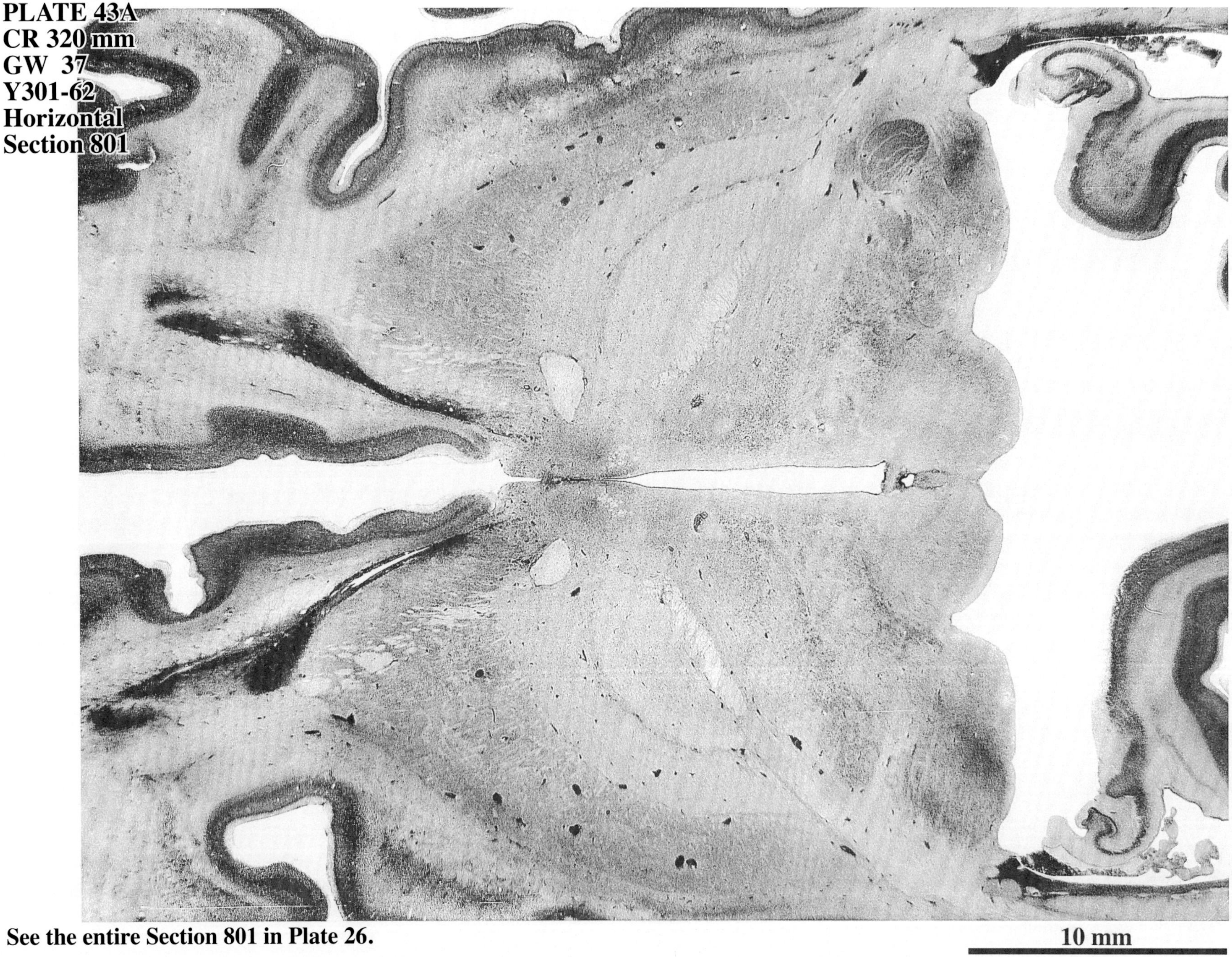

See the entire Section 801 in Plate 26.

Germinal and transitional structures in *italics*

PLATE 43B

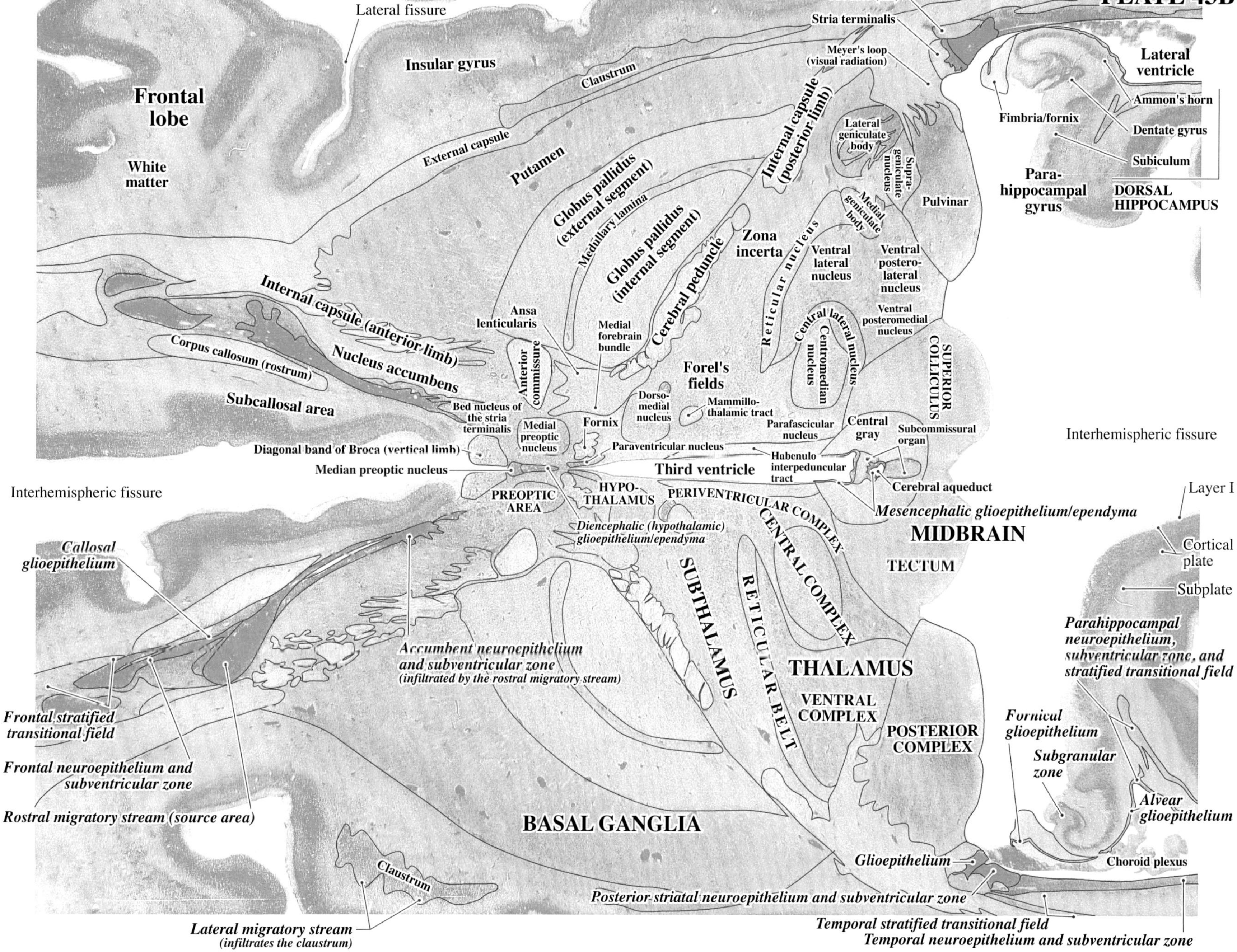

PLATE 44A
CR 320 mm
GW 37
Y301-62
Horizontal
Section 851

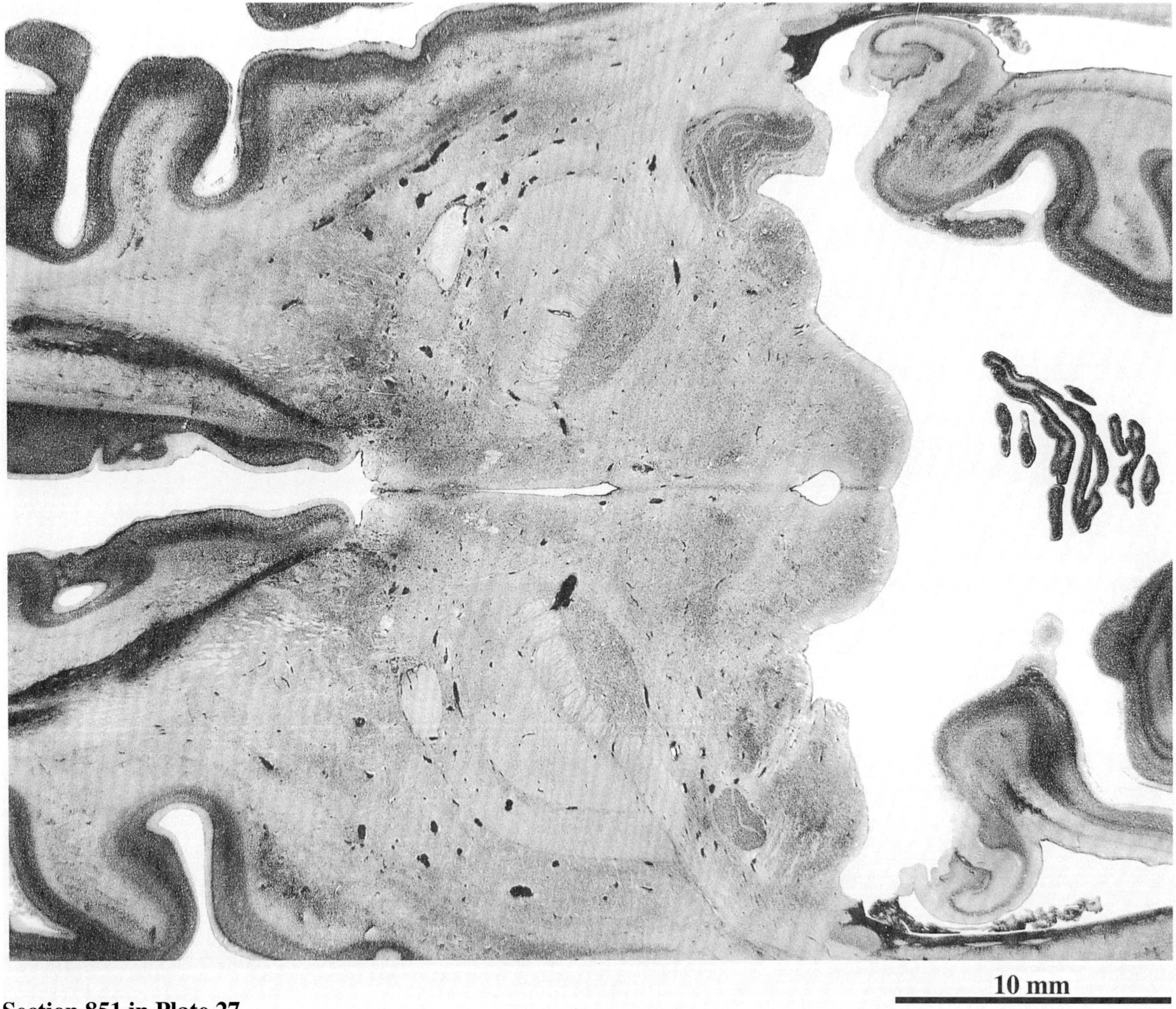

See the entire Section 851 in Plate 27.

Germinal and transitional structures in *italics*

PLATE 44B

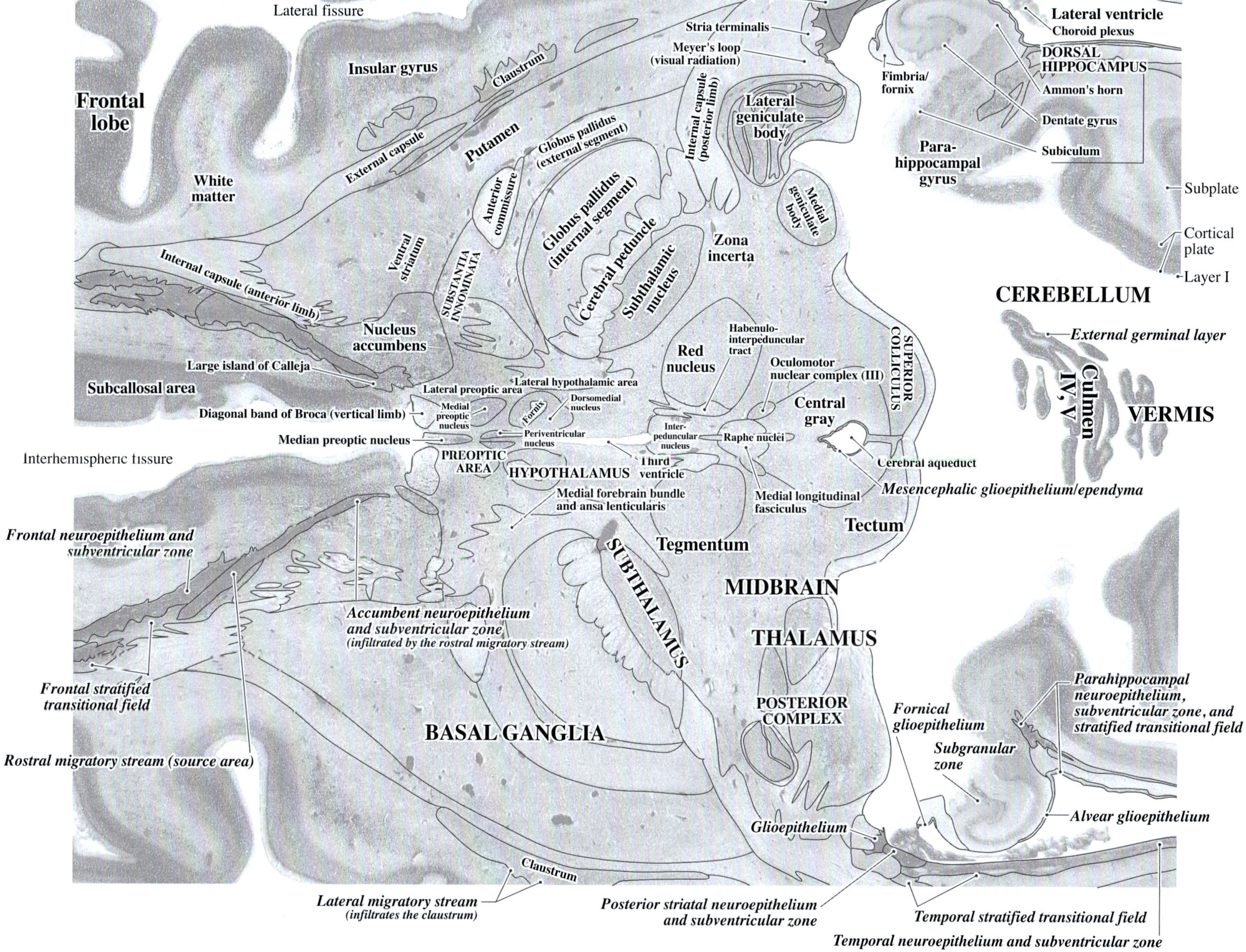

PLATE 45A
CR 320 mm
GW 37
Y301-62
Horizontal
Section 921

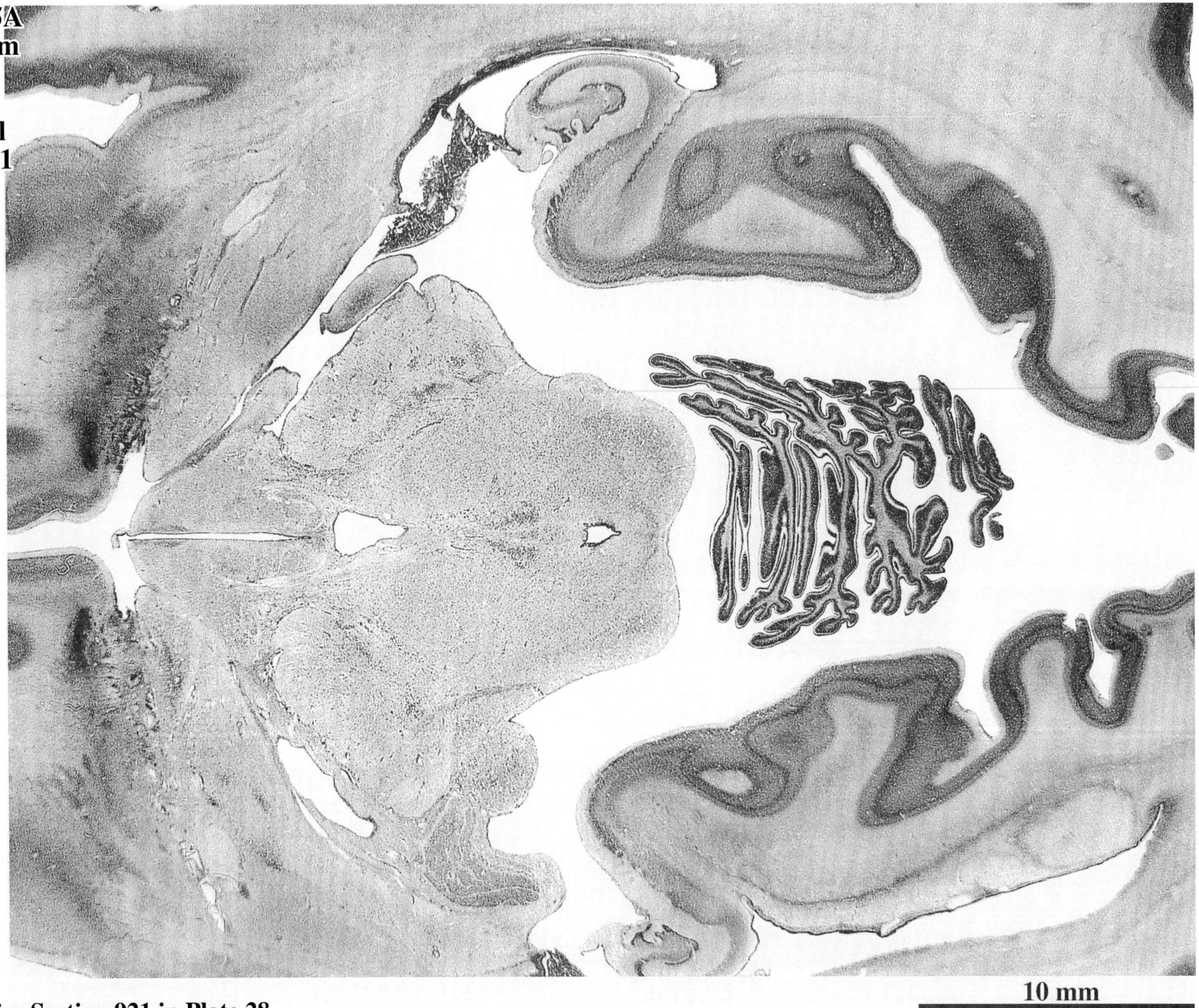

See the entire Section 921 in Plate 28.

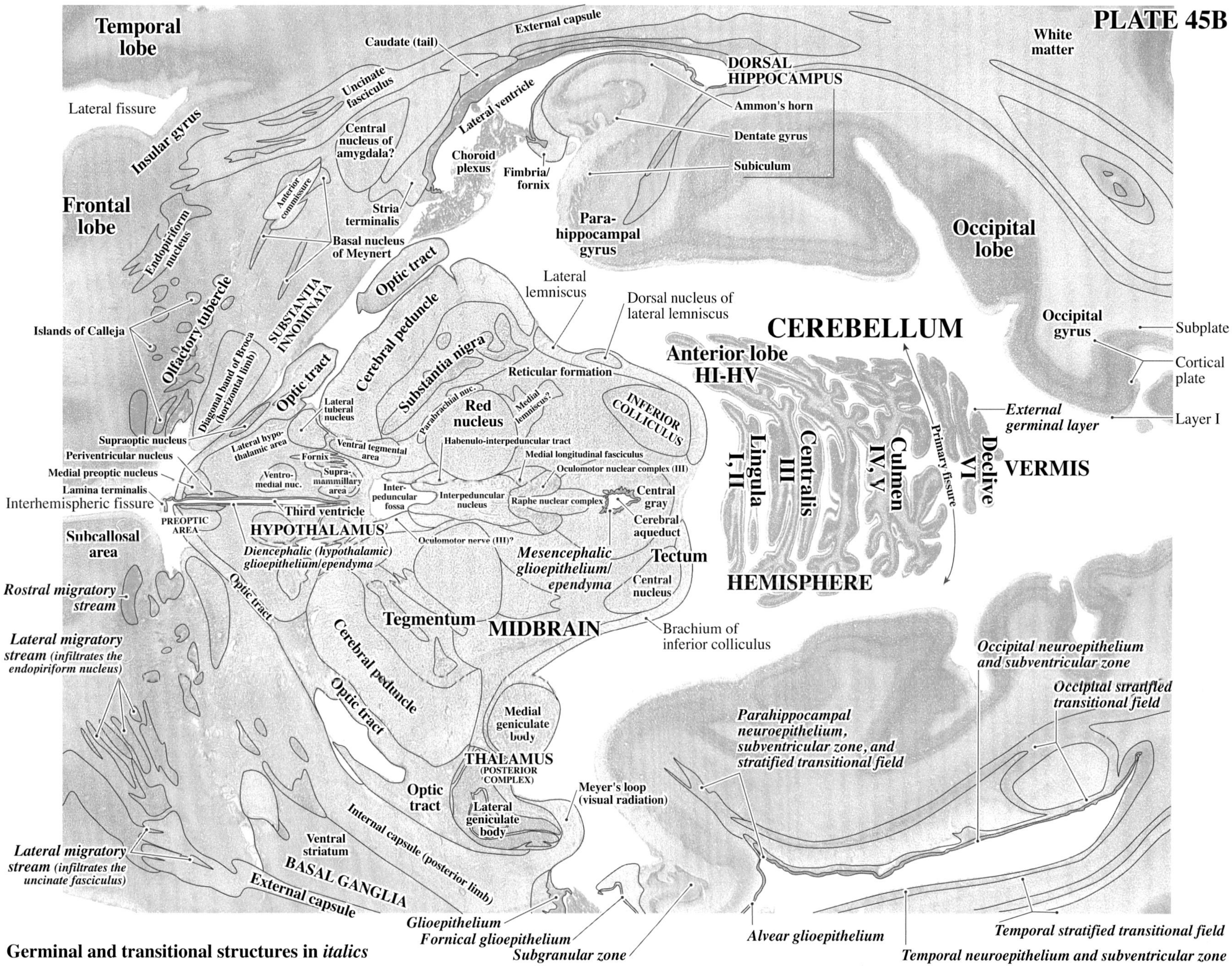

Germinal and transitional structures in *italics*

PLATE 46A
CR 320 mm
GW 37
Y301-62
Horizontal
Section 981

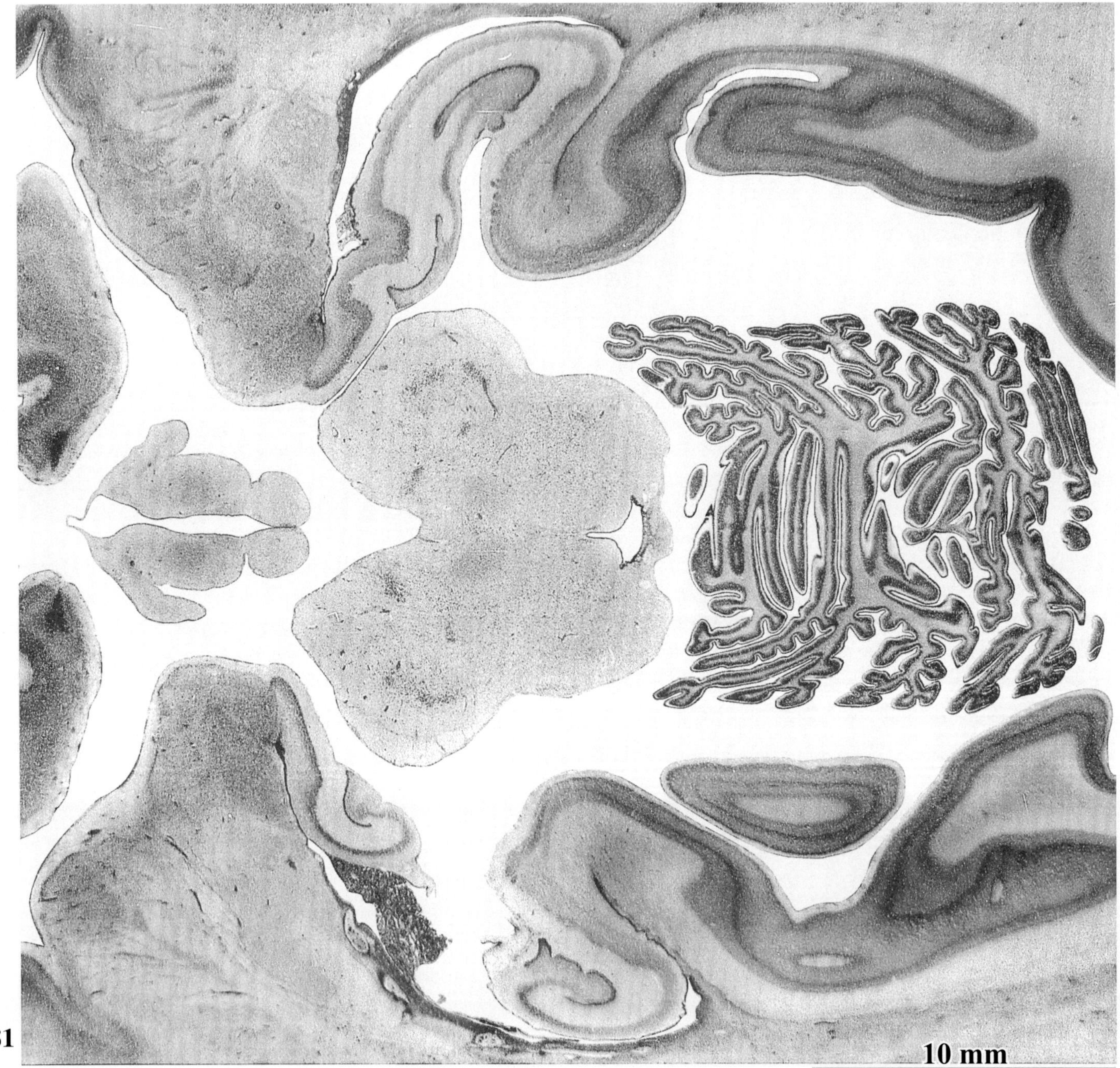

See the entire Section 981 in Plate 29.

PLATE 46B

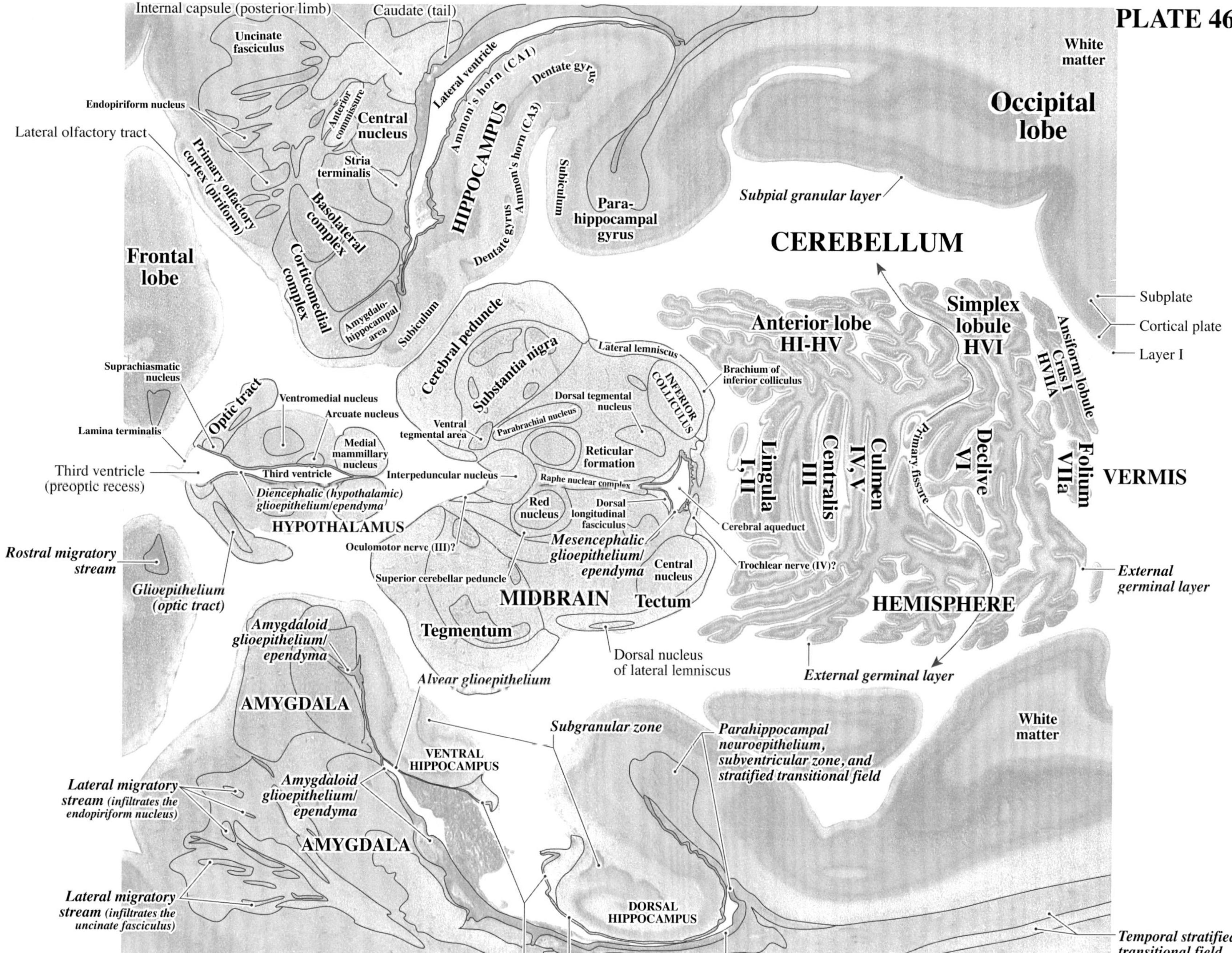

Germinal and transitional structures in *italics*

PLATE 47A
CR 320 mm
GW 37
Y301-62
Horizontal
Section 1051

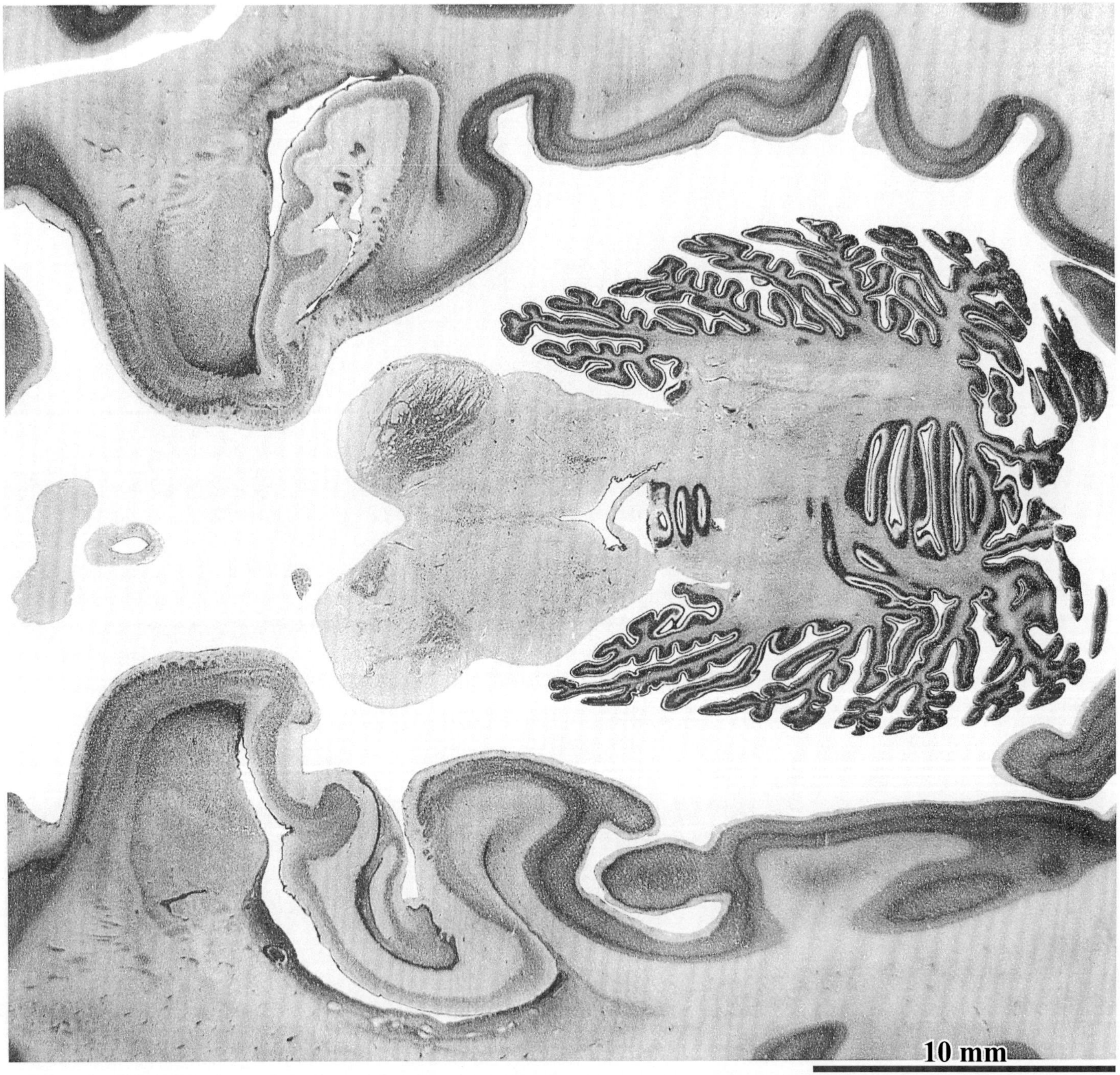

See the entire Section 1051 in Plate 30.

PLATE 47B

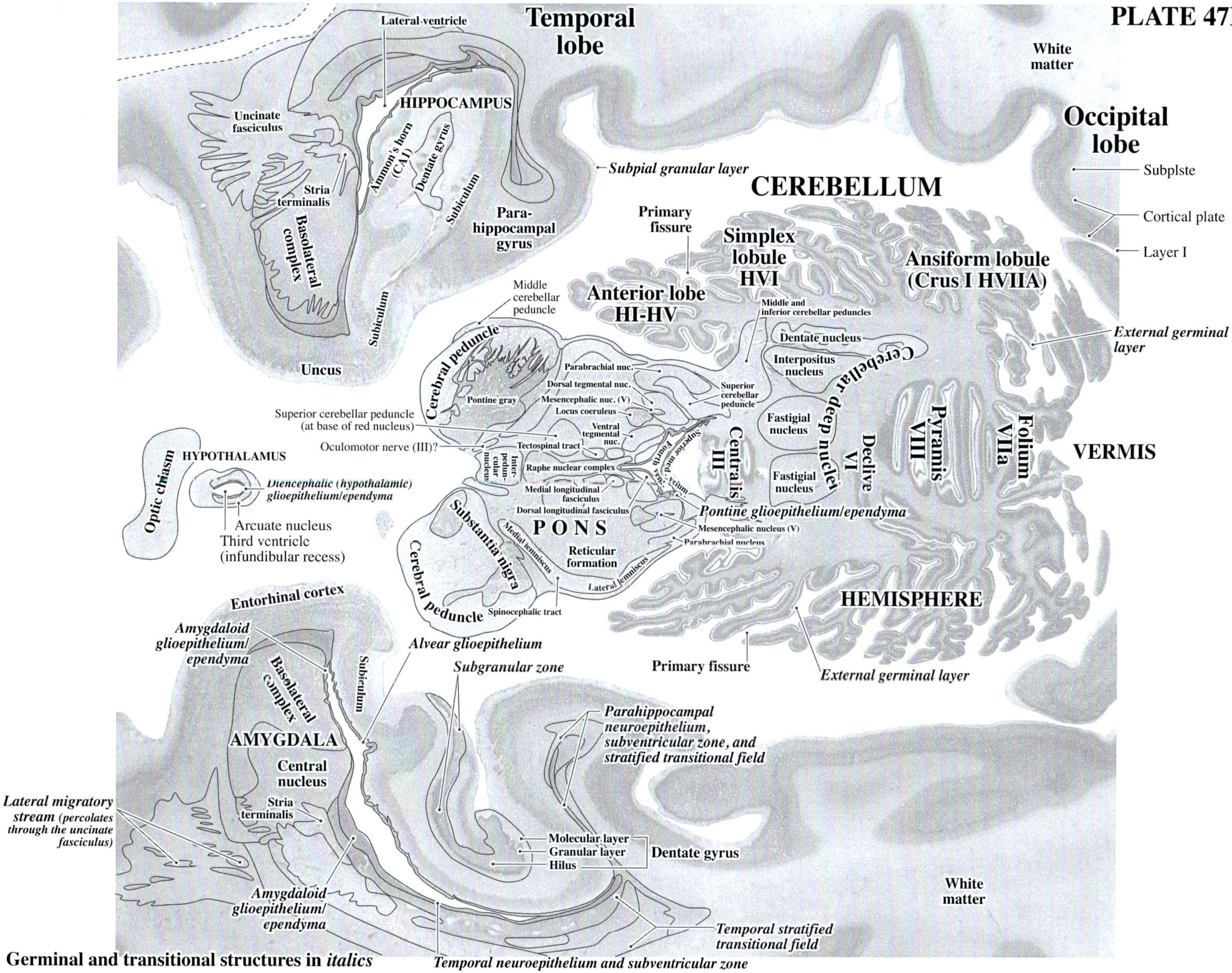

Germinal and transitional structures in *italics*

PLATE 48A
CR 320 mm
GW 37
Y301-62
Horizontal
Section 1111

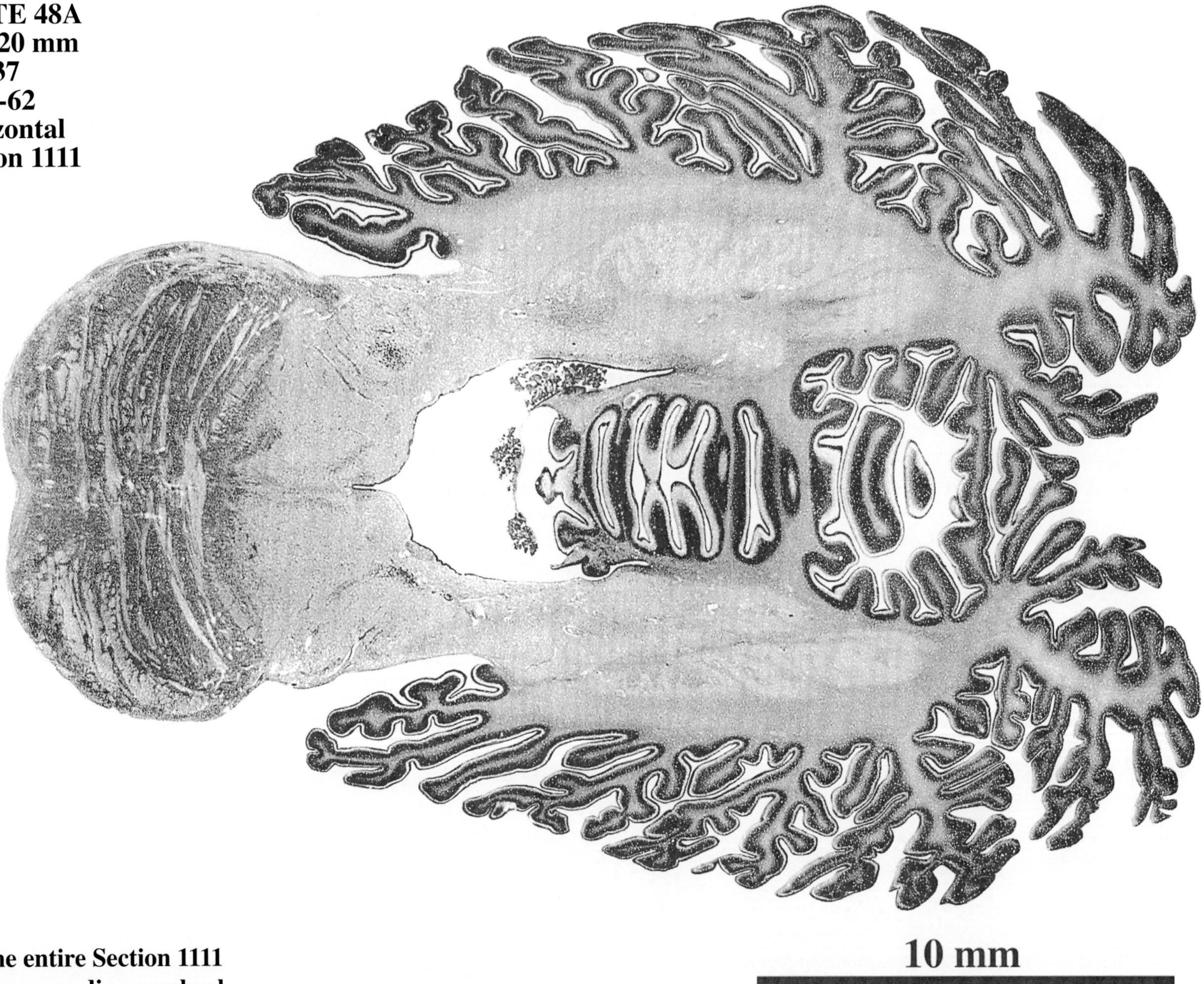

See the entire Section 1111 with surrounding cerebral cortex in Plate 31.

PLATE 48B

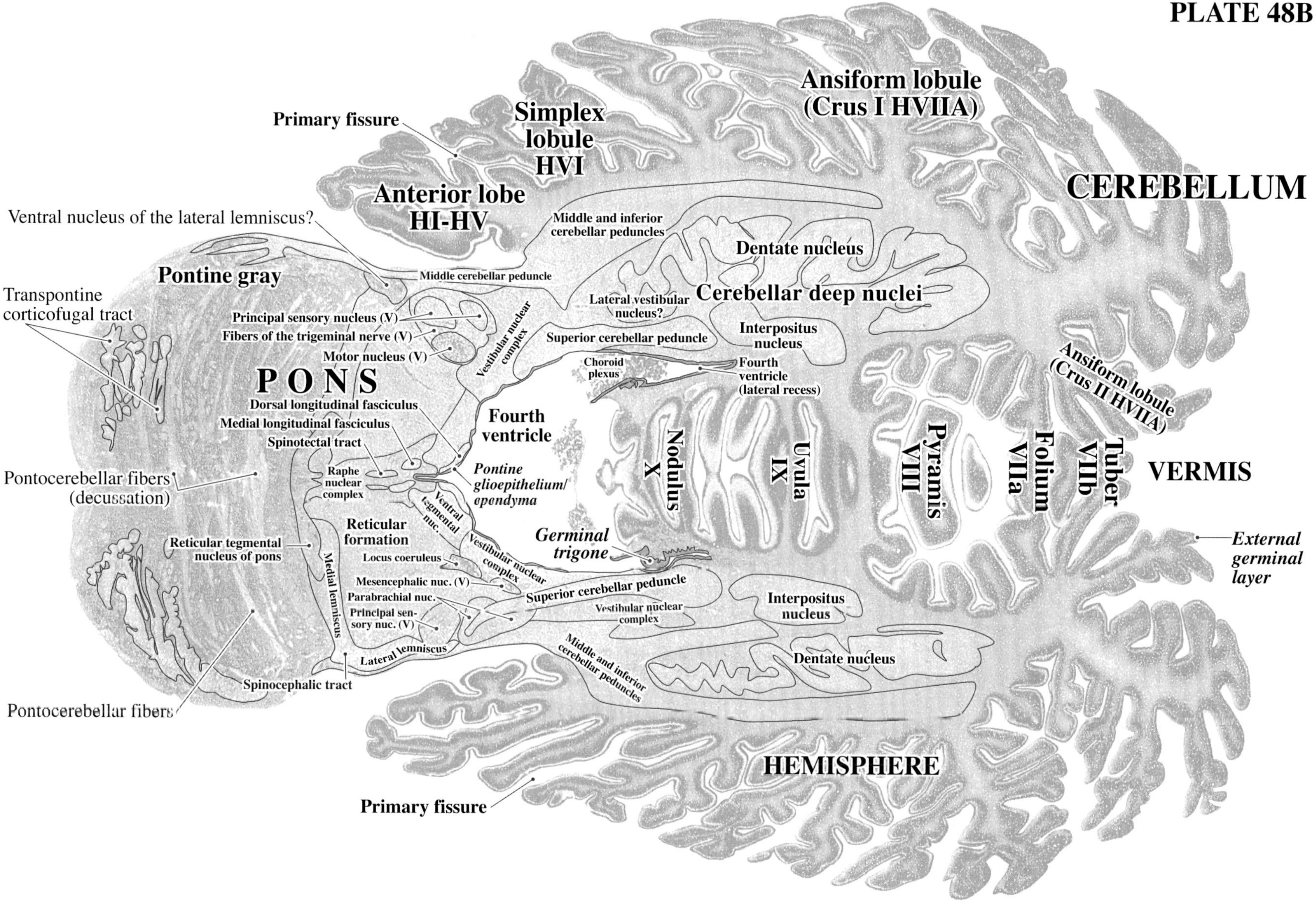

Germinal and transitional structures in *italics*

PLATE 49A
CR 320 mm
GW 37
Y301-62
Horizontal
Section 1161

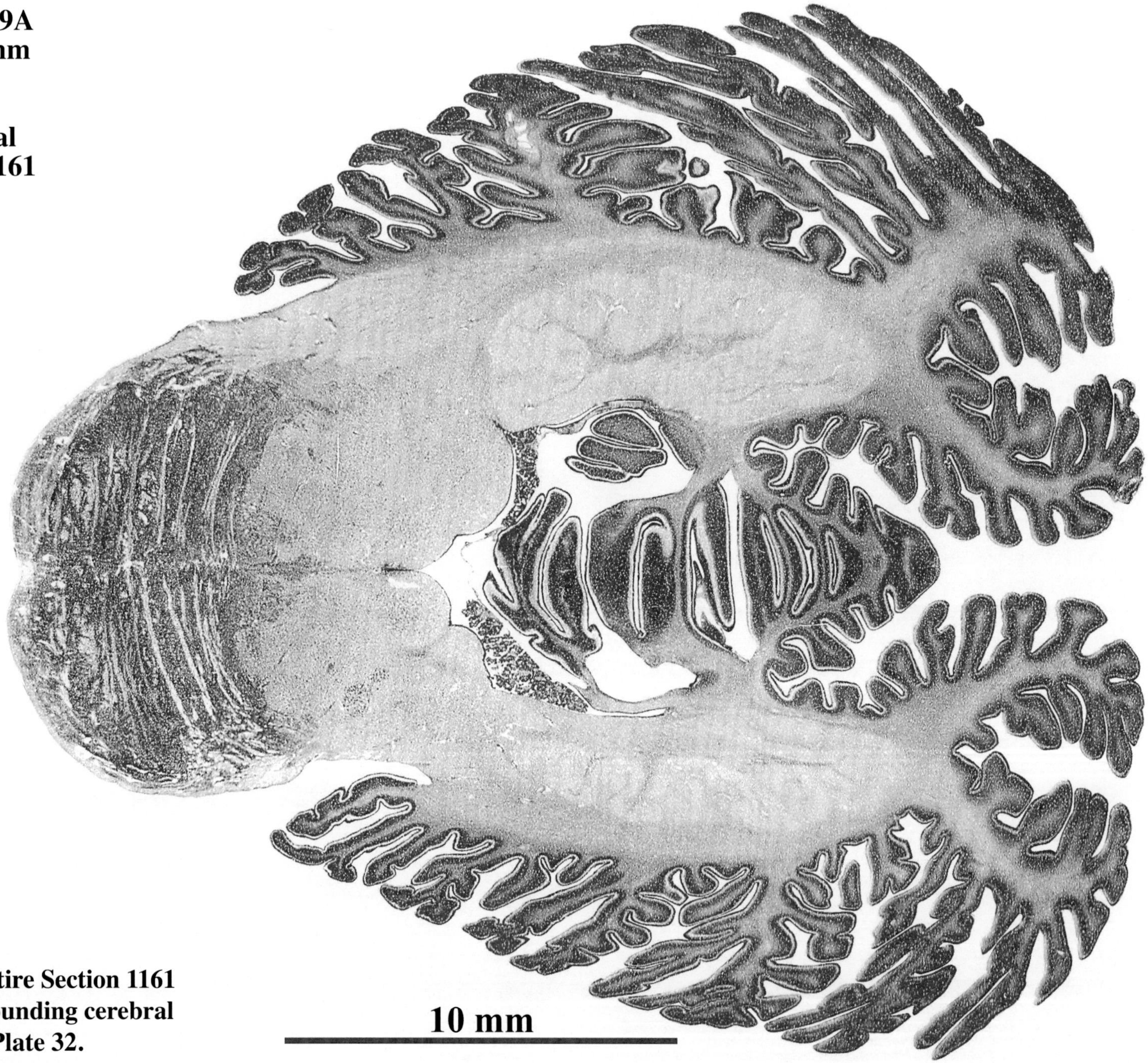

See the entire Section 1161 with surrounding cerebral cortex in Plate 32.

PLATE 49B

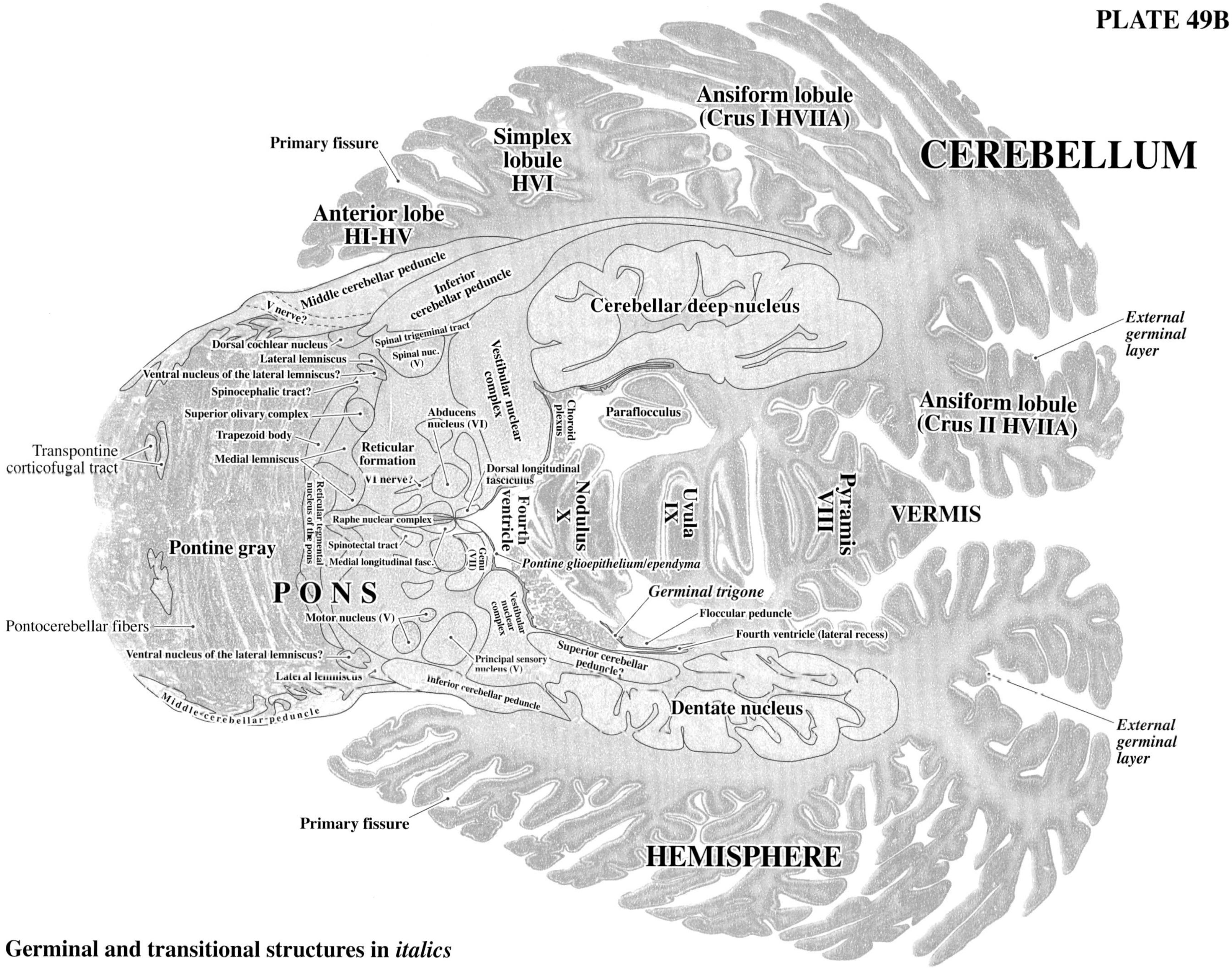

Germinal and transitional structures in *italics*

PART IV: Y217-65 CR 350 mm (GW 37) Frontal

This specimen is case number W-217-65 (Perinatal RPSL) in the Yakovlev Collection. A male infant, weighing 3,057 grams, was born alive on August 10, 1965, and survived for 44 hours. Death occurred because a hyaline membrane obstructed the airway to the lungs. Upon autopsy, a subarachnoid hemmorhage in the left temporal/parietal area of the cerebral cortex was noted. The brain itself, cut in the frontal plane in 35-µm thick sections, is classified as a Normative Control in the Yakovlev Collection (Haleem, 1990). Although there is no photograph of this brain before it was embedded and cut, the photograph of the lateral view of another GW 37 brain that Larroche published in 1967 (**Figure 7**) is similar to the features of the brain in Y217-65. The approximate cutting plane of this brain is indicated in **Figure 8** (facing page) with lines superimposed on the GW 37 brain from the Larroche (1967) series. The dorsal part of each section is posterior to the ventral part. This brain is nearly even in mediolateral orientation. For example, the temporal poles on right vs. left sides of the brain appear at the same time in Section 611 (**Plate 54**). The sections chosen for illustration are more closely spaced to show small structures in the diencephalon, midbrain, pons, and medulla. Illustrated sections are spaced farther apart when they contain only large brain structures, such as the cerebral cortex, basal ganglia, and cerebellum. Low-magnification photographs of 21 Nissl-stained sections are shown in **Plates 50–70**. Different areas of the cerebral cortex are shown at very high magnification in **Plates 71–80**. The core of the brain and the cerebellum are shown at high magnification in **Plates 81-97**.

In the cortical regions of the telencephalon, remnants of the germinal matrices are present in all lobes of the cerebral cortex where the ***neuroepithelium/subventric-***

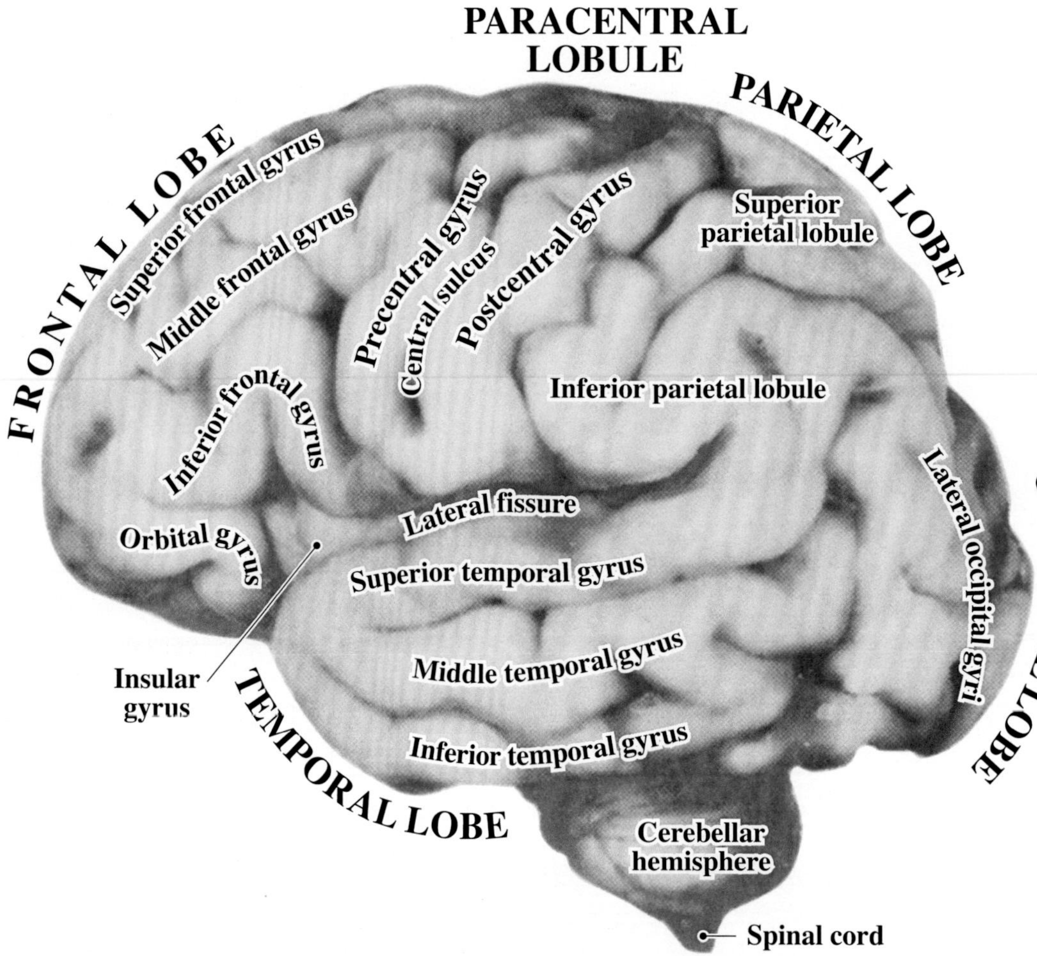

Figure 7. Lateral view of a GW 37 brain with major structures in the cerebral hemispheres labeled. (From the photographic series of J. C. Larroche (1967) Maturation morphologique du système nerveux central: ses rapports avec le développement pondéral du foetus et son age gestationnel. In: *Regional Development of the Brain in Early Life*, A. Minkowski (ed.), London: Blackwell, page 248.)

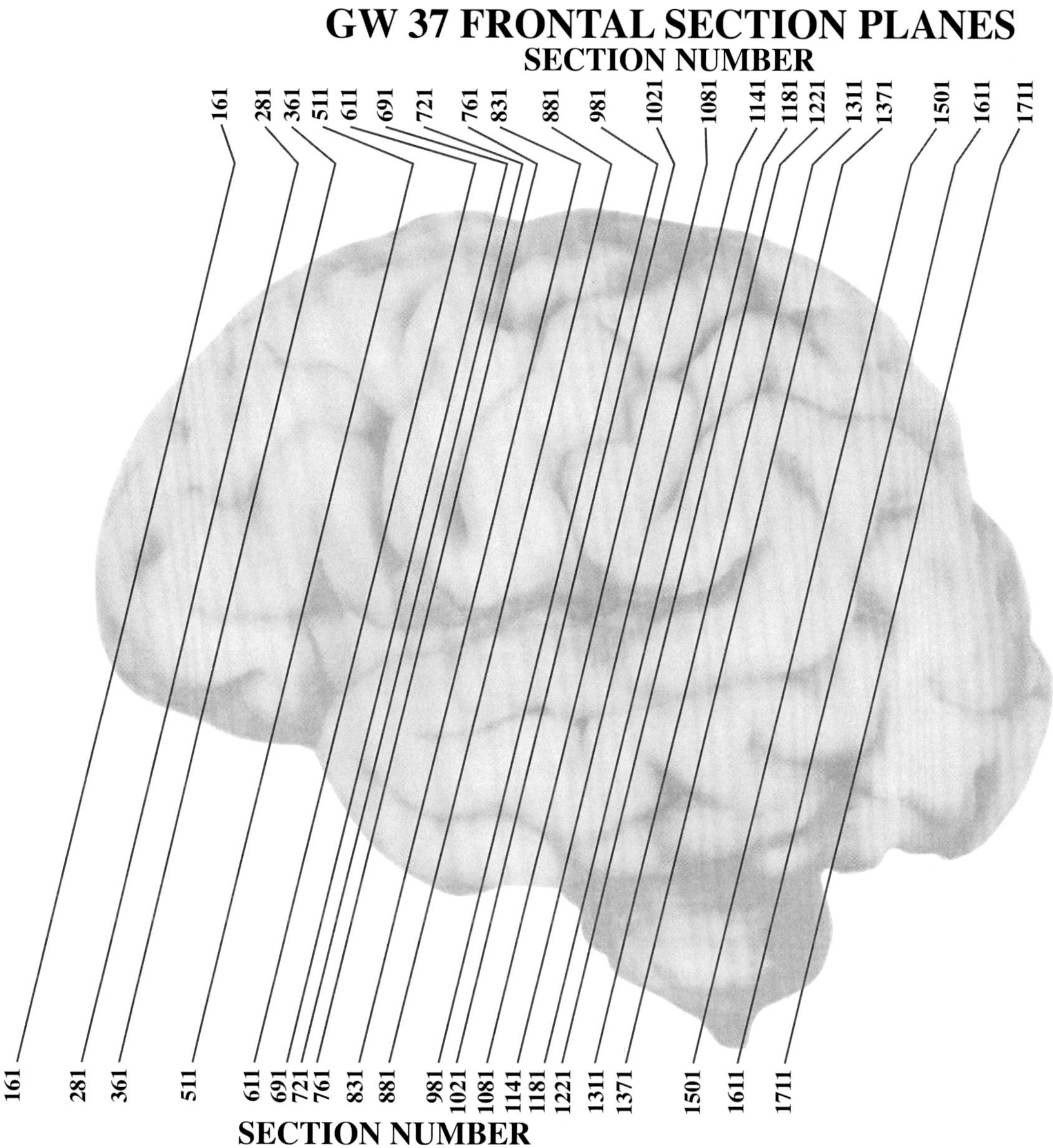

Figure 8. Lateral view of the same GW37 brain shown in **Figure 1** with the approximate locations and cutting angle of the sections of Y217-65. (From the photographic series of J. C. Larroche (1967) Maturation morphologique du système nerveux central: ses rapports avec le développement pondéral du foetus et son age gestationnel. In: *Regional Development of the Brain in Early Life*, A. Minkowski (ed.), London: Blackwell, page 248.)

ular zone is still generating neocortical interneurons. Remnants of migrating and sojourning neurons and/or glia are visible in all lobes of the cerebral cortex in the dwindling ***stratified transitional fields***. Many neurons, glia, and their mitotic precursor cells are still migrating in the ***rostral migratory stream*** from a presumed source area in the germinal matrix at the junction between the cerebral cortex, striatum, and nucleus accumbens. Within the lateral parts of the cerebral cortex, streams of neurons and glia are in the ***lateral migratory stream*** that percolates through the claustrum, endopiriform nucleus, external capsule, and uncinate fasciculus. The cells appear to be heading toward the insular cortex, primary olfactory cortex, temporal cortex, and basolateral parts of the amygdaloid complex. There is active neurogenesis in the ***subgranular zone*** in the hilus of the dentate gyrus that is generating granule cells.

In the basal ganglia, there is a prominent ***neuroepithelium/subventricular zone*** overlying the striatum and nucleus accumbens where neurons are being generated. The striatal portion can be subdivided into anterolateral, anteromedial, and posterior parts. Other structures in the telencephalon, such as the septum, fornix, and Ammon's horn part of the hippocampus, have only a thin, darkly staining layer at the ventricle, and these are presumed to be generating glia, cells of the choroid plexus, and the ependymal lining of the ventricle

Most of the structures in the diencephalon appear to be settled and are maturing, and the third ventricle is lined by a thin ***glioepithelium/ependyma***. In the midbrain and anterior pons, there is a slightly thicker and more convoluted ***glioepithelium/ependyma*** lining the posterior cerebral aqueduct and anterior fourth ventricle. The posterior pons and entire medulla have a thin ***glioepithelium/ependyma*** lining the rest of the fourth ventricle. The ***external germinal layer*** is prominent over the entire surface of the cerebellar cortex and is producing basket, stellate, and granule cells. The ***germinal trigone*** is visible at the base of the nodulus and along the floccular peduncle; choroid plexus cells and glia may be originating here.

PLATE 50A
CR 350 mm
GW 37
Y217-65
Coronal
Section 161

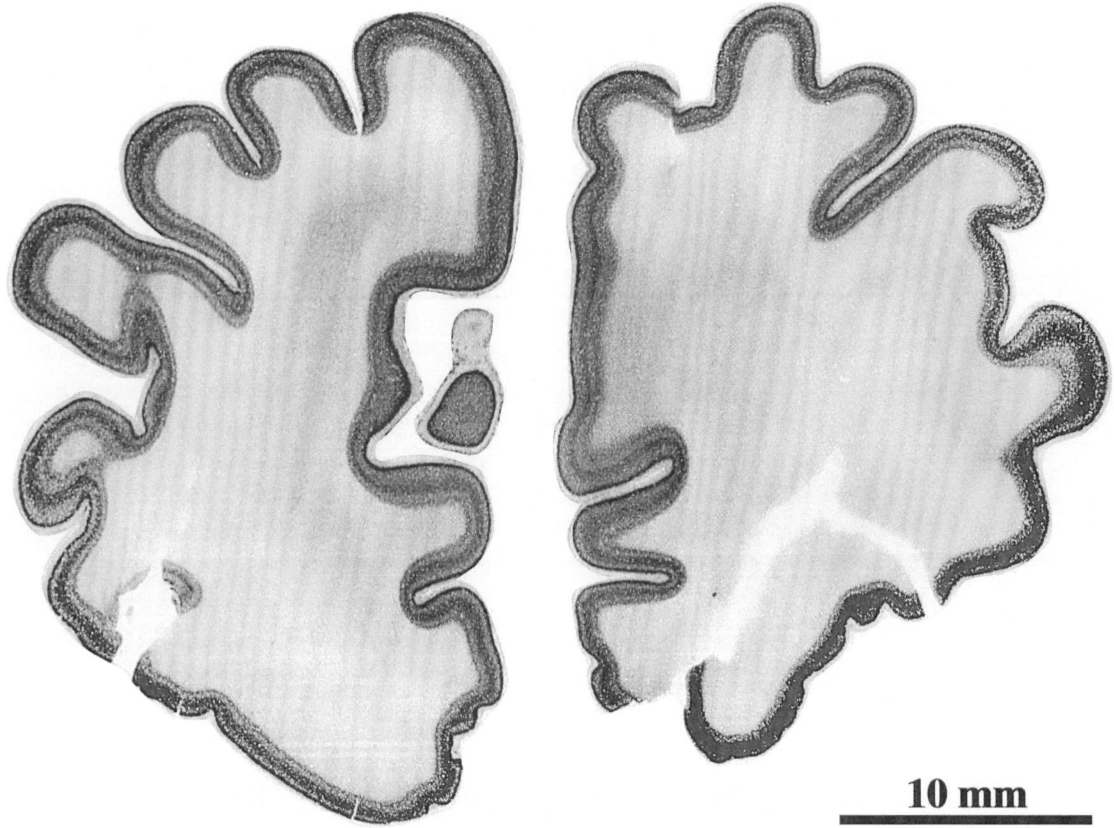

Remnants of the germinal matrix migratory streams, and transitional fields

1 ***Subpial granular layer (cortical)***

PLATE 50B

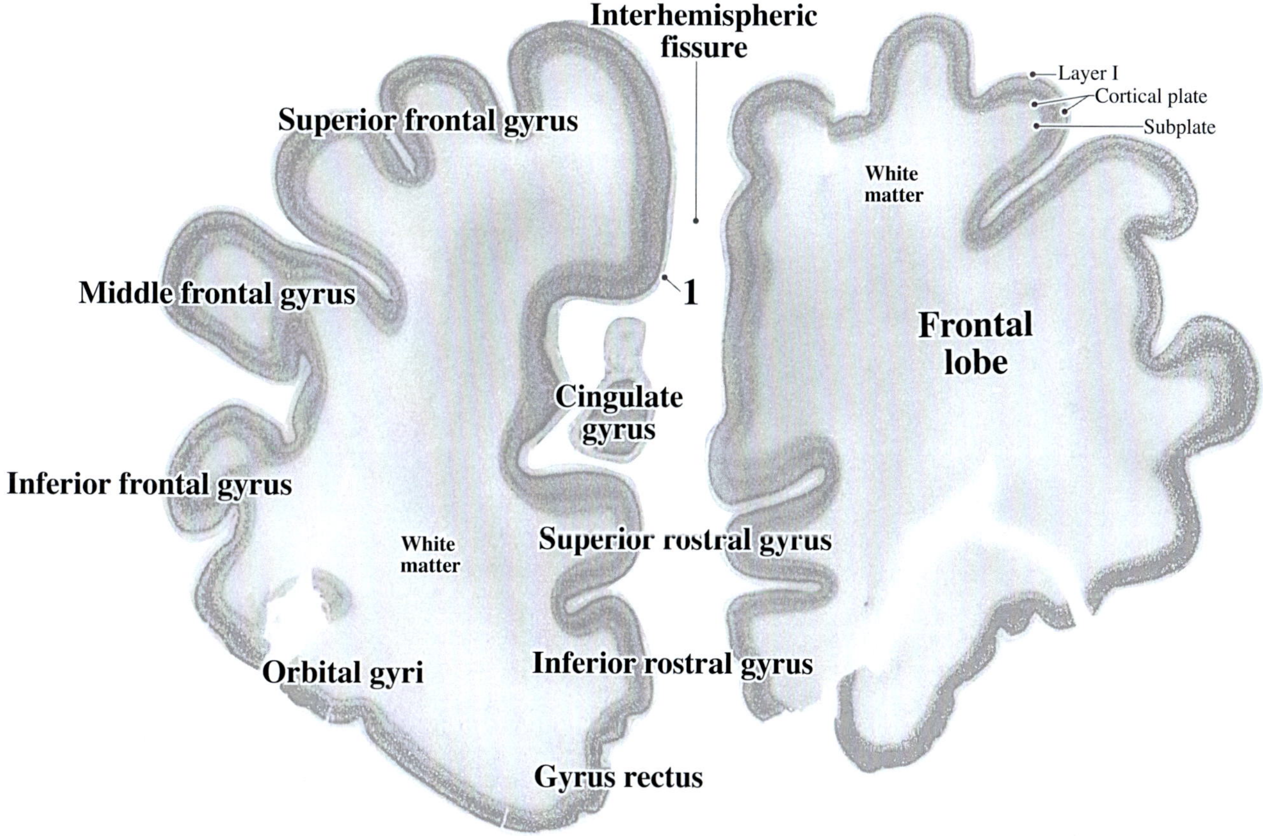
Interhemispheric fissure
Layer I
Cortical plate
Subplate
Superior frontal gyrus
White matter
Middle frontal gyrus
1
Frontal lobe
Cingulate gyrus
Inferior frontal gyrus
White matter
Superior rostral gyrus
Orbital gyri
Inferior rostral gyrus
Gyrus rectus

PLATE 51A
CR 350 mm
GW 37
Y217-65
Coronal
Section 281

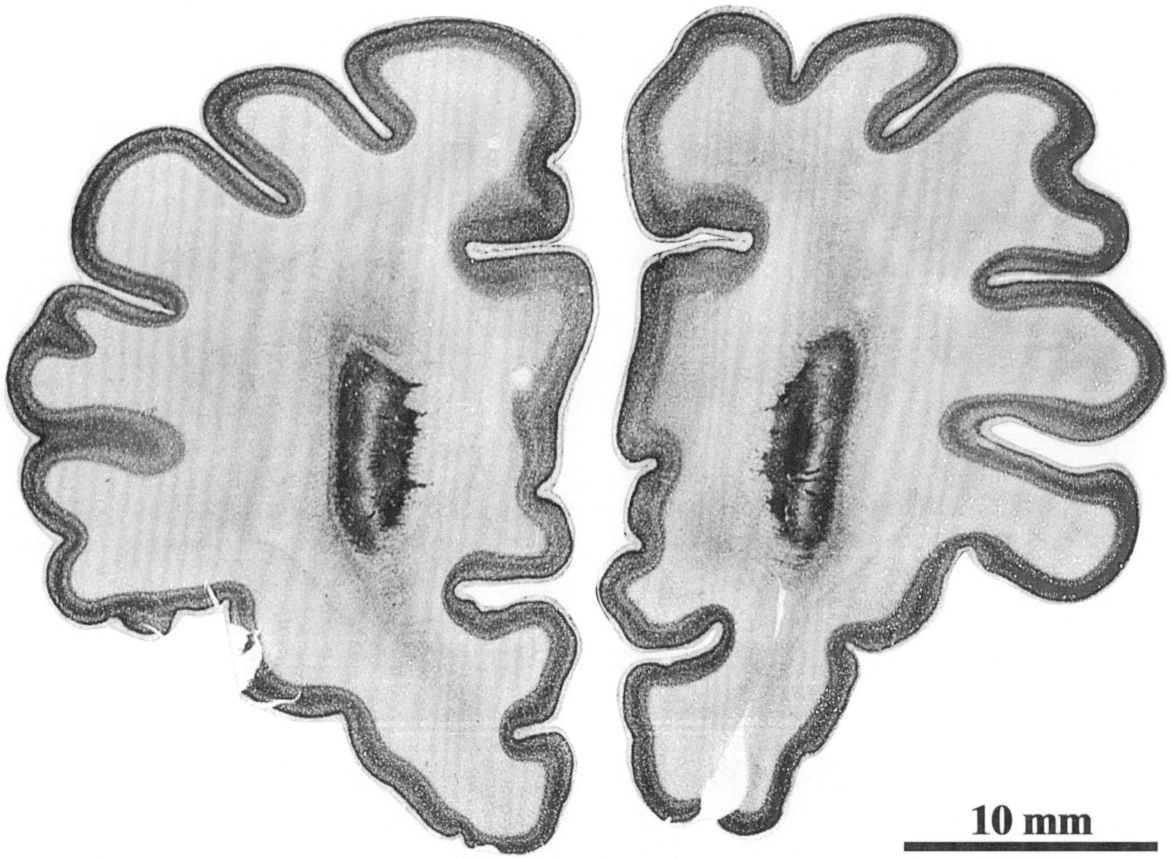

Remnants of the germinal matrix migratory streams, and transitional fields

1 *Frontal neuroepithelium and subventricular zone*

2 *Frontal stratified transitional field*

3 *Subpial granular layer (cortical)*

PLATE 51B

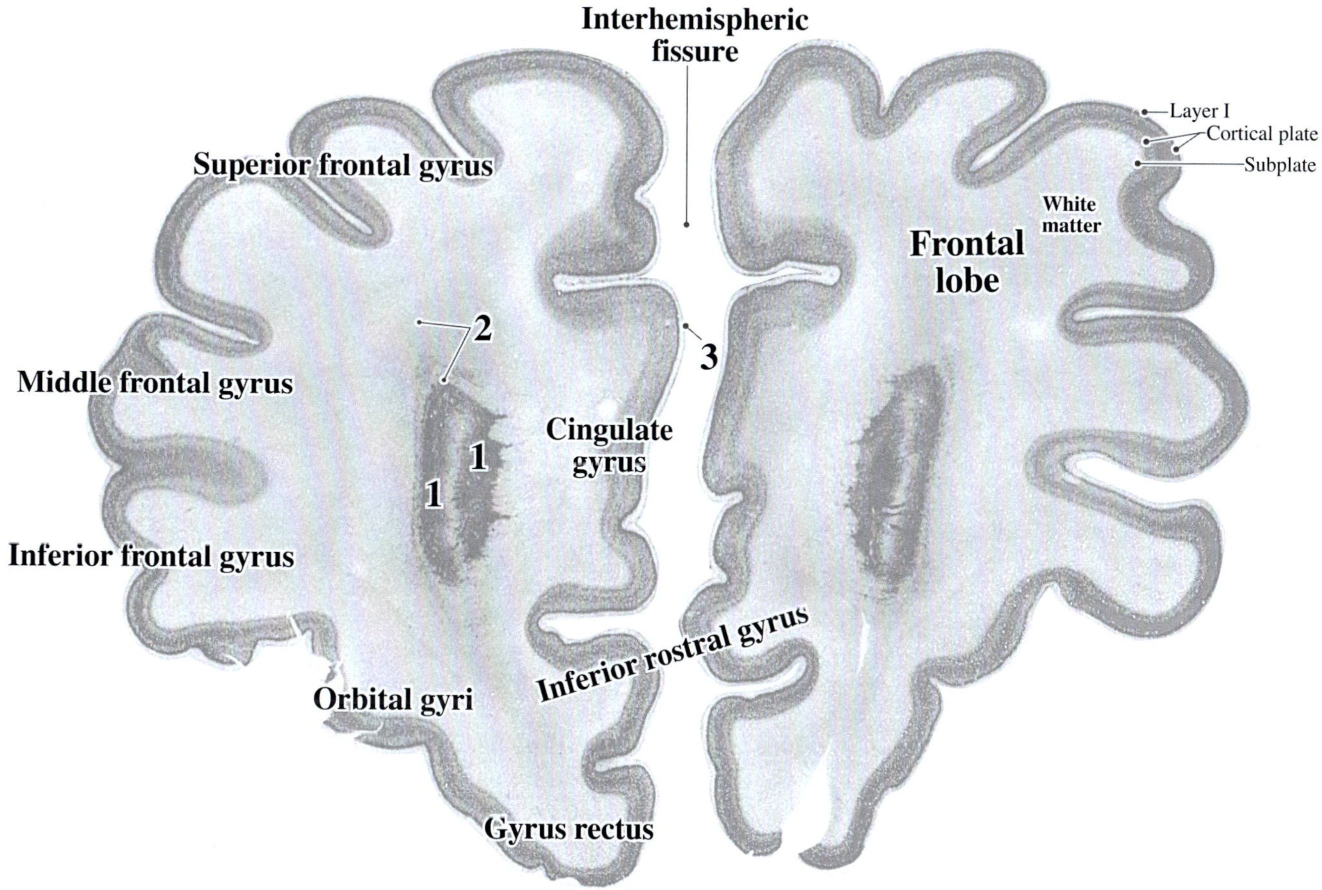
Interhemispheric fissure
Superior frontal gyrus
Layer I
Cortical plate
Subplate
White matter
Frontal lobe
2
3
Middle frontal gyrus
Cingulate gyrus
1
1
Inferior frontal gyrus
Inferior rostral gyrus
Orbital gyri
Gyrus rectus

PLATE 52A
CR 350 mm
GW 37
Y217-65
Coronal
Section 361

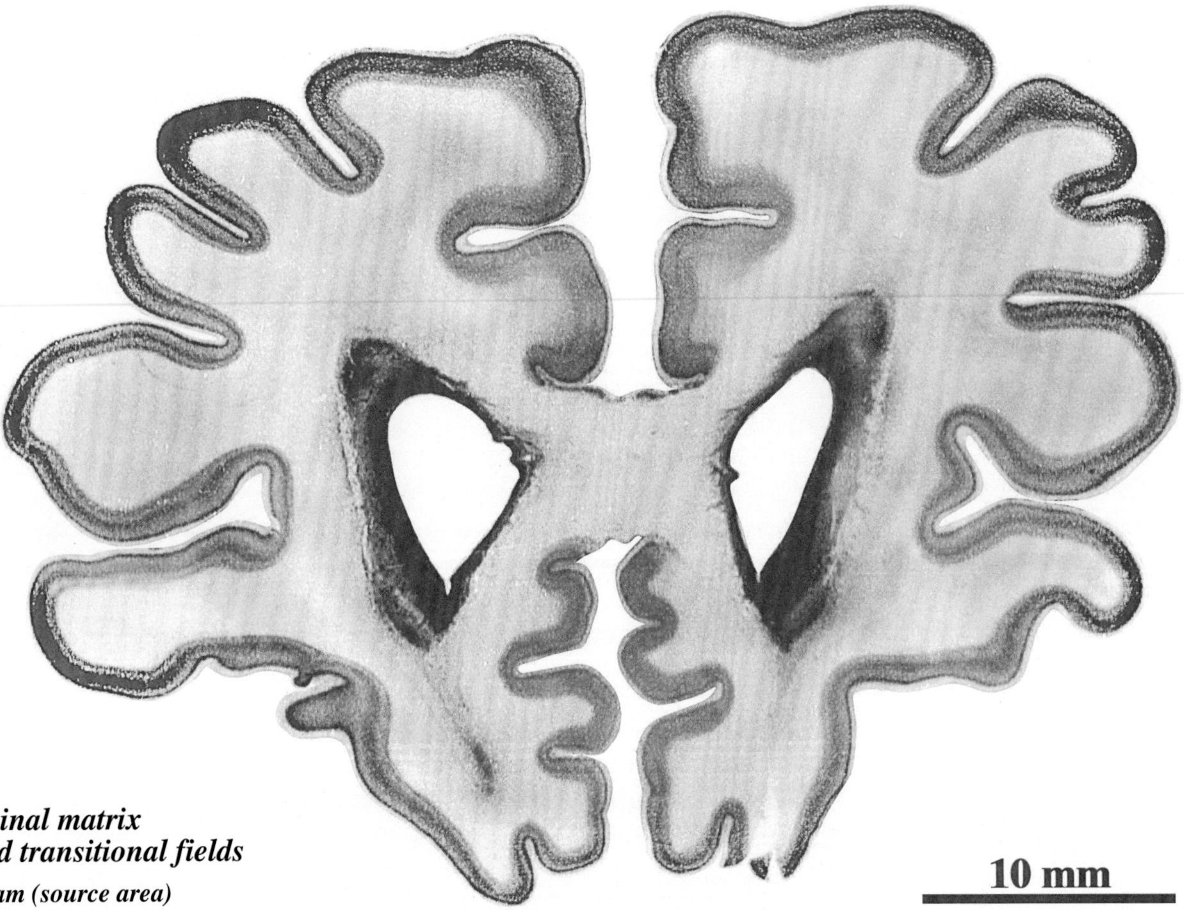

Remnants of the germinal matrix migratory streams, and transitional fields

1 *Rostral migratory stream (source area)*
2 *Frontal NEP and SVZ*
3 *Frontal stratified transitional field*
4 *Callosal GEP*
5 *Anterolateral striatal NEP and SVZ*
6 *Subpial granular layer (cortical)*

GEP - Glioepithelium
G/EP - Glioepithelium/ependyma
NEP - Neuroepithelium
STF - Stratified transitional field
SVZ - Subventricular zone

PLATE 52B

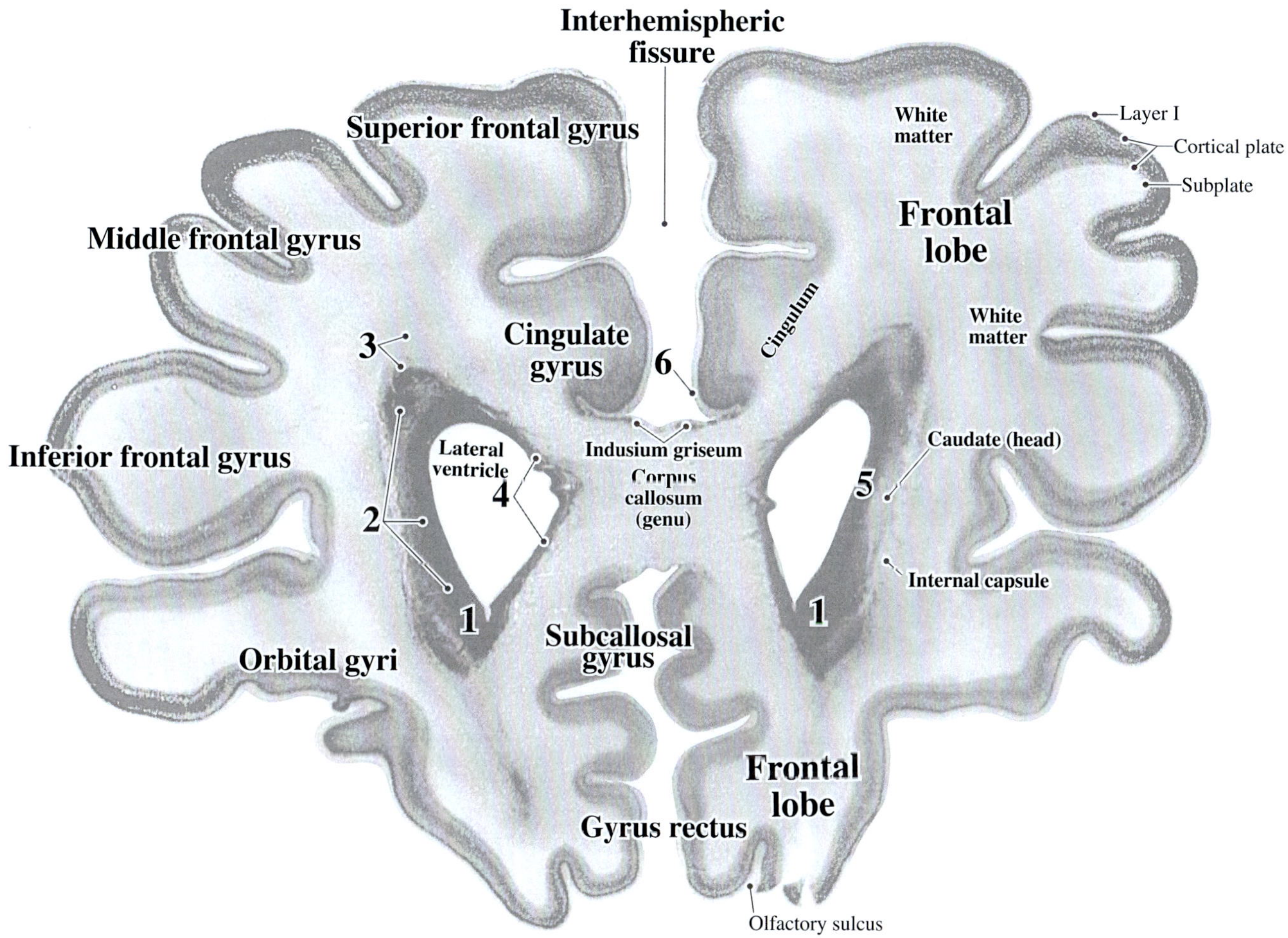
Interhemispheric fissure
Superior frontal gyrus
White matter
Layer I
Cortical plate
Subplate
Frontal lobe
Middle frontal gyrus
Cingulum
White matter
Cingulate gyrus
3
6
Indusium griseum
Lateral ventricle
Inferior frontal gyrus
4
Corpus callosum (genu)
Caudate (head)
5
2
Internal capsule
1
1
Subcallosal gyrus
Orbital gyri
Frontal lobe
Gyrus rectus
Olfactory sulcus

PLATE 53A
CR 350 mm
GW 37
Y217-65
Coronal
Section 511

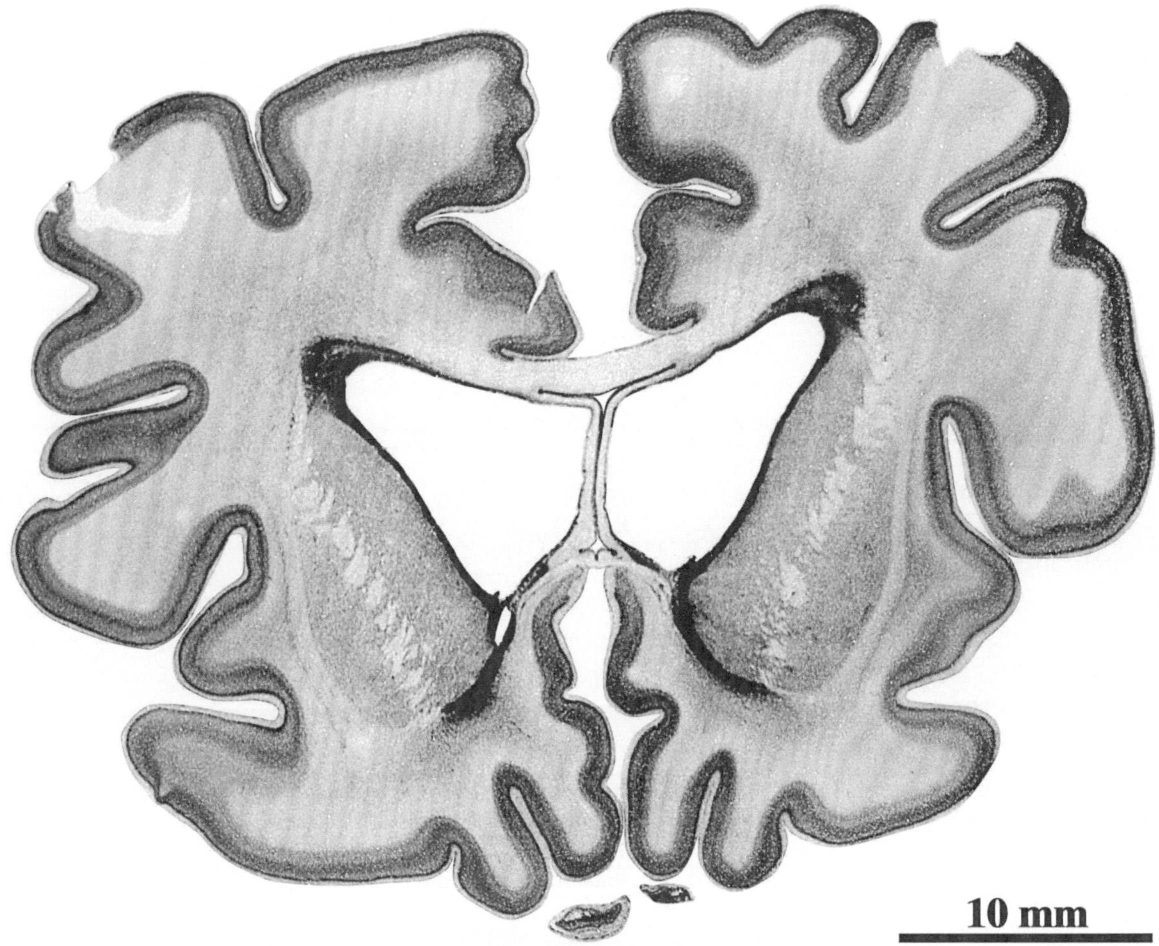

Remnants of the germinal matrix migratory streams, and transitional fields

1 ***Rostral migratory stream***

2 ***Frontal NEP and SVZ***

3 ***Frontal STF***

4 ***Frontal NEP and SVZ (intermingled with the source of the rostral migratory stream)***

5 ***Callosal GEP***

6 ***Callosal sling***

7 ***Fornical GEP***

8 ***Anterolateral striatal NEP and SVZ***

9 ***Accumbent NEP and SVZ (intermingled with the source of the rostral migratory stream)***

10 ***Subpial granular layer (cortical)***

GEP - Glioepithelium
NEP - Neuroepithelium
STF - Stratified transitional field
SVZ - Subventricular zone

PLATE 53B

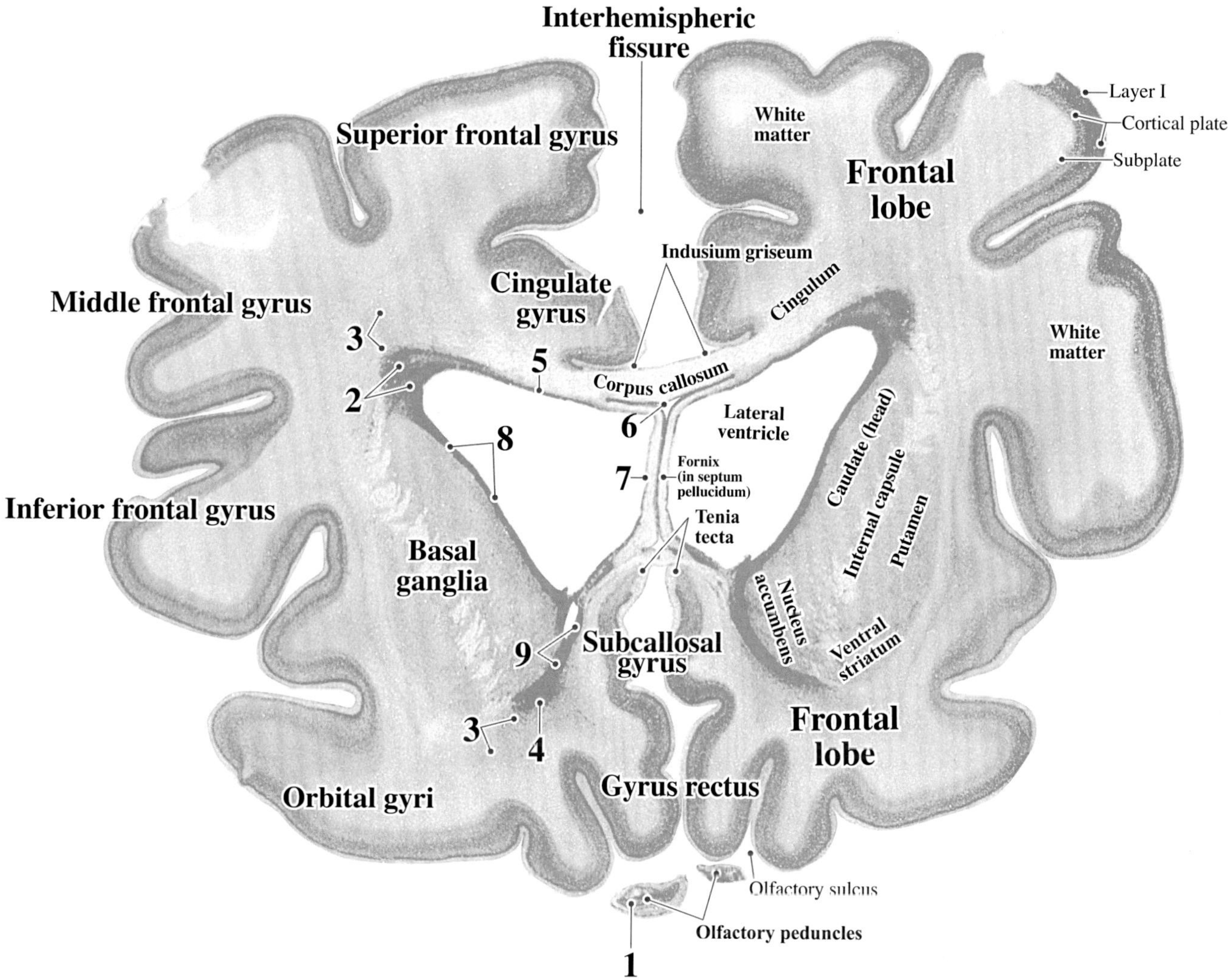
Interhemispheric fissure
Superior frontal gyrus
White matter
Frontal lobe
Layer I
Cortical plate
Subplate
Indusium griseum
Cingulate gyrus
Cingulum
Middle frontal gyrus
White matter
3
2
5
Corpus callosum
Lateral ventricle
6
8
Caudate (head)
7
Fornix (in septum pellucidum)
Internal capsule
Putamen
Inferior frontal gyrus
Tenia tecta
Basal ganglia
Nucleus accumbens
Ventral striatum
9
Subcallosal gyrus
Frontal lobe
3
4
Gyrus rectus
Orbital gyri
Olfactory sulcus
Olfactory peduncles
1

PLATE 54A
CR 350 mm
GW 37
Y217-65
Coronal
Section 611

See detail of the brain core in Plates 81A and B.

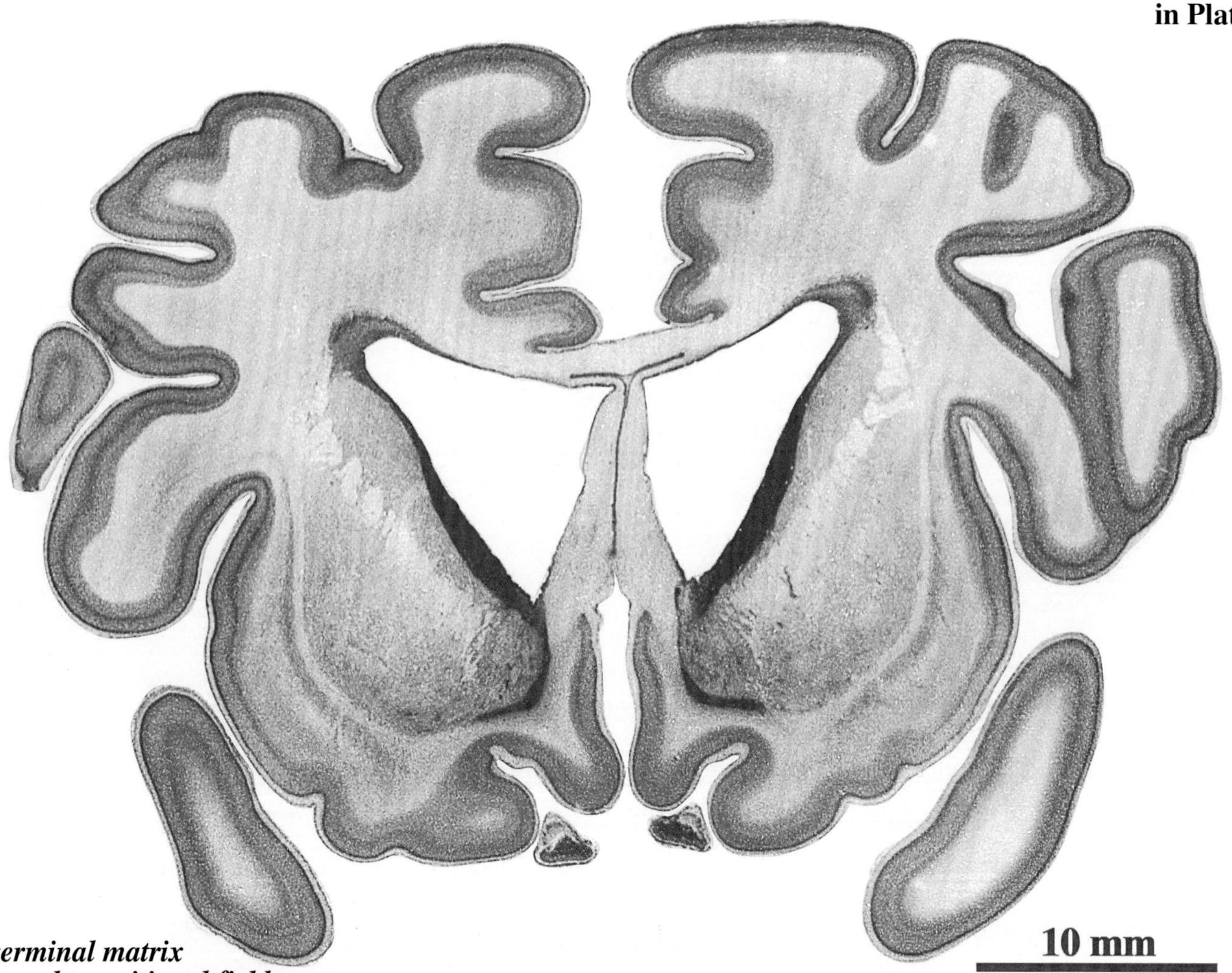

Remnants of the germinal matrix migratory streams, and transitional fields

1 *Rostral migratory stream*
2 *Frontal NEP and SVZ*
3 *Frontal STF*
4 *Callosal GEP*
5 *Callosal sling*
6 *Fornical GEP*
7 *Anterolateral striatal NEP and SVZ*
8 *Anteromedial striatal NEP and SVZ*
9 *Accumbent NEP and SVZ (intermingled with the source of the rostral migratory stream)*
10 *Septal G/EP*
11 *Subpial granular layer (cortical)*

GEP - Glioepithelium
G/EP - Glioepithelium/ependyma
NEP - Neuroepithelium
STF - Stratified transitional field
SVZ - Subventricular zone

PLATE 54B

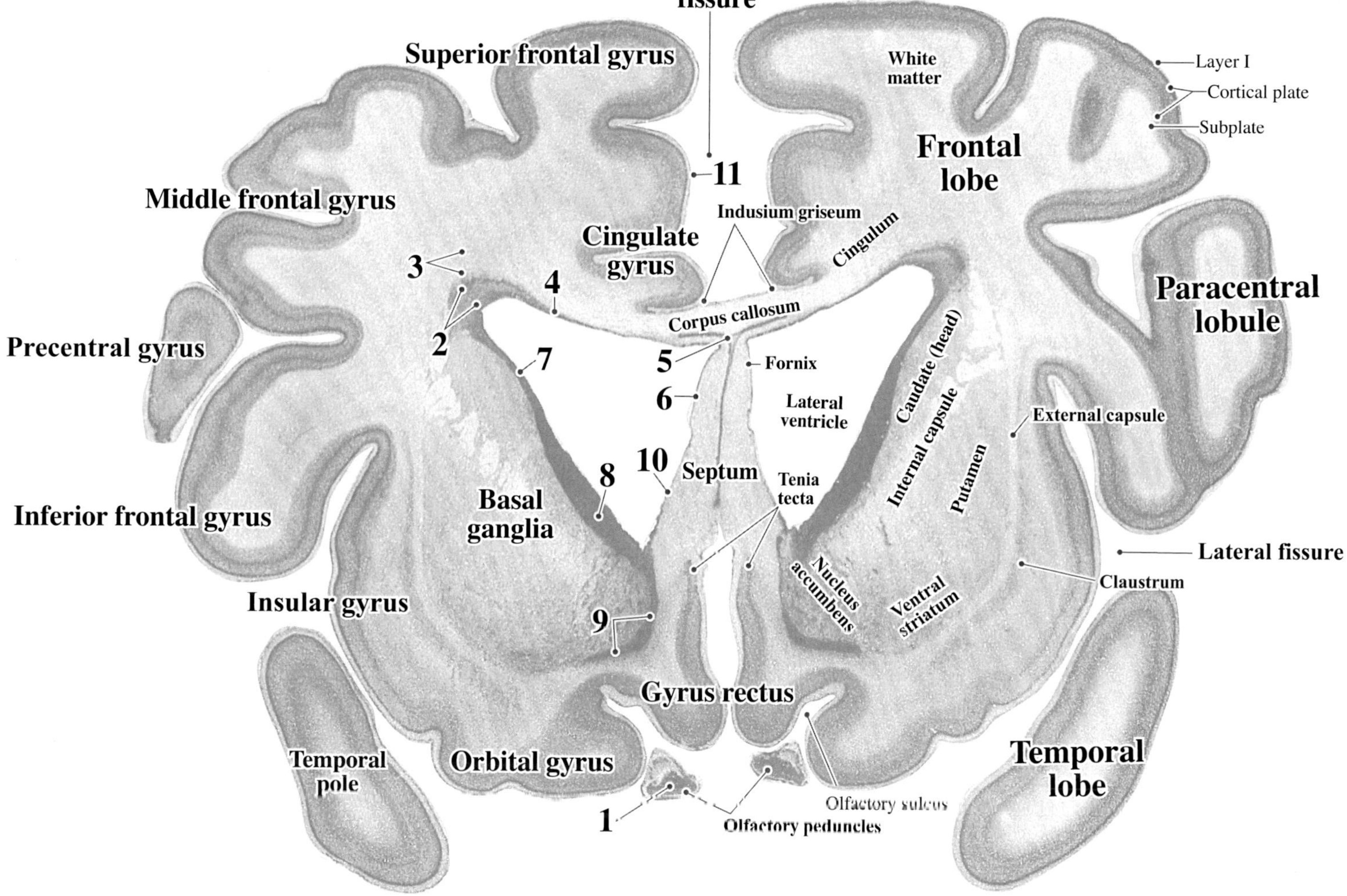
Interhemispheric fissure
Superior frontal gyrus
White matter
Layer I
Cortical plate
Subplate
Frontal lobe
11
Middle frontal gyrus
Indusium griseum
Cingulate gyrus
Cingulum
3
4
Corpus callosum
Paracentral lobule
2
Precentral gyrus
7
5
Fornix
6
Lateral ventricle
Caudate (head)
Internal capsule
External capsule
Putamen
10
Septum
8
Basal ganglia
Tenia tecta
Inferior frontal gyrus
Lateral fissure
Nucleus accumbens
Claustrum
Insular gyrus
Ventral striatum
9
Gyrus rectus
Temporal pole
Orbital gyrus
Temporal lobe
Olfactory sulcus
1
Olfactory peduncles

PLATE 55A
CR 350 mm
GW 37
Y217-65
Coronal
Section 691

See detail of the brain core in Plates 82A and B.

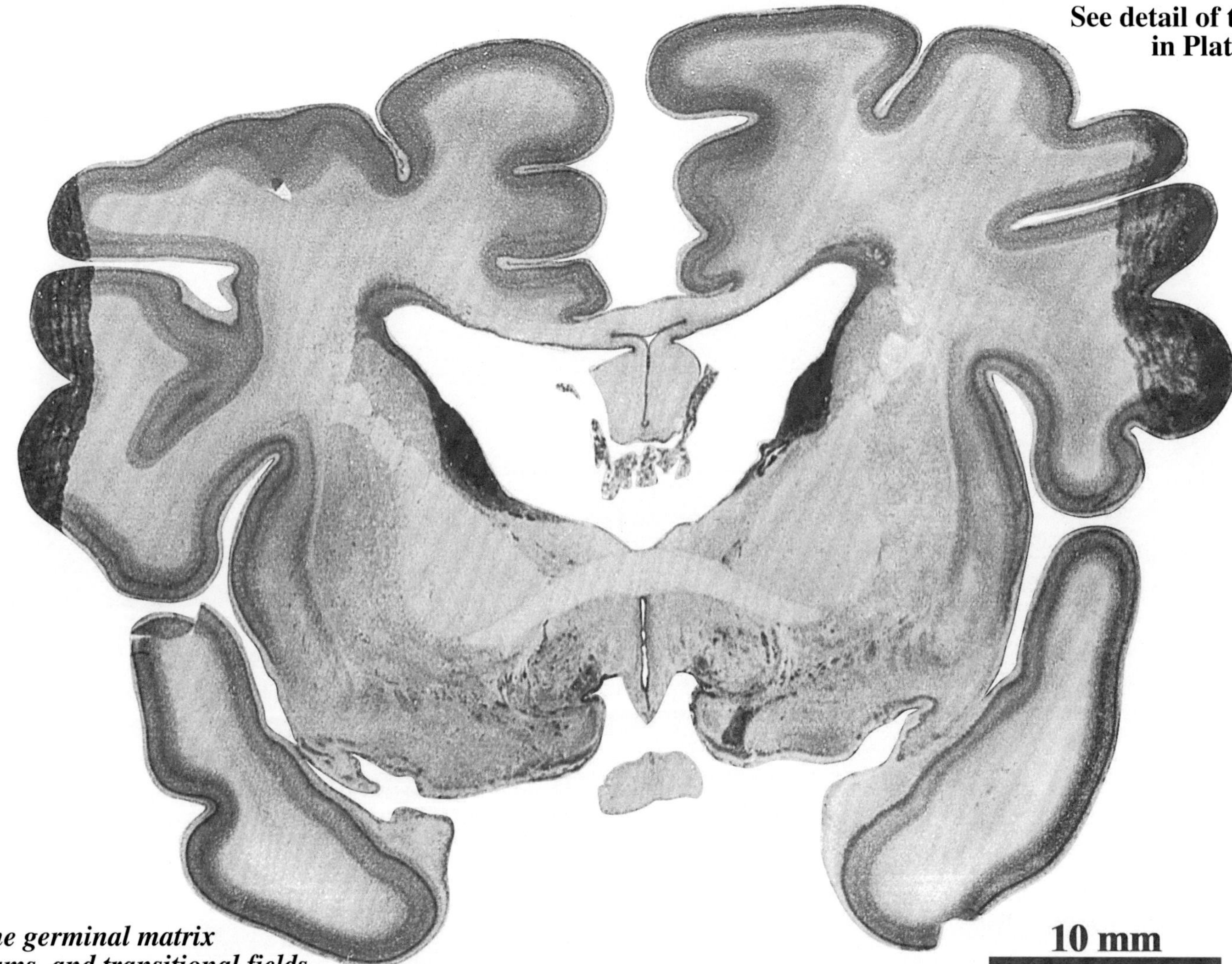

Remnants of the germinal matrix migratory streams, and transitional fields

1 *Rostral migratory stream*
2 *Frontal NEP and SVZ*
3 *Frontal STF*
4 *Callosal GEP*
5 *Callosal sling*
6 *Fornical GEP*
7 *Lateral migratory stream (cortical)*
8 *Anterolateral striatal NEP and SVZ*
9 *Anteromedial striatal NEP and SVZ*
10 *Strionuclear GEP*
11 *Diencephalic (preoptic) G/EP*
12 *Subpial granular layer (cortical)*

GEP - Glioepithelium
G/EP - Glioepithelium/ependyma
NEP - Neuroepithelium
STF - Stratified transitional field
SVZ - Subventricular zone

PLATE 55B

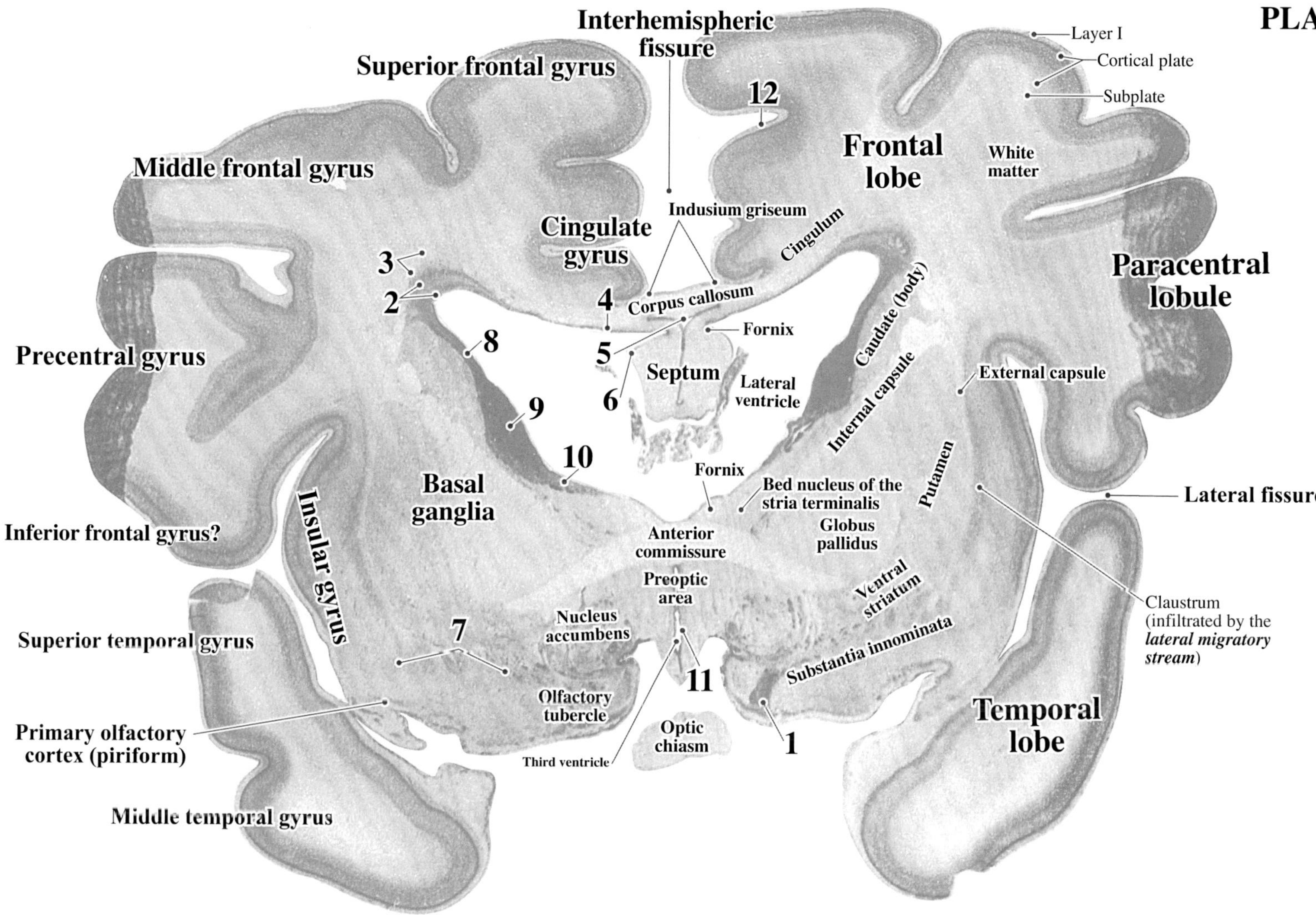
Interhemispheric fissure
Superior frontal gyrus
Layer I
Cortical plate
Subplate
12
Frontal lobe
White matter
Middle frontal gyrus
Indusium griseum
Cingulate gyrus
Cingulum
3
2
4
Corpus callosum
Caudate (body)
Paracentral lobule
Fornix
8
5
Septum
Precentral gyrus
Lateral ventricle
External capsule
6
9
Internal capsule
10
Putamen
Fornix
Basal ganglia
Bed nucleus of the stria terminalis
Lateral fissure
Insular gyrus
Inferior frontal gyrus?
Anterior commissure
Globus pallidus
Preoptic area
Ventral striatum
Claustrum (infiltrated by the *lateral migratory stream*)
Superior temporal gyrus
Nucleus accumbens
7
Substantia innominata
11
Olfactory tubercle
Temporal lobe
Primary olfactory cortex (piriform)
Optic chiasm
1
Third ventricle
Middle temporal gyrus

PLATE 56A
CR 350 mm
GW 37
Y217-65
Coronal
Section 721

See detail of the brain core in Plates 83A and B.

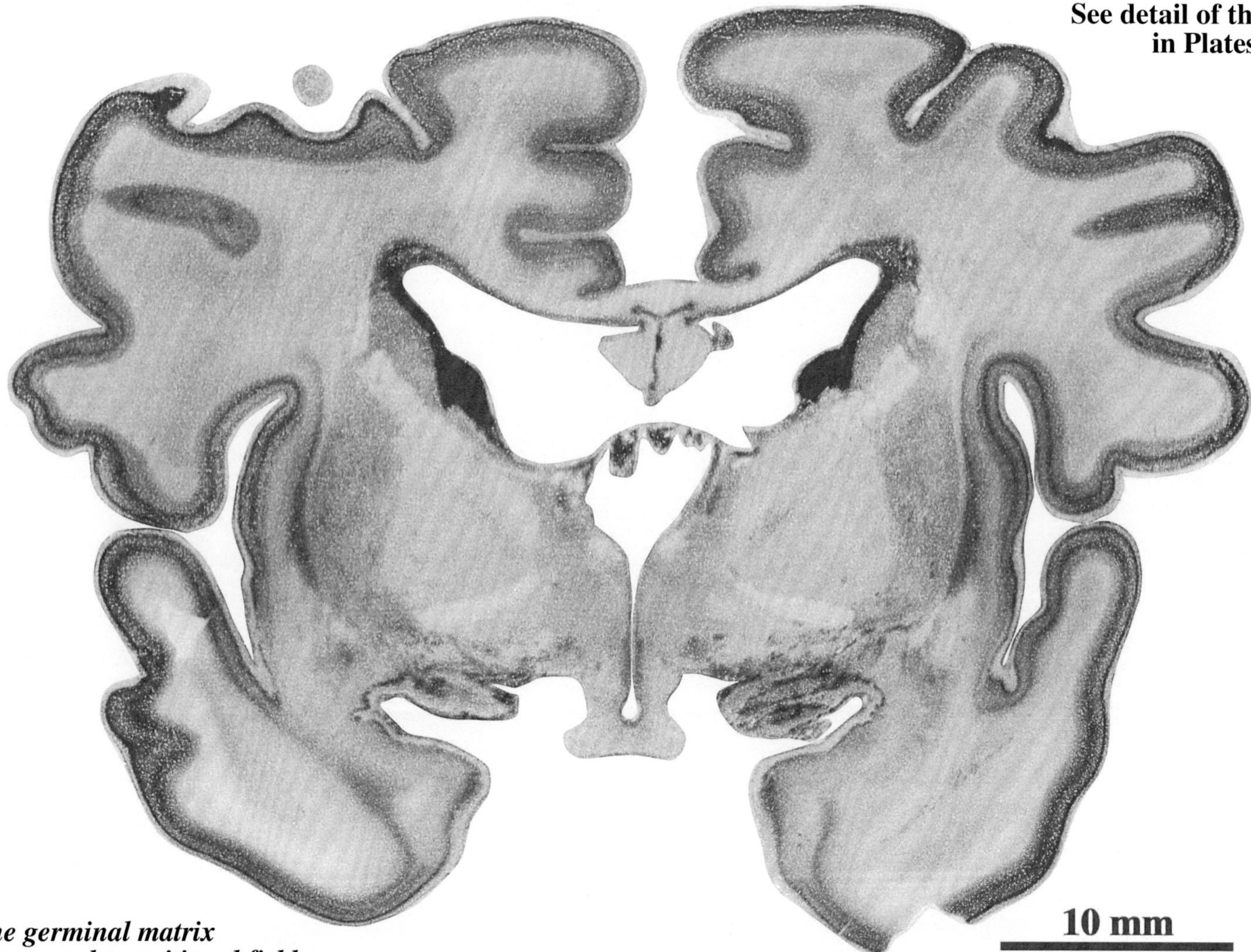

Remnants of the germinal matrix migratory streams, and transitional fields

1 *Frontal neuroepithelium and subventricular zone*
2 *Frontal stratified transitional field*
3 *Callosal glioepithelium*
4 *Callosal sling*
5 *Fornical glioepithelium*
6 *Lateral migratory stream (cortical)*
7 *Anterolateral striatal neuroepithelium and subventricular zone*
8 *Anteromedial striatal neuroepithelium and subventricular zone*
9 *Strionuclear glioepithelium*
10 *Diencephalic (epithalamic) glioepithelium/ependyma*
11 *Diencephalic (preoptic) glioepithelium/ependyma*
12 *Subpial granular layer (cortical)*

PLATE 56B

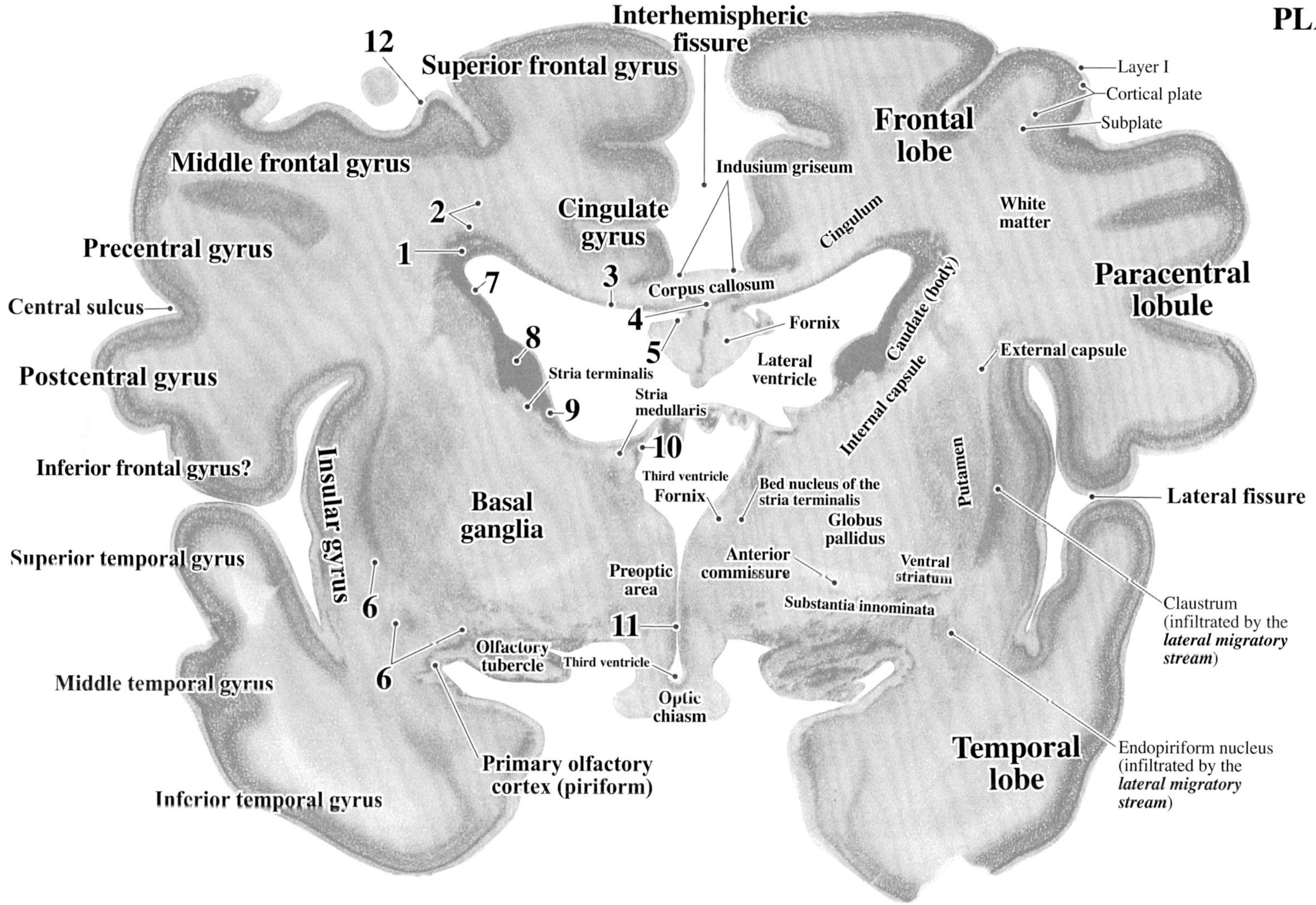

Interhemispheric fissure
12
Superior frontal gyrus
Layer I
Cortical plate
Subplate
Frontal lobe
Middle frontal gyrus
Indusium griseum
White matter
2
Cingulate gyrus
Cingulum
Precentral gyrus
1
7
3
Corpus callosum
Paracentral lobule
Central sulcus
4
Caudate (body)
Fornix
8
5
Lateral ventricle
External capsule
Postcentral gyrus
Stria terminalis
Stria medullaris
9
Internal capsule
10
Putamen
Inferior frontal gyrus?
Third ventricle
Insular gyrus
Fornix
Bed nucleus of the stria terminalis
Lateral fissure
Basal ganglia
Globus pallidus
Superior temporal gyrus
Anterior commissure
Preoptic area
Ventral striatum
6
Substantia innominata
Claustrum (infiltrated by the *lateral migratory stream*)
11
Olfactory tubercle
Third ventricle
6
Middle temporal gyrus
Optic chiasm
Endopiriform nucleus (infiltrated by the *lateral migratory stream*)
Temporal lobe
Primary olfactory cortex (piriform)
Inferior temporal gyrus

PLATE 57A
CR 350 mm
GW 37
Y217-65
Coronal
Section 761

✳ See this area of cortex in Plates 71 and 72

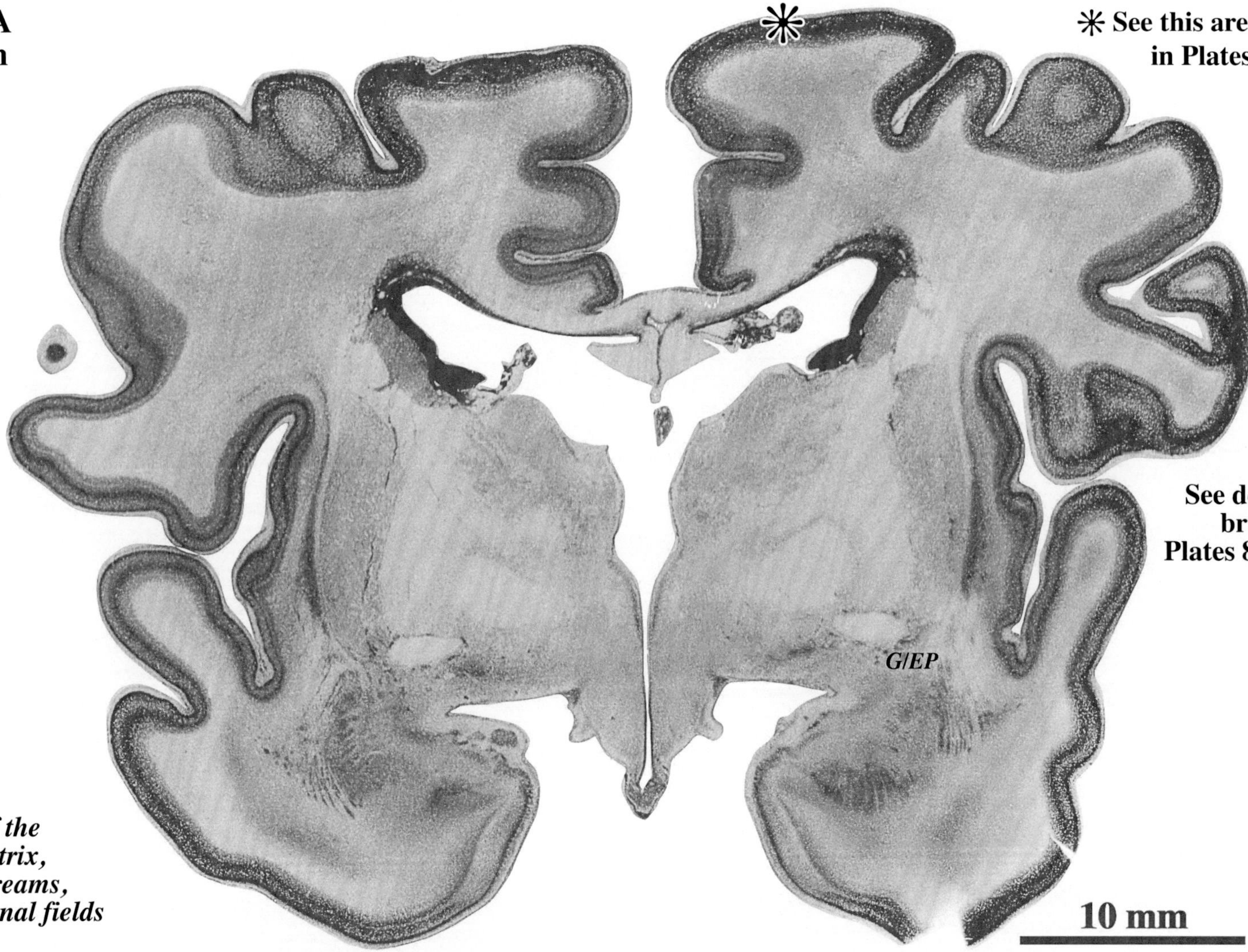

See detail of the brain core in Plates 84A and B.

Remnants of the germinal matrix, migratory streams, and transitional fields

1 *Frontal/paracentral NEP and SVZ*
2 *Frontal STF*
3 *Callosal GEP*
4 *Callosal sling*
5 *Fornical GEP*
6 *Lateral migratory stream (cortical)*
7 *Anterolateral striatal NEP and SVZ*
8 *Anteromedial striatal NEP and SVZ*
9 *Strionuclear GEP*
10 *Diencephalic (thalamic) G/EP*
11 *Diencephalic (hypothalamic) G/EP*
12 *Subpial granular layer (cortical)*

GEP - Glioepithelium
NEP - Neuroepithelium
STF - Stratified transitional field
SVZ - Subventricular zone

PLATE 57B

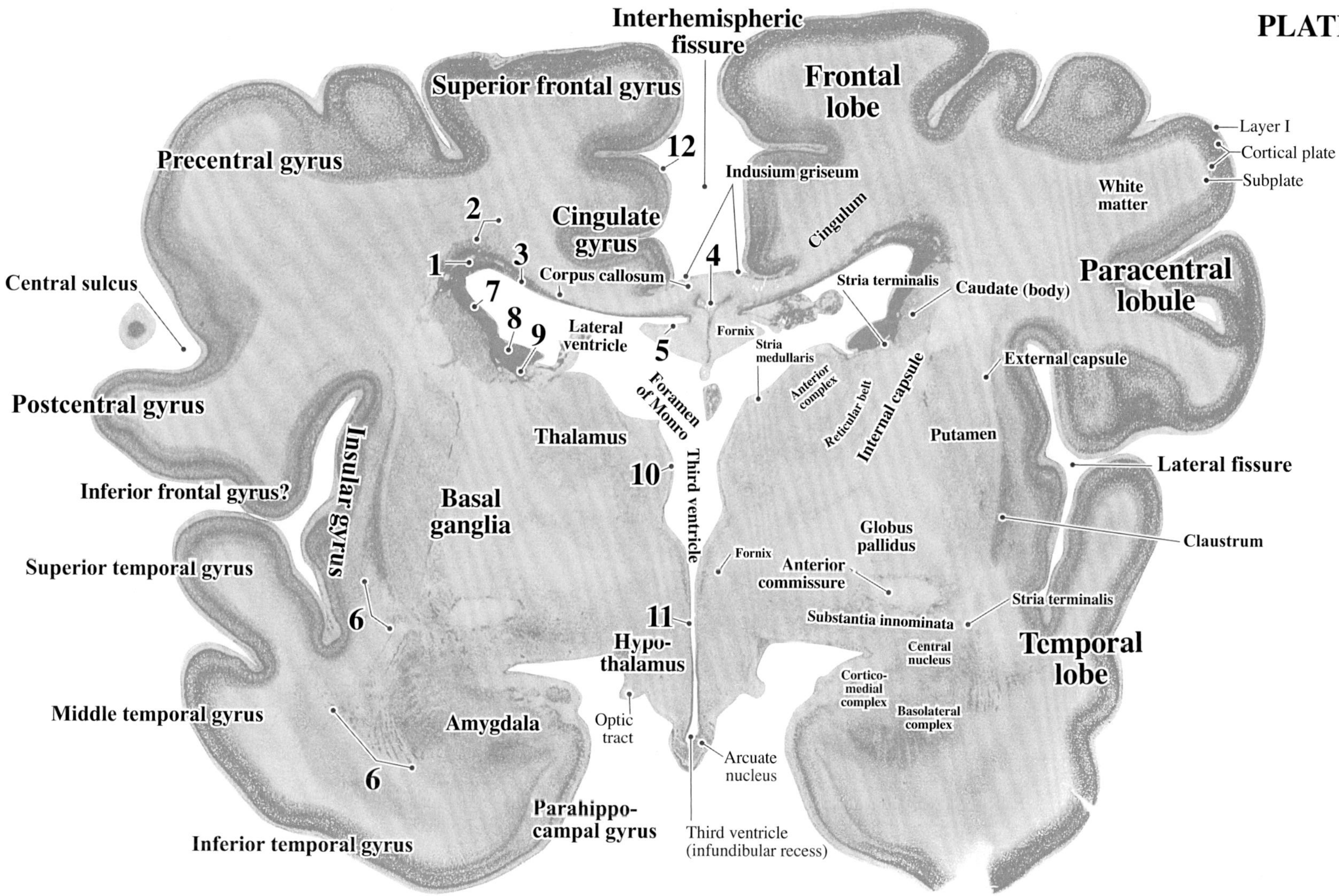
Interhemispheric fissure
Superior frontal gyrus
Frontal lobe
Layer I
Cortical plate
Subplate
White matter
Precentral gyrus
12
Indusium griseum
Cingulum
2
Cingulate gyrus
1
3
4
Central sulcus
Corpus callosum
Stria terminalis
Caudate (body)
Paracentral lobule
7
8
9
Lateral ventricle
5
Fornix
Stria medullaris
Anterior complex
External capsule
Foramen of Monro
Reticular belt
Internal capsule
Postcentral gyrus
Thalamus
Putamen
Insular gyrus
Third ventricle
10
Lateral fissure
Inferior frontal gyrus?
Basal ganglia
Globus pallidus
Claustrum
Fornix
Anterior commissure
Superior temporal gyrus
Stria terminalis
6
11
Substantia innominata
Hypo-thalamus
Central nucleus
Temporal lobe
Cortico-medial complex
Basolateral complex
Middle temporal gyrus
Amygdala
Optic tract
Arcuate nucleus
6
Parahippo-campal gyrus
Inferior temporal gyrus
Third ventricle (infundibular recess)

PLATE 58A
CR 350 mm
GW 37
Y217-65
Coronal
Section 831

See detail of the brain core in Plates 85A and B.

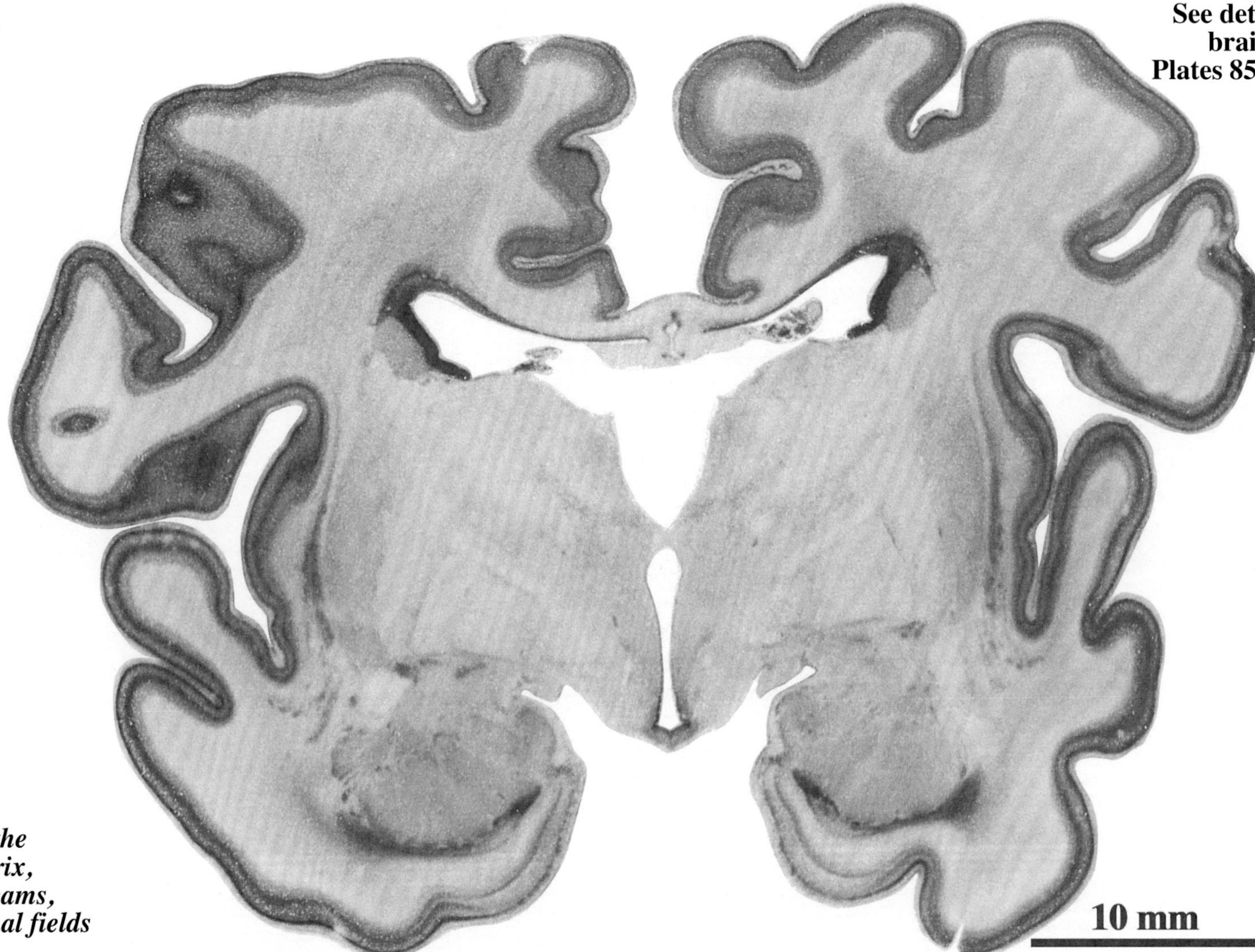

Remnants of the germinal matrix, migratory streams, and transitional fields

1 *Frontal/paracentral NEP and SVZ*
2 *Frontal/paracentral STF*
3 *Callosal GEP*
4 *Callosal sling*
5 *Fornical GEP*
6 *Parahippocampal NEP, SVZ, and STF*
7 *Lateral migratory stream (cortical)*
8 *Amygdaloid G/EP*
9 *Posterior striatal NEP and SVZ*
10 *Strionuclear GEP*
11 *Diencephalic (thalamic) G/EP*
12 *Diencephalic (hypothalamic) G/EP*
13 *Subpial granular layer*

GEP - Glioepithelium
G/EP - Glioepithelium/ependyma
NEP - Neuroepithelium
STF - Stratified transitional field
SVZ - Subventricular zone

PLATE 58B

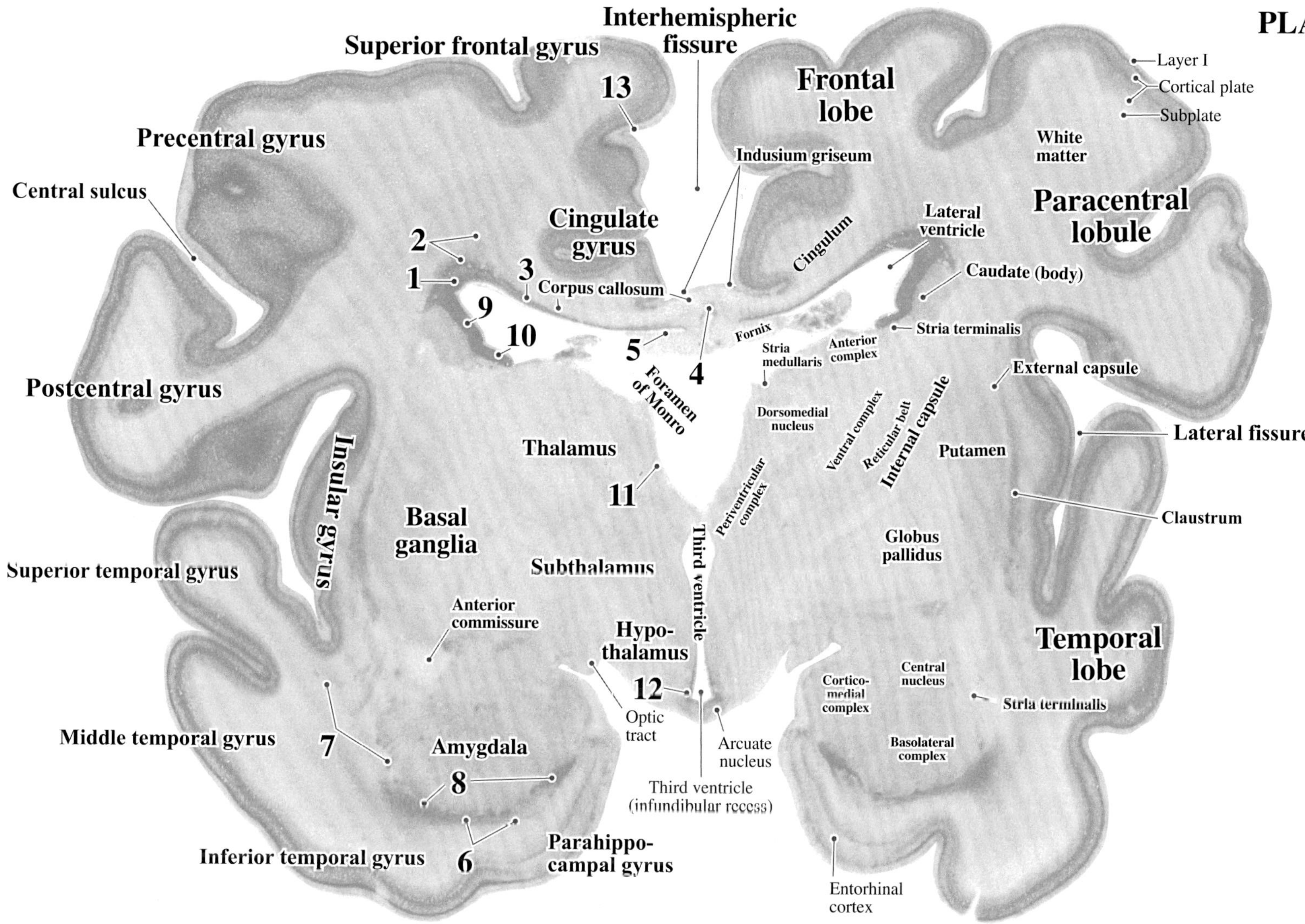
Interhemispheric fissure
Superior frontal gyrus
13
Frontal lobe
Layer I
Cortical plate
Subplate
Precentral gyrus
Indusium griseum
White matter
Central sulcus
Cingulate gyrus
Lateral ventricle
Paracentral lobule
Cingulum
2
1
3
Corpus callosum
Caudate (body)
9
10
Fornix
Stria terminalis
5
Stria medullaris
Anterior complex
4
Foramen of Monro
External capsule
Postcentral gyrus
Dorsomedial nucleus
Ventral complex
Reticular belt
Internal capsule
Lateral fissure
Thalamus
Putamen
Insular gyrus
11
Periventricular complex
Claustrum
Basal ganglia
Third ventricle
Globus pallidus
Superior temporal gyrus
Subthalamus
Anterior commissure
Hypo-thalamus
Temporal lobe
Central nucleus
Cortico-medial complex
12
Stria terminalis
Optic tract
Middle temporal gyrus
7
Amygdala
Arcuate nucleus
Basolateral complex
8
Third ventricle (infundibular recess)
Inferior temporal gyrus
6
Parahippo-campal gyrus
Entorhinal cortex

PLATE 59A
CR 350 mm
GW 37
Y217-65
Coronal
Section 881

See detail of the brain core in Plates 86A and B.

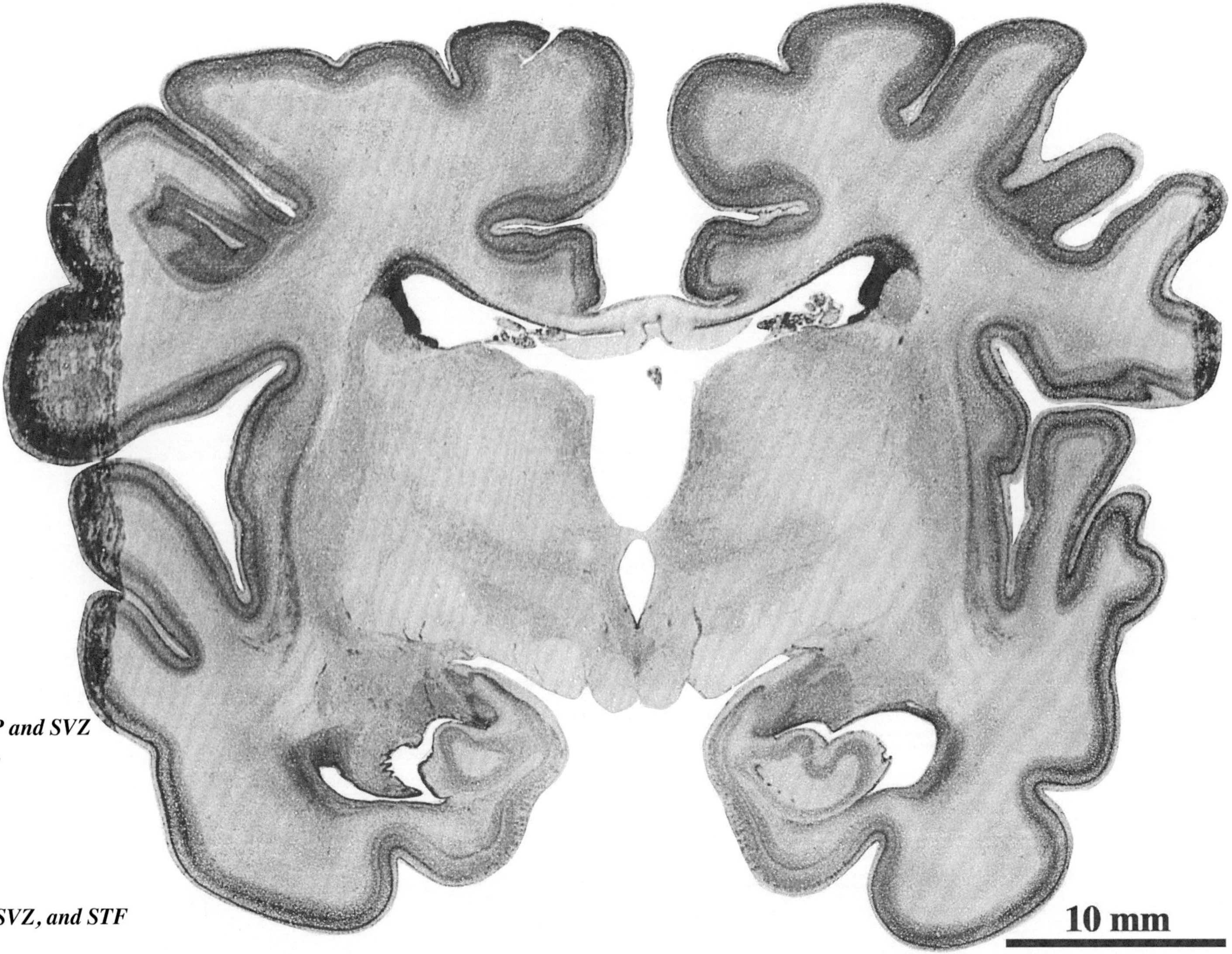

Remnants of the germinal matrix, migratory streams, and transitional fields

1 *Frontal/paracentral NEP and SVZ*
2 *Frontal/paracentral STF*
3 *Callosal GEP*
4 *Callosal sling*
5 *Fornical GEP*
6 *Parahippocampal NEP, SVZ, and STF*
7 *Temporal NEP and SVZ*
8 *Temporal STF*
9 *Alvear GEP*
10 *Lateral migratory stream (cortical)*
11 *Amygdaloid G/EP*
12 *Posterior striatal NEP and SVZ*
13 *Strionuclear GEP*
14 *Diencephalic (thalamic) G/EP*
15 *Diencephalic (hypothalamic) G/EP*
16 *Subpial granular layer*

GEP - Glioepithelium
G/EP - Glioepithelium/ependyma
NEP - Neuroepithelium
STF - Stratified transitional field
SVZ - Subventricular zone

PLATE 59B

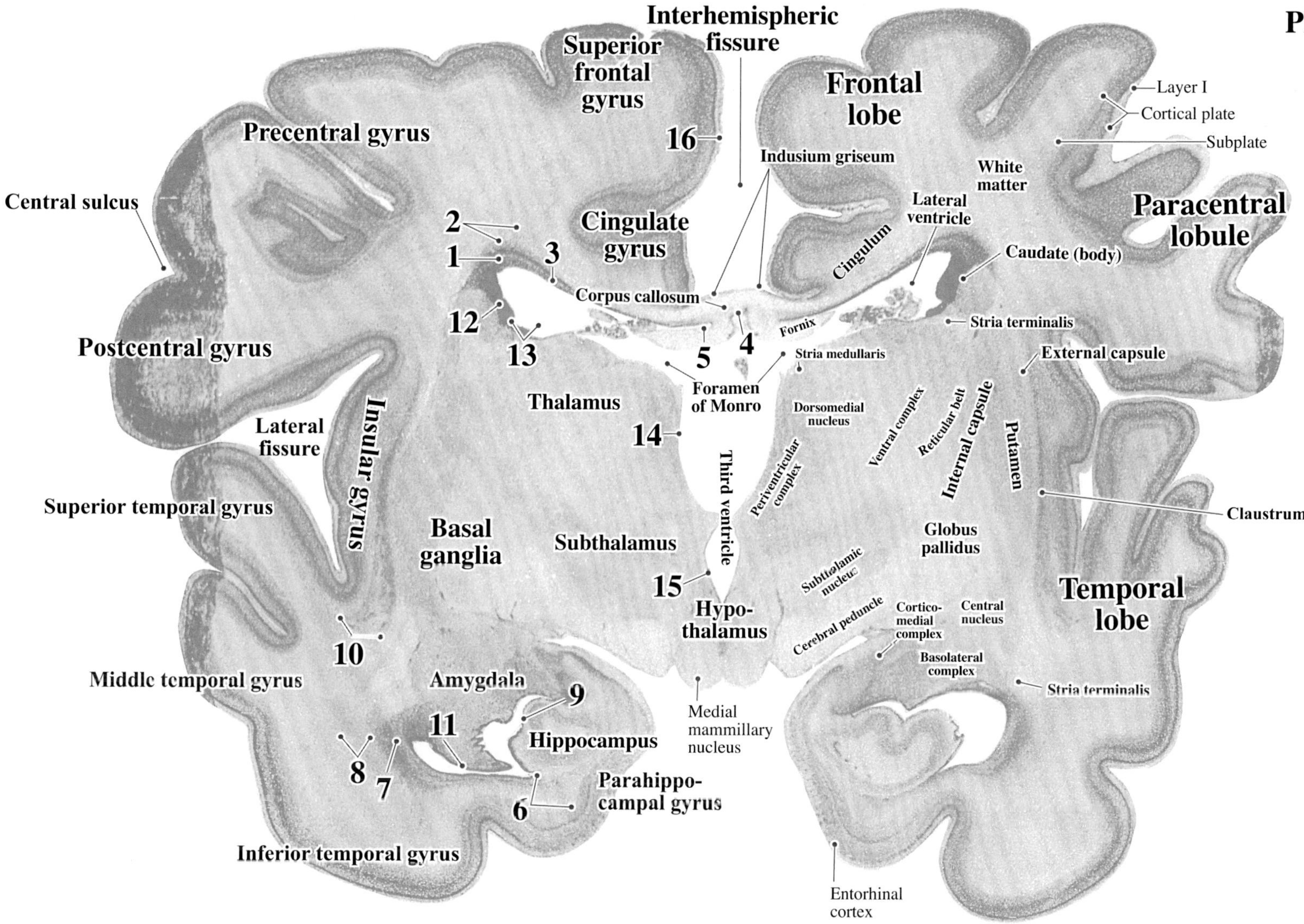
Interhemispheric fissure
Superior frontal gyrus
Frontal lobe
Layer I
Cortical plate
Subplate
Precentral gyrus
16
Indusium griseum
White matter
Central sulcus
Lateral ventricle
Paracentral lobule
2
Cingulate gyrus
1
3
Cingulum
Caudate (body)
Corpus callosum
12
Fornix
Stria terminalis
13
5
4
Postcentral gyrus
Stria medullaris
External capsule
Foramen of Monro
Thalamus
Dorsomedial nucleus
Ventral complex
Reticular belt
Internal capsule
Lateral fissure
14
Putamen
Insular gyrus
Third ventricle
Periventricular complex
Superior temporal gyrus
Claustrum
Basal ganglia
Subthalamus
Globus pallidus
Subthalamic nucleus
15
Temporal lobe
Hypo-thalamus
Cerebral peduncle
Cortico-medial complex
Central nucleus
10
Basolateral complex
Middle temporal gyrus
Amygdala
Stria terminalis
9
Medial mammillary nucleus
11
Hippocampus
8
7
Parahippo-campal gyrus
6
Inferior temporal gyrus
Entorhinal cortex

PLATE 60A
CR 350 mm
GW 37
Y217-65
Coronal
Section 981

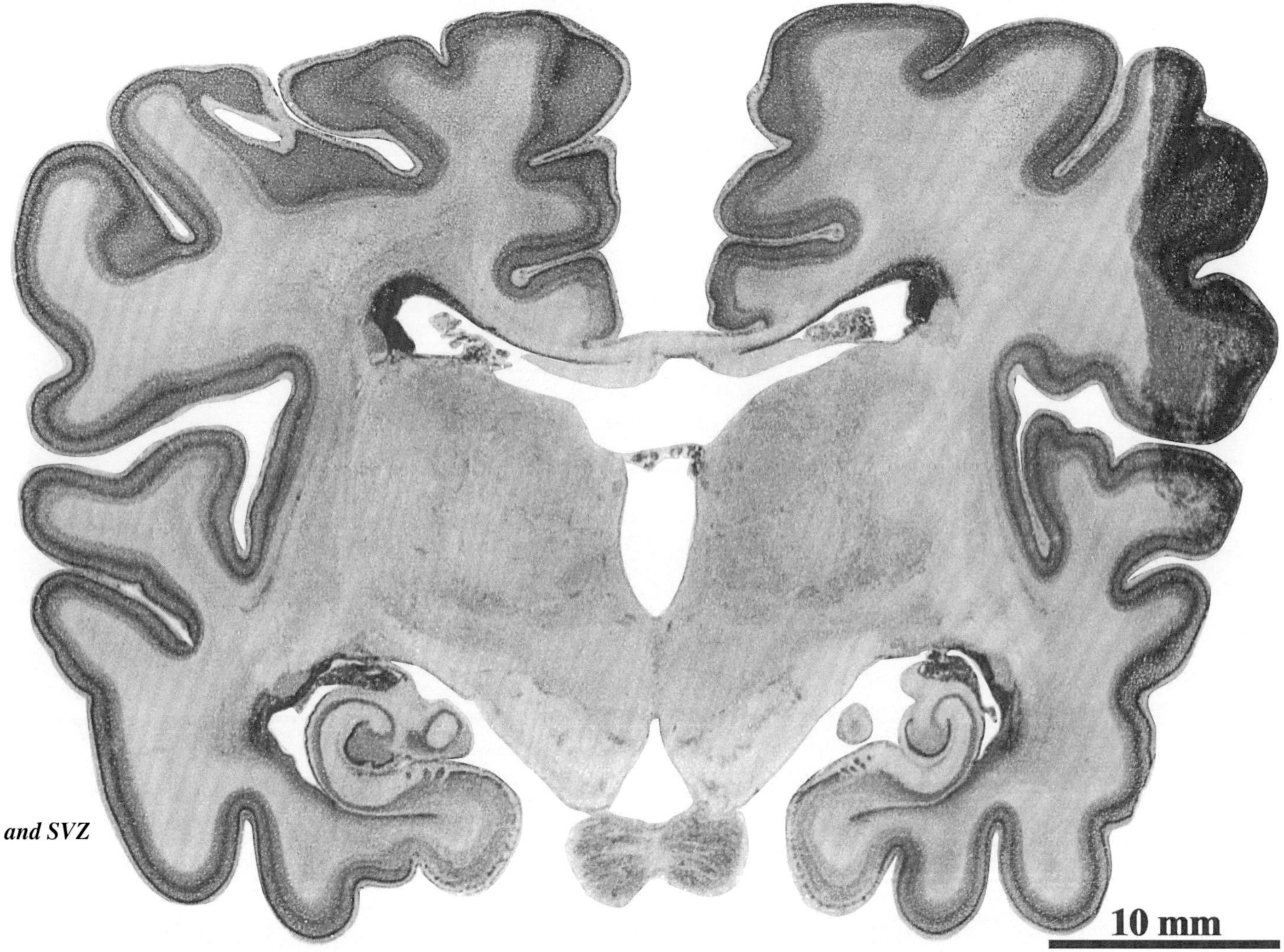

See detail of the brain core in Plates 87A and B.

Remnants of the germinal matrix, migratory streams, and transitional fields

1 *Frontal/paracentral NEP and SVZ*
2 *Frontal/paracentral STF*
3 *Callosal GEP*
4 *Callosal sling*
5 *Fornical GEP*
6 *Parahippocampal NEP, SVZ, and STF*
7 *Temporal NEP and SVZ*
8 *Temporal STF*
9 *Alvear GEP*
10 *Subgranular zone (dentate)*
11 *Lateral migratory stream (cortical)*
12 *Posterior striatal NEP and SVZ*
13 *Strionuclear GEP*
14 *Diencephalic (thalamic) G/EP*
15 *Subpial granular layer (cortical)*

GEP - Glioepithelium
G/EP - Glioepithelium/ependyma
NEP - Neuroepithelium
STF - Stratified transitional field
SVZ - Subventricular zone

PLATE 60B

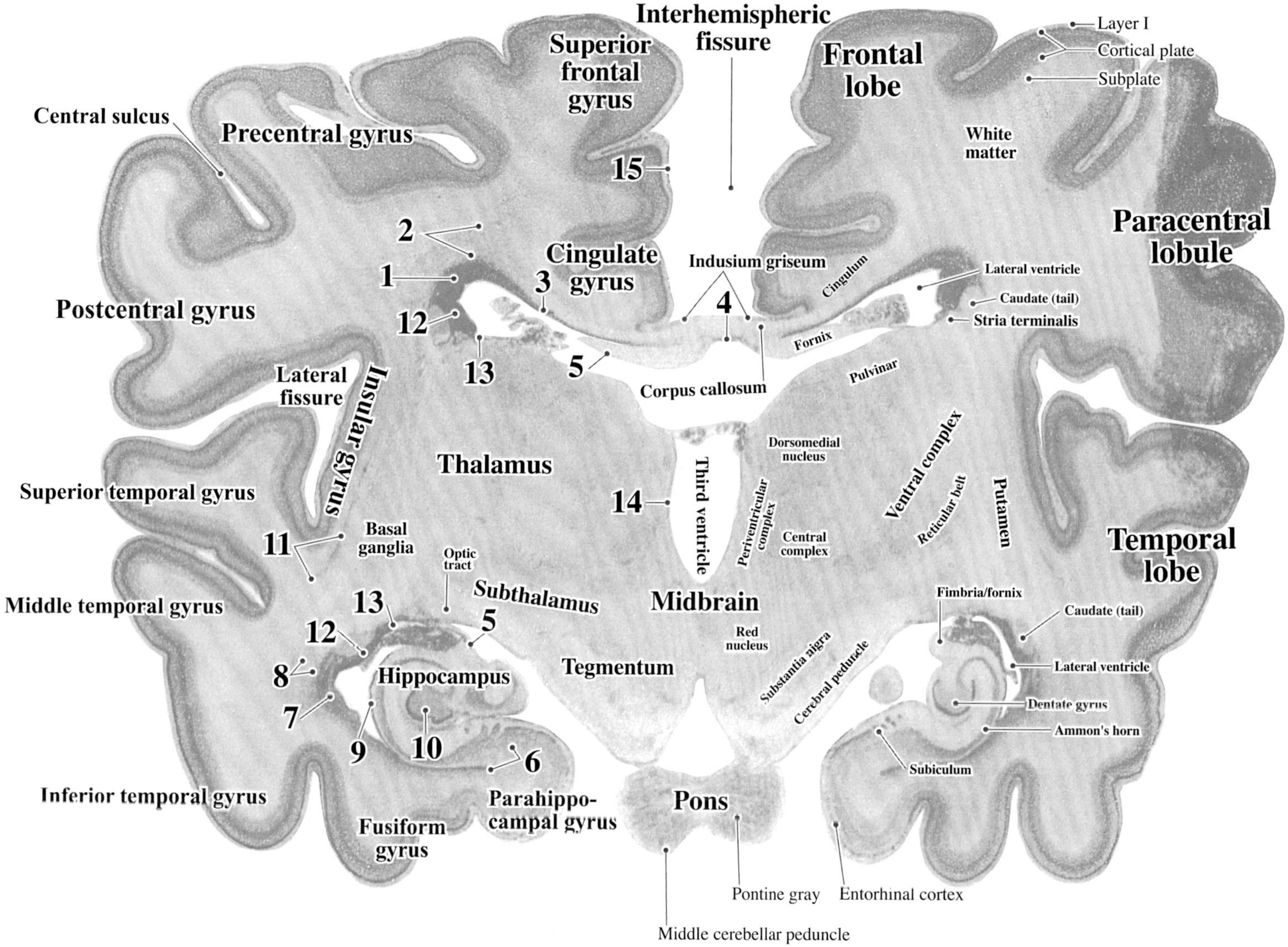
Interhemispheric fissure
Superior frontal gyrus
Frontal lobe
Layer I
Cortical plate
Subplate
Central sulcus
Precentral gyrus
White matter
15
Paracentral lobule
2
Cingulate gyrus
1
3
Indusium griseum
4
Cingulum
Lateral ventricle
Caudate (tail)
Stria terminalis
Postcentral gyrus
12
13
5
Fornix
Corpus callosum
Pulvinar
Lateral fissure
Insular gyrus
Thalamus
Dorsomedial nucleus
Third ventricle
Ventral complex
Superior temporal gyrus
14
Periventricular complex
Reticular belt
Putamen
Central complex
Temporal lobe
11
Basal ganglia
Optic tract
Subthalamus
Midbrain
Fimbria/fornix
Middle temporal gyrus
13
12
5
Caudate (tail)
Red nucleus
8
Hippocampus
Tegmentum
Substantia nigra
Cerebral peduncle
Lateral ventricle
7
9
10
Dentate gyrus
Ammon's horn
6
Subiculum
Inferior temporal gyrus
Fusiform gyrus
Parahippocampal gyrus
Pons
Pontine gray
Entorhinal cortex
Middle cerebellar peduncle

PLATE 61A
CR 350 mm
GW 37
Y217-65
Coronal
Section 1021

See detail of the brain core in Plates 88A and B.

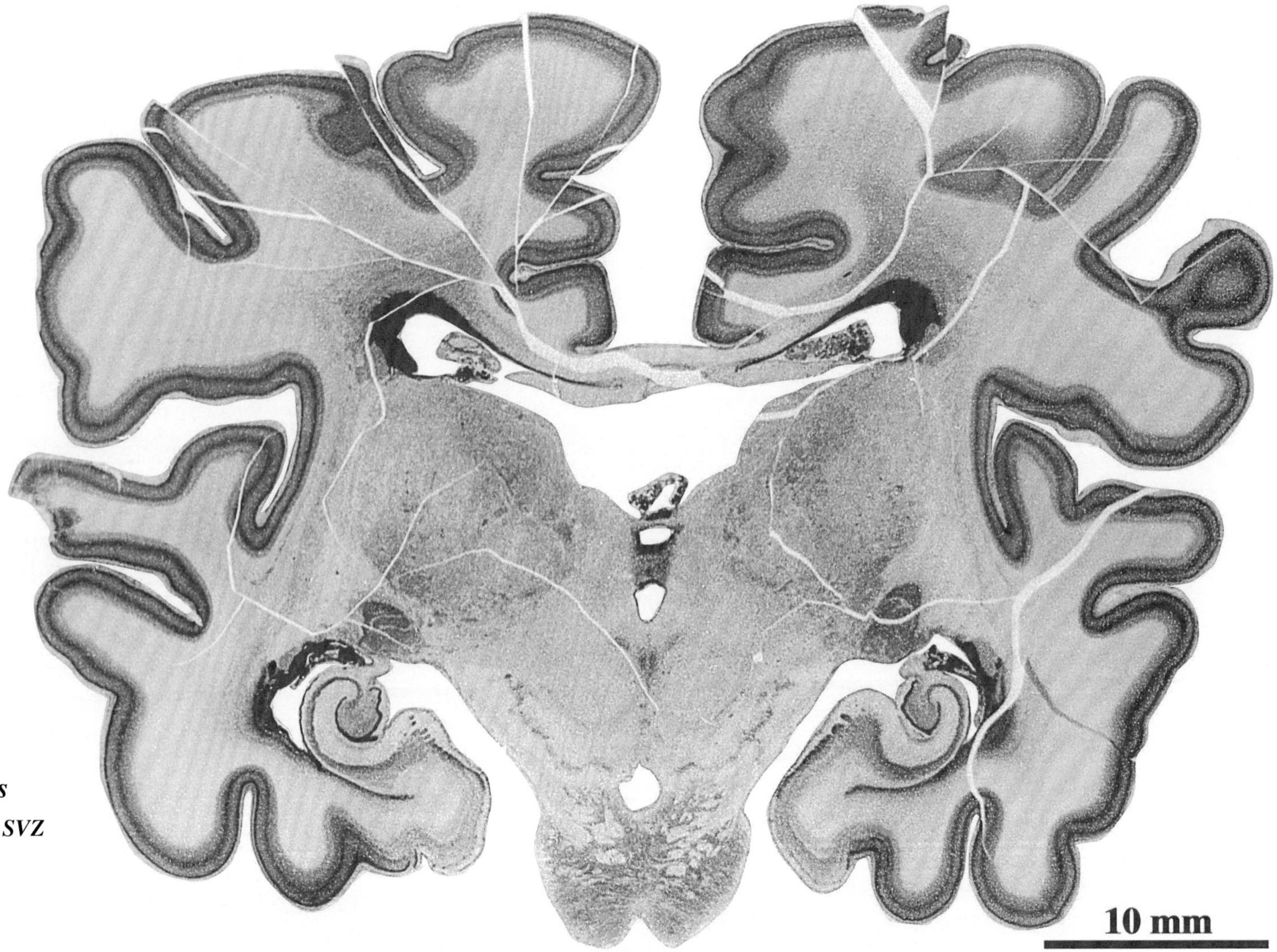

Remnants of the germinal matrix, migratory streams, and transitional fields

1 *Paracentral NEP and SVZ*
2 *Paracentral STF*
3 *Callosal GEP*
4 *Callosal sling*
5 *Fornical GEP*
6 *Parahippocampal NEP, SVZ, and STF*
7 *Temporal NEP and SVZ*
8 *Temporal STF*
9 *Alvear GEP*
10 *Subgranular zone (dentate)*
11 *Lateral migratory stream (cortical)*
12 *Posterior striatal NEP and SVZ*
13 *Strionuclear GEP*
14 *Mesencephalic G/EP*
15 *Subpial granular layer (cortical)*

GEP - Glioepithelium
G/EP - Glioepithelium/ependyma
NEP - Neuroepithelium
STF - Stratified transitional field
SVZ - Subventricular zone

PLATE 61B

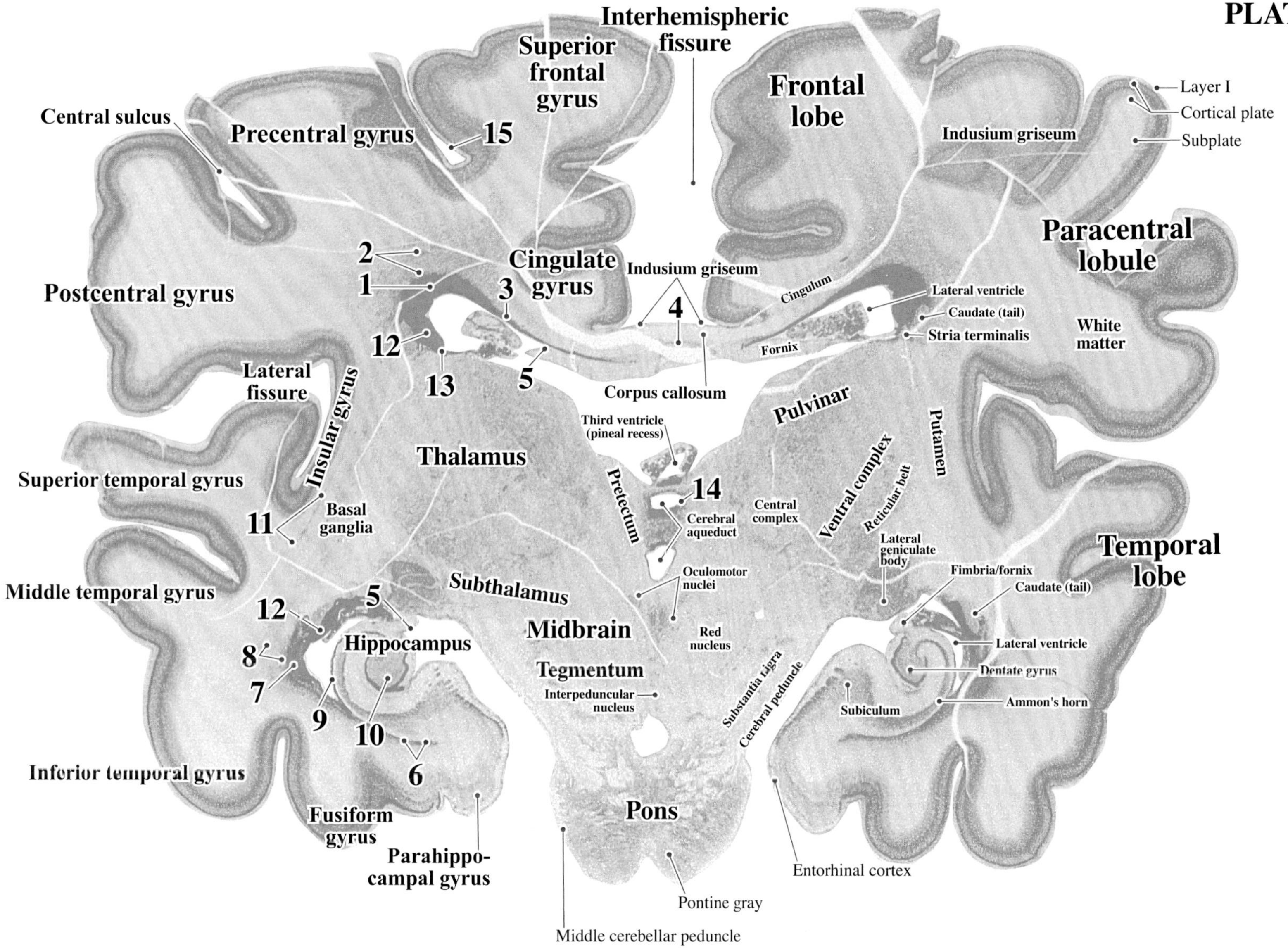
Interhemispheric fissure
Superior frontal gyrus
Frontal lobe
Central sulcus
Precentral gyrus
15
Indusium griseum
Layer I
Cortical plate
Subplate
Paracentral lobule
2
1
Cingulate gyrus
3
Indusium griseum
4
Postcentral gyrus
Cingulum
Lateral ventricle
Caudate (tail)
Stria terminalis
White matter
Fornix
12
13
5
Corpus callosum
Lateral fissure
Insular gyrus
Pulvinar
Putamen
Third ventricle (pineal recess)
Thalamus
Ventral complex
Reticular belt
Superior temporal gyrus
Pretectum
14
Cerebral aqueduct
Central complex
11
Basal ganglia
Lateral geniculate body
Temporal lobe
Fimbria/fornix
Oculomotor nuclei
Subthalamus
Middle temporal gyrus
Caudate (tail)
12
5
Midbrain
Red nucleus
Hippocampus
Lateral ventricle
8
Tegmentum
Substantia nigra
Cerebral peduncle
Dentate gyrus
7
Interpeduncular nucleus
Subiculum
Ammon's horn
9
10
6
Inferior temporal gyrus
Pons
Fusiform gyrus
Parahippocampal gyrus
Entorhinal cortex
Pontine gray
Middle cerebellar peduncle

PLATE 62A
CR 350 mm
GW 37
Y217-65
Coronal
Section 1081

✳ See this area of cortex in Plates 73 and 74.

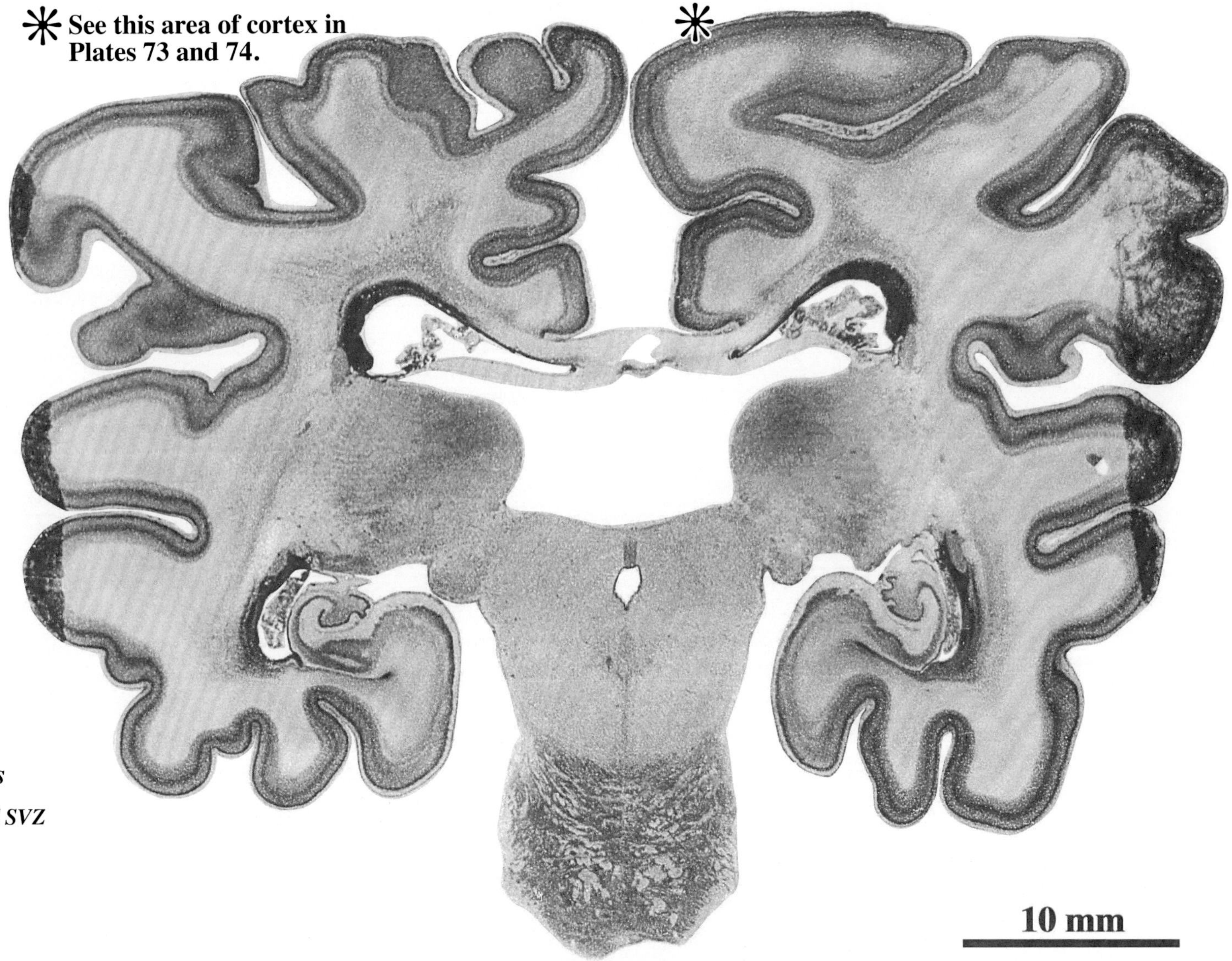

See detail of the brain core in Plates 89A and B.

Remnants of the germinal matrix, migratory streams, and transitional fields

1. *Paracentral NEP and SVZ*
2. *Paracentral STF*
3. *Callosal GEP*
4. *Callosal sling*
5. *Fornical GEP*
6. *Parahippocampal NEP, SVZ, and STF*
7. *Temporal NEP and SVZ*
8. *Temporal STF*
9. *Alvear GEP*
10. *Subgranular zone (dentate)*
11. *Posterior striatal NEP and SVZ*
12. *Strionuclear GEP*
13. *Mesencephalic G/EP*
14. *Subpial granular layer (cortical)*

GEP - Glioepithelium
G/EP - Glioepithelium/ependyma
NEP - Neuroepithelium
STF - Stratified transitional field
SVZ - Subventricular zone

PLATE 62B

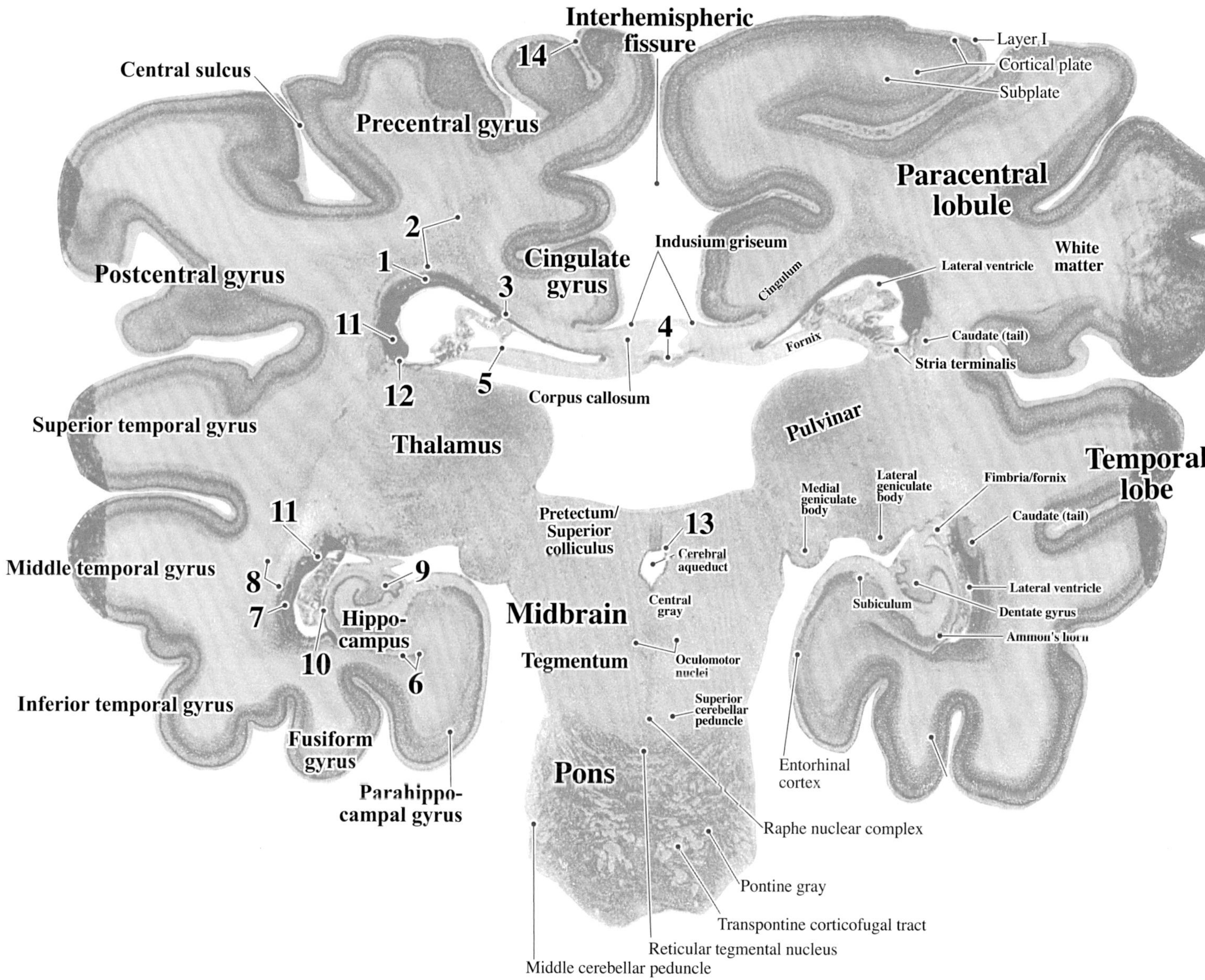
Interhemispheric fissure
14
Central sulcus
Precentral gyrus
Layer I
Cortical plate
Subplate
Paracentral lobule
2
1
Cingulate gyrus
3
Indusium griseum
White matter
Lateral ventricle
Cingulum
Postcentral gyrus
11
4
Caudate (tail)
Fornix
Stria terminalis
5
12
Corpus callosum
Superior temporal gyrus
Thalamus
Pulvinar
Temporal lobe
Medial geniculate body
Lateral geniculate body
Fimbria/fornix
11
Pretectum/ Superior colliculus
13
Caudate (tail)
Cerebral aqueduct
Middle temporal gyrus
8
9
Lateral ventricle
Central gray
Subiculum
Dentate gyrus
7
Hippo-campus
Midbrain
Ammon's horn
10
Tegmentum
Oculomotor nuclei
6
Inferior temporal gyrus
Superior cerebellar peduncle
Fusiform gyrus
Entorhinal cortex
Pons
Parahippo-campal gyrus
Raphe nuclear complex
Pontine gray
Transpontine corticofugal tract
Reticular tegmental nucleus
Middle cerebellar peduncle

PLATE 63A
CR 350 mm
GW 37
Y217-65
Coronal
Section 1141

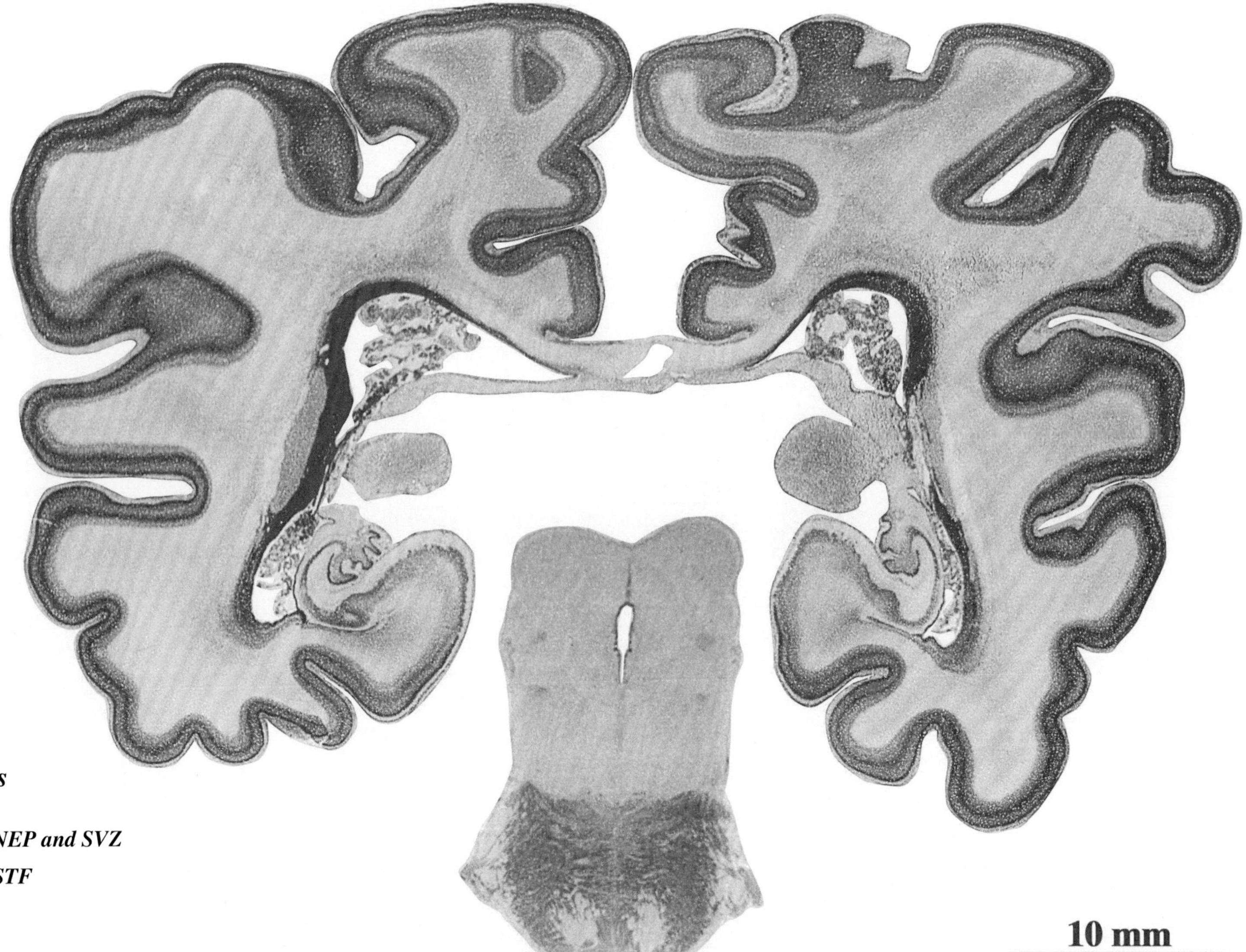

See detail of the brain core in Plates 90A and B.

Remnants of the germinal matrix, migratory streams, and transitional fields

1 *Paracentral/parietal NEP and SVZ*
2 *Paracentral/parietal STF*
3 *Callosal GEP*
4 *Fornical GEP*
5 *Parahippocampal NEP, SVZ, and STF*
6 *Temporal NEP and SVZ*
7 *Temporal STF*
8 *Alvear GEP*
9 *Subgranular zone (dentate)*
10 *Posterior striatal NEP and SVZ*
11 *Strionuclear GEP*
12 *Mesencephalic G/EP*
13 *Subpial granular layer (cortical)*

GEP - Glioepithelium
G/EP - Glioepithelium/ependyma
NEP - Neuroepithelium
STF - Stratified transitional field
SVZ - Subventricular zone

10 mm

PLATE 63B

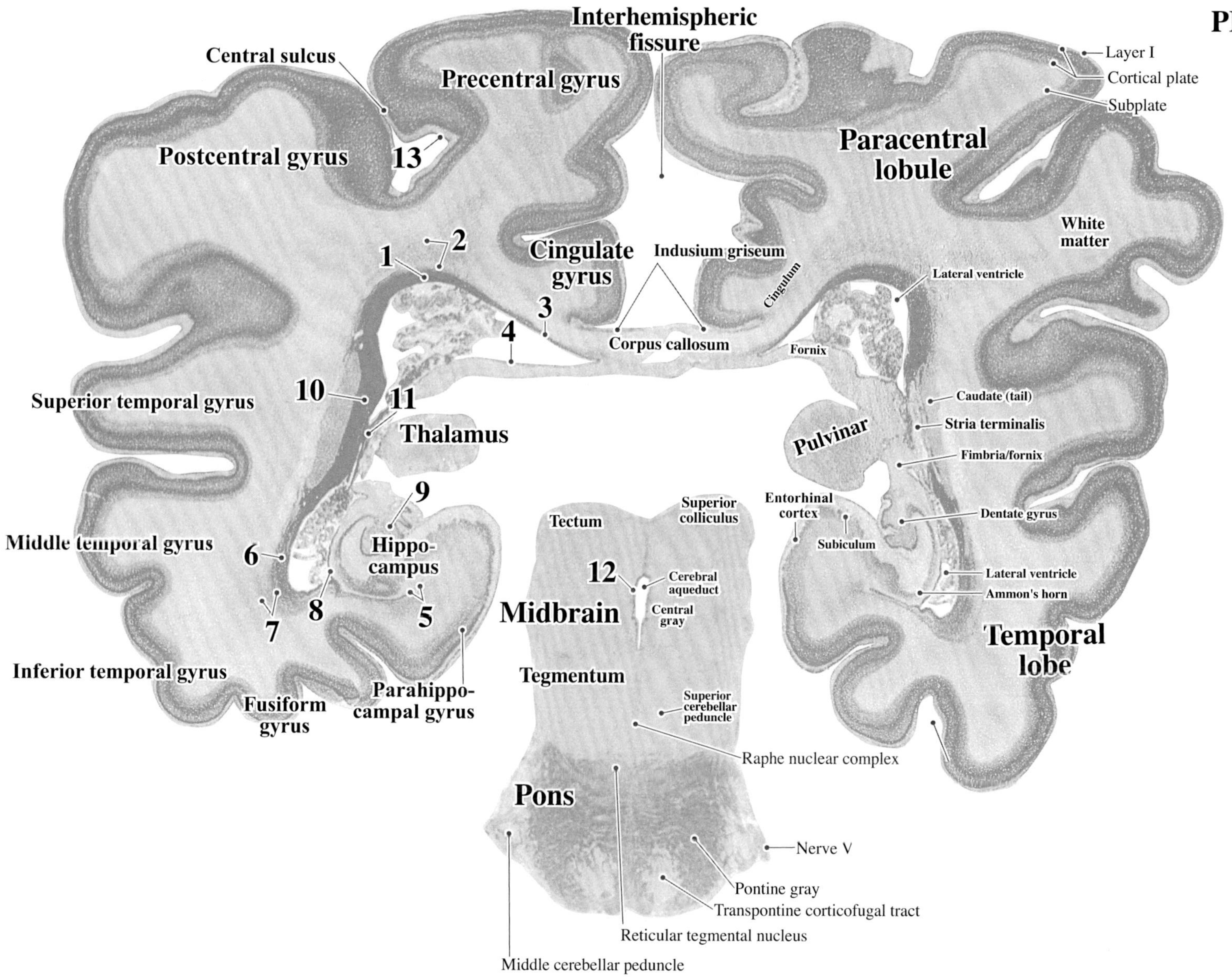
Interhemispheric fissure
Central sulcus
Precentral gyrus
Postcentral gyrus
13
Layer I
Cortical plate
Subplate
Paracentral lobule
White matter
1
2
Cingulate gyrus
Indusium griseum
Cingulum
Lateral ventricle
3
4
Corpus callosum
Fornix
Superior temporal gyrus
10
11
Thalamus
Caudate (tail)
Stria terminalis
Pulvinar
Fimbria/fornix
9
Hippo-campus
Entorhinal cortex
Subiculum
Dentate gyrus
Middle temporal gyrus
6
Tectum
Superior colliculus
12
Cerebral aqueduct
Central gray
Lateral ventricle
Ammon's horn
7
8
5
Midbrain
Temporal lobe
Inferior temporal gyrus
Tegmentum
Fusiform gyrus
Parahippo-campal gyrus
Superior cerebellar peduncle
Raphe nuclear complex
Pons
Nerve V
Pontine gray
Transpontine corticofugal tract
Reticular tegmental nucleus
Middle cerebellar peduncle

PLATE 64A
CR 350 mm
GW 37
Y217-65
Coronal
Section 1181

✳ See this area of cortex in Plates 75 and 76.

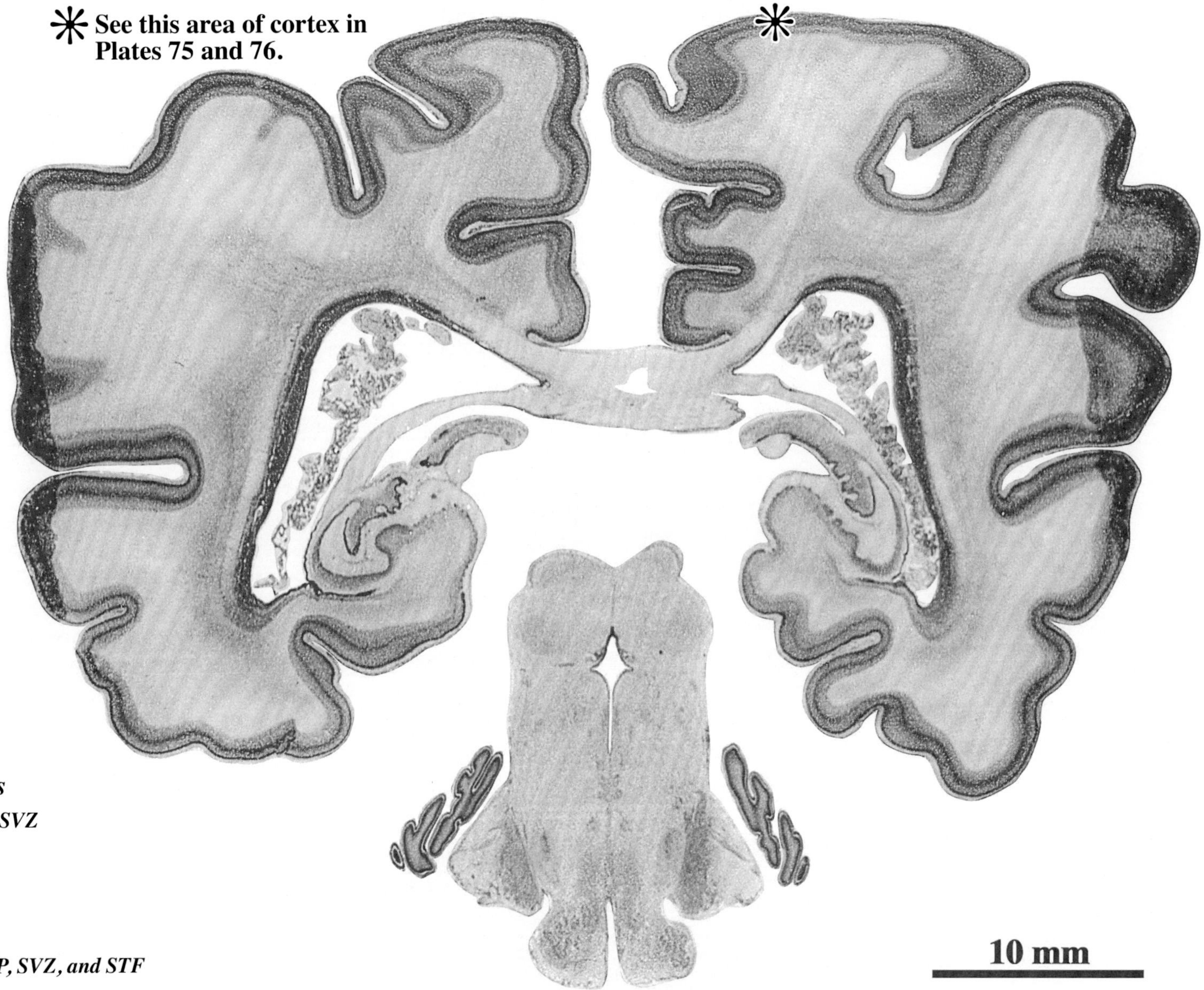

See detail of the brain core in Plates 91A and B.

Remnants of the germinal matrix, migratory streams, and transitional fields

1 *Paracentral NEP and SVZ*
2 *Paracentral STF*
3 *Callosal GEP*
4 *Fornical GEP*
5 *Parahippocampal NEP, SVZ, and STF*
6 *Temporal NEP and SVZ*
7 *Temporal STF*
8 *Alvear GEP*
9 *Subgranular zone (dentate)*
10 *Mesencephalic G/EP*
11 *External germinal layer (cerebellum)*
12 *Subpial granular layer (cortical)*

GEP - Glioepithelium
G/EP - Glioepithelium/ependyma
NEP - Neuroepithelium
STF - Stratified transitional field
SVZ - Subventricular zone

PLATE 64B

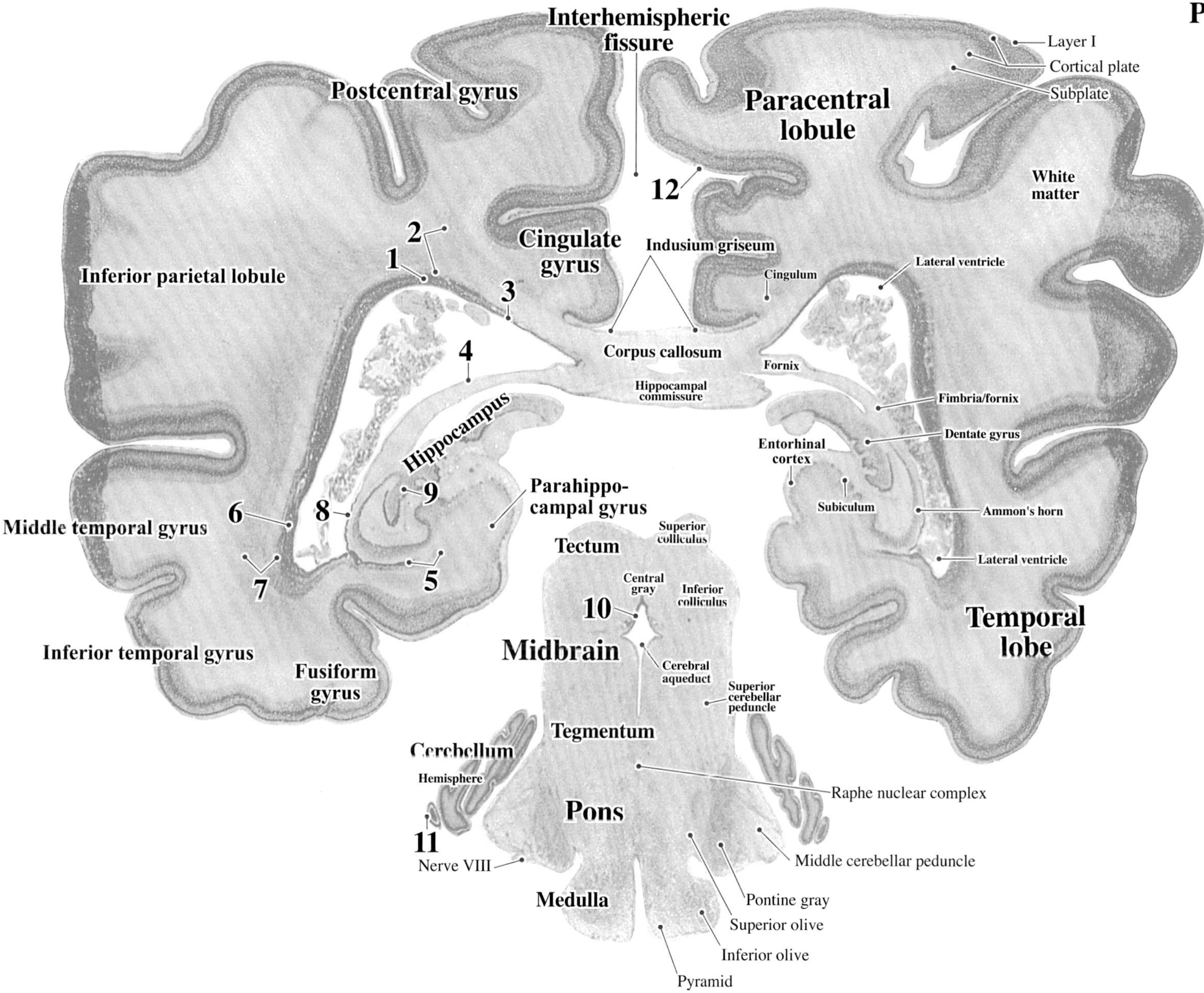
Interhemispheric fissure
Layer I
Cortical plate
Subplate
Postcentral gyrus
Paracentral lobule
White matter
12
2
1
Cingulate gyrus
Indusium griseum
Inferior parietal lobule
3
Cingulum
Lateral ventricle
4
Corpus callosum
Fornix
Hippocampal commissure
Fimbria/fornix
Hippocampus
Dentate gyrus
Entorhinal cortex
9
Parahippocampal gyrus
8
6
Middle temporal gyrus
Subiculum
Ammon's horn
Superior colliculus
Tectum
Lateral ventricle
7
5
Central gray
Inferior colliculus
10
Temporal lobe
Midbrain
Inferior temporal gyrus
Fusiform gyrus
Cerebral aqueduct
Superior cerebellar peduncle
Tegmentum
Cerebellum
Hemisphere
Raphe nuclear complex
Pons
11
Nerve VIII
Middle cerebellar peduncle
Medulla
Pontine gray
Superior olive
Inferior olive
Pyramid

PLATE 65A
CR 350 mm
GW 37
Y217-65
Coronal
Section 1221

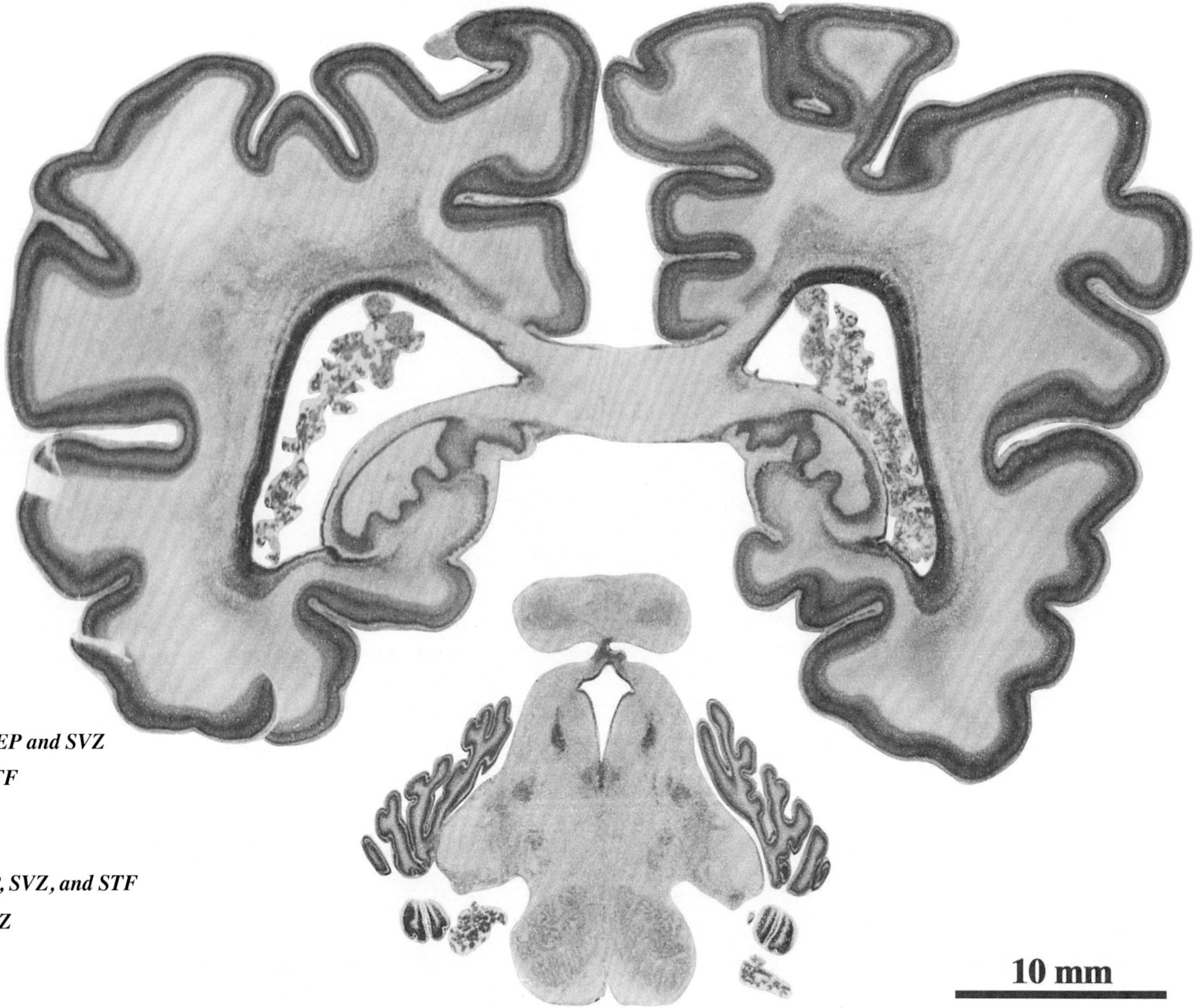

See detail of the brain core in Plates 92A and B.

Remnants of the germinal matrix, migratory streams, and transitional fields

1 ***Paracentral/parietal NEP and SVZ***
2 ***Paracentral/parietal STF***
3 ***Callosal GEP***
4 ***Fornical GEP***
5 ***Parahippocampal NEP, SVZ, and STF***
6 ***Temporal NEP and SVZ***
7 ***Temporal STF***
8 ***Alvear GEP***
9 ***Mesencephalic G/EP***
10 ***External germinal layer (cerebellum)***
11 ***Subpial granular layer (cortical)***

GEP - Glioepithelium
G/EP - Glioepithelium/ependyma
NEP - Neuroepithelium
STF - Stratified transitional field
SVZ - Subventricular zone

PLATE 65B

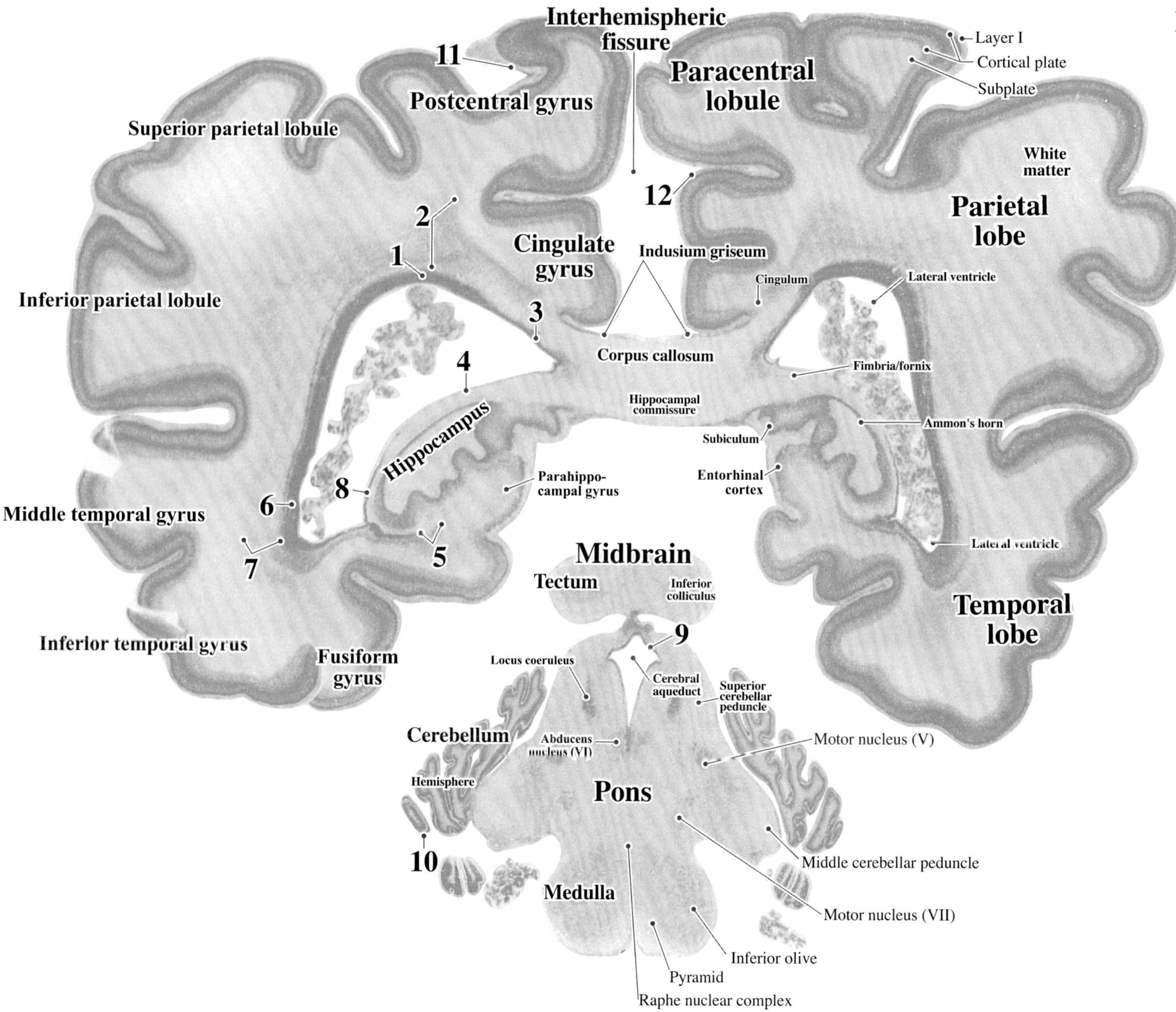
Interhemispheric fissure
11
Postcentral gyrus
Paracentral lobule
Layer I
Cortical plate
Subplate
Superior parietal lobule
White matter
12
Parietal lobe
2
1
Cingulate gyrus
Indusium griseum
Inferior parietal lobule
Cingulum
Lateral ventricle
3
Corpus callosum
4
Fimbria/fornix
Hippocampal commissure
Hippocampus
Ammon's horn
Subiculum
Parahippo-campal gyrus
Entorhinal cortex
8
6
Middle temporal gyrus
5
7
Lateral ventricle
Midbrain
Tectum
Inferior colliculus
Temporal lobe
9
Inferior temporal gyrus
Fusiform gyrus
Locus coeruleus
Cerebral aqueduct
Superior cerebellar peduncle
Cerebellum
Abducens nucleus (VI)
Motor nucleus (V)
Hemisphere
Pons
10
Middle cerebellar peduncle
Medulla
Motor nucleus (VII)
Inferior olive
Pyramid
Raphe nuclear complex

PLATE 66A
CR 350 mm
GW 37
Y217-65
Coronal
Section 1311

See detail of the brain core in Plates 93A and B.

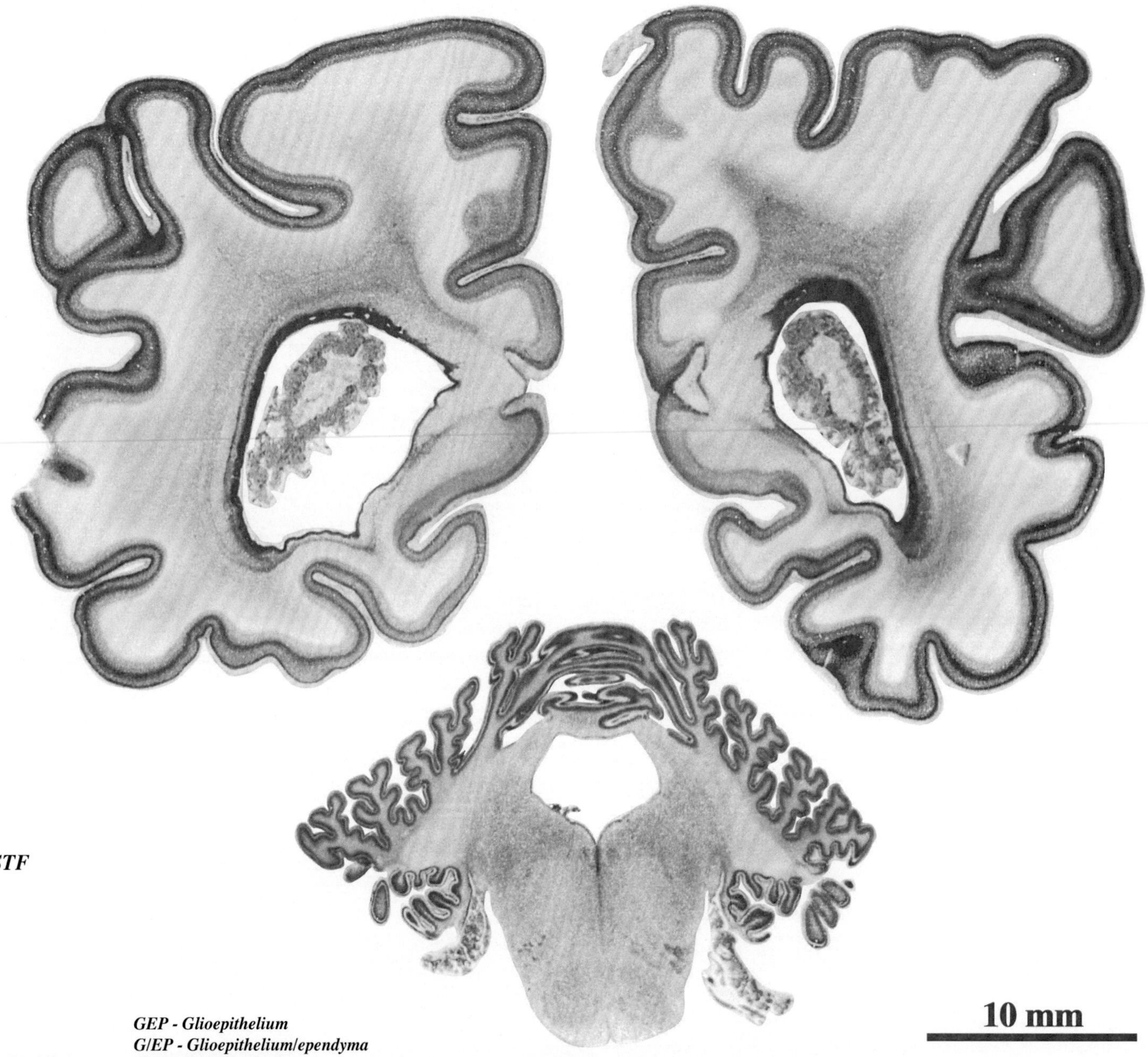

Remnants of the germinal matrix, migratory streams, and transitional fields

1 *Parietal NEP and SVZ*
2 *Parietal STF*
3 *Callosal GEP*
4 *Cingulate NEP, SVZ, and STF*
5 *Occipital NEP and SVZ*
6 *Occipital STF*
7 *Temporal NEP and SVZ*
8 *Temporal STF*
9 *Medullary G/EP*
10 *Pontine G/EP*
11 *External germinal layer (cerebellum)*
12 *Subpial granular layer (cortical)*

GEP - Glioepithelium
G/EP - Glioepithelium/ependyma
NEP - Neuroepithelium
STF - Stratified transitional field
SVZ - Subventricular zone

PLATE 66B

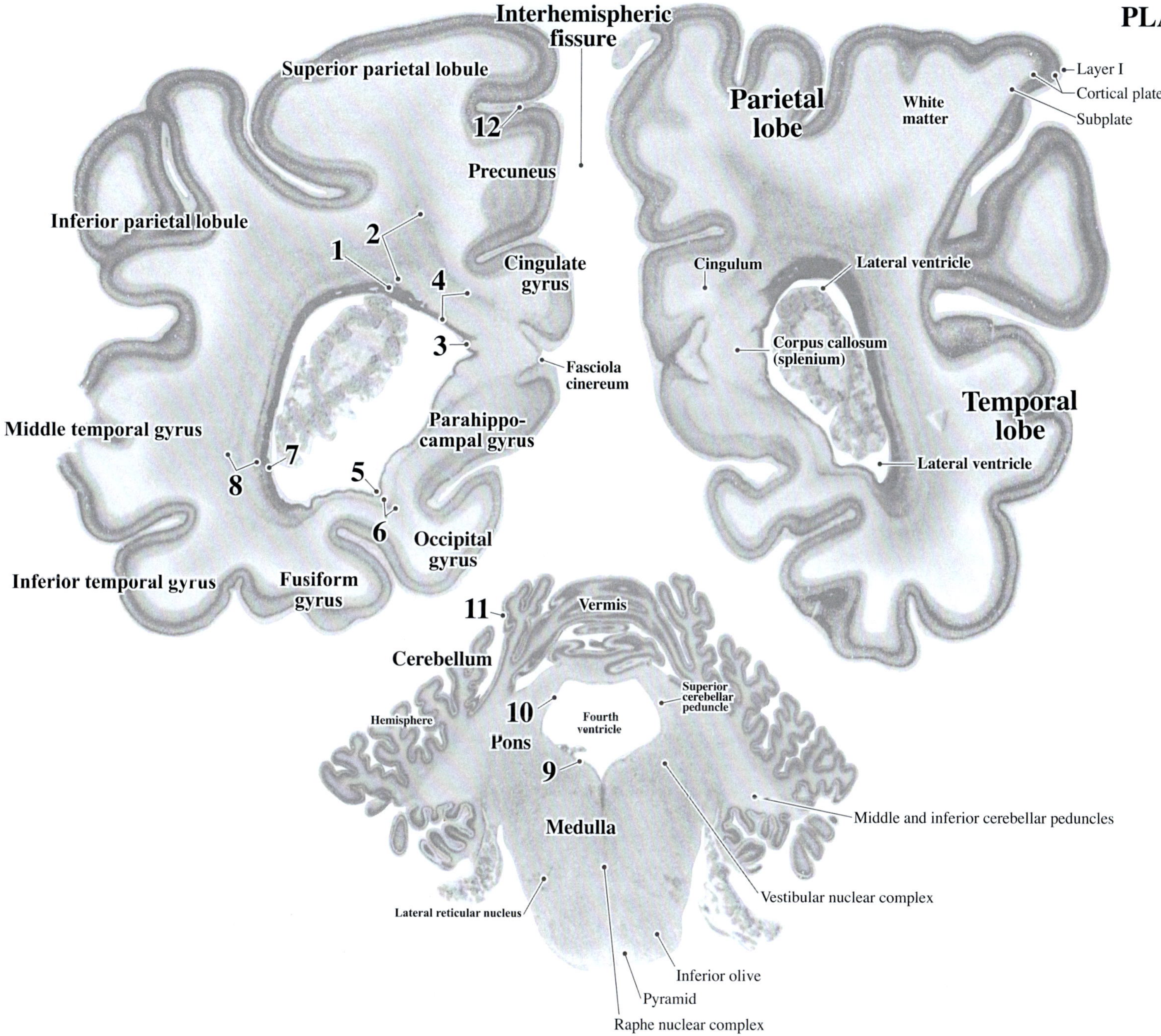
Interhemispheric fissure
Superior parietal lobule
12
Precuneus
Inferior parietal lobule
2
1
4
Cingulate gyrus
3
Fasciola cinereum
Parahippo-campal gyrus
Middle temporal gyrus
7
8
5
6
Occipital gyrus
Inferior temporal gyrus
Fusiform gyrus
Parietal lobe
White matter
Layer I
Cortical plate
Subplate
Cingulum
Lateral ventricle
Corpus callosum (splenium)
Temporal lobe
Lateral ventricle
11
Vermis
Cerebellum
10
Fourth ventricle
Superior cerebellar peduncle
Hemisphere
Pons
9
Medulla
Middle and inferior cerebellar peduncles
Lateral reticular nucleus
Vestibular nuclear complex
Inferior olive
Pyramid
Raphe nuclear complex

PLATE 67A
CR 350 mm
GW 37
Y217-65
Coronal
Section 1371

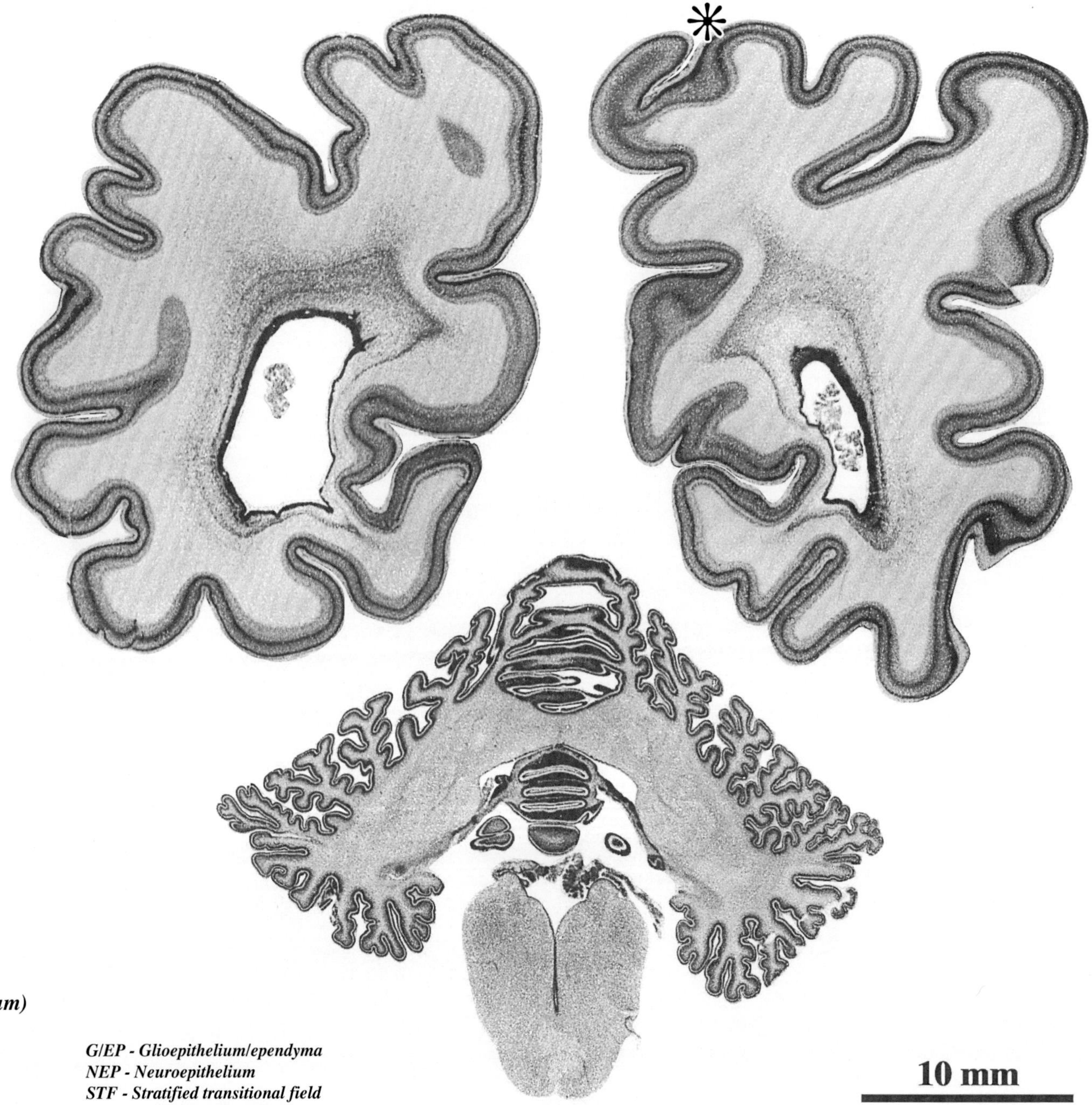

*** See this area of cortex in Plates 77 and 78**

See detail of the brain core and cerebellum in Plates 94A and B.

Remnants of the germinal matrix, migratory streams, and transitional fields

1 *Parietal NEP and SVZ*
2 *Parietal STF*
3 *Cingulate NEP and SVZ*
4 *Cingulate STF*
5 *Occipital NEP and SVZ*
6 *Occipital STF*
7 *Temporal NEP and SVZ*
8 *Temporal STF*
9 *External germinal layer (cerebellum)*
10 *Germinal trigone (cerebellum)*
11 *Medullary G/EP*
12 *Subpial granular layer (cortical)*

G/EP - Glioepithelium/ependyma
NEP - Neuroepithelium
STF - Stratified transitional field
SVZ - Subventricular zone

PLATE 67B

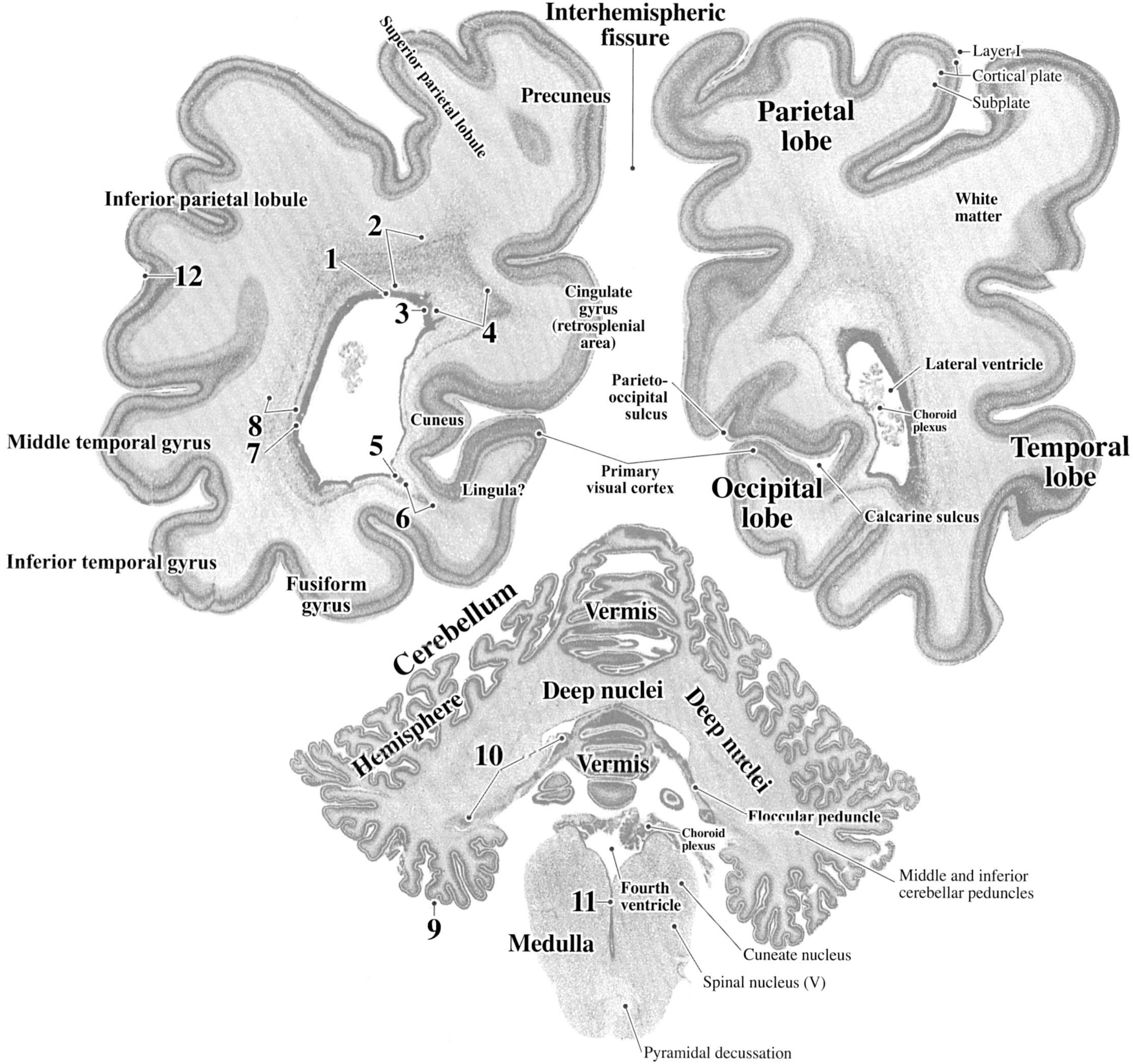
Interhemispheric fissure
Superior parietal lobule
Precuneus
Parietal lobe
Layer I
Cortical plate
Subplate
White matter
Inferior parietal lobule
2
1
12
3
4
Cingulate gyrus (retrosplenial area)
Lateral ventricle
Choroid plexus
Parieto-occipital sulcus
8
7
Middle temporal gyrus
5
Cuneus
Primary visual cortex
Occipital lobe
Temporal lobe
Calcarine sulcus
6
Lingula?
Inferior temporal gyrus
Fusiform gyrus
Cerebellum
Vermis
Hemisphere
Deep nuclei
Deep nuclei
10
Vermis
Floccular peduncle
Choroid plexus
Middle and inferior cerebellar peduncles
Fourth ventricle
11
9
Medulla
Cuneate nucleus
Spinal nucleus (V)
Pyramidal decussation

PLATE 68A
CR 350 mm
GW 37
Y217-65
Coronal
Section 1501

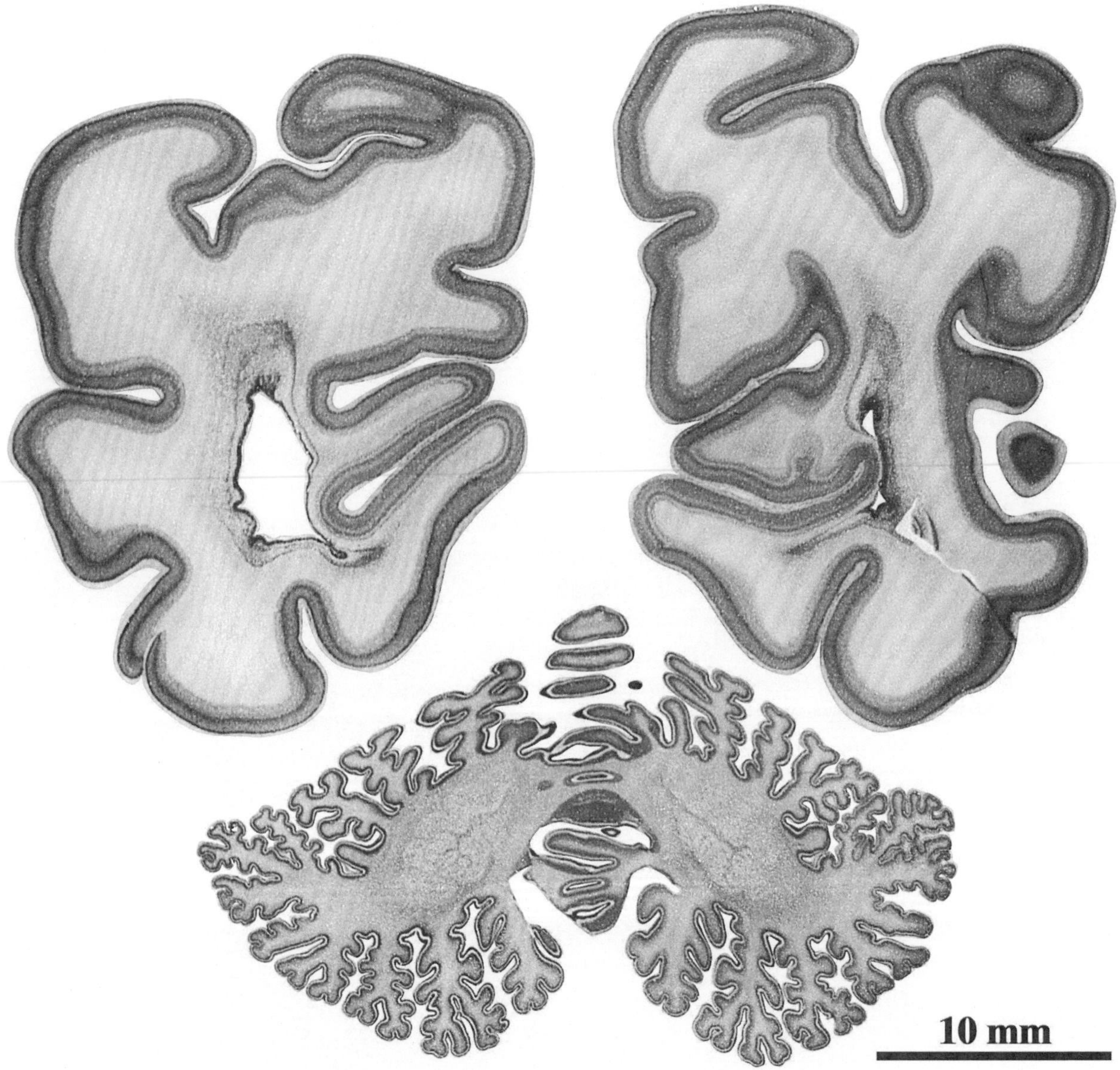

See detail of the cerebellum in Plates 95A and B.

Remnants of the germinal matrix, migratory streams, and transitional fields

1 *Parietal NEP and SVZ*
2 *Parietal STF*
3 *Occipital NEP and SVZ*
4 *Occipital STF*
5 *Temporal NEP and SVZ*
6 *Temporal STF*
7 *External germinal layer (cerebellum)*
8 *Subpial granular layer (cortical)*

NEP - Neuroepithelium
STF - Stratified transitional field
SVZ - Subventricular zone

PLATE 68B

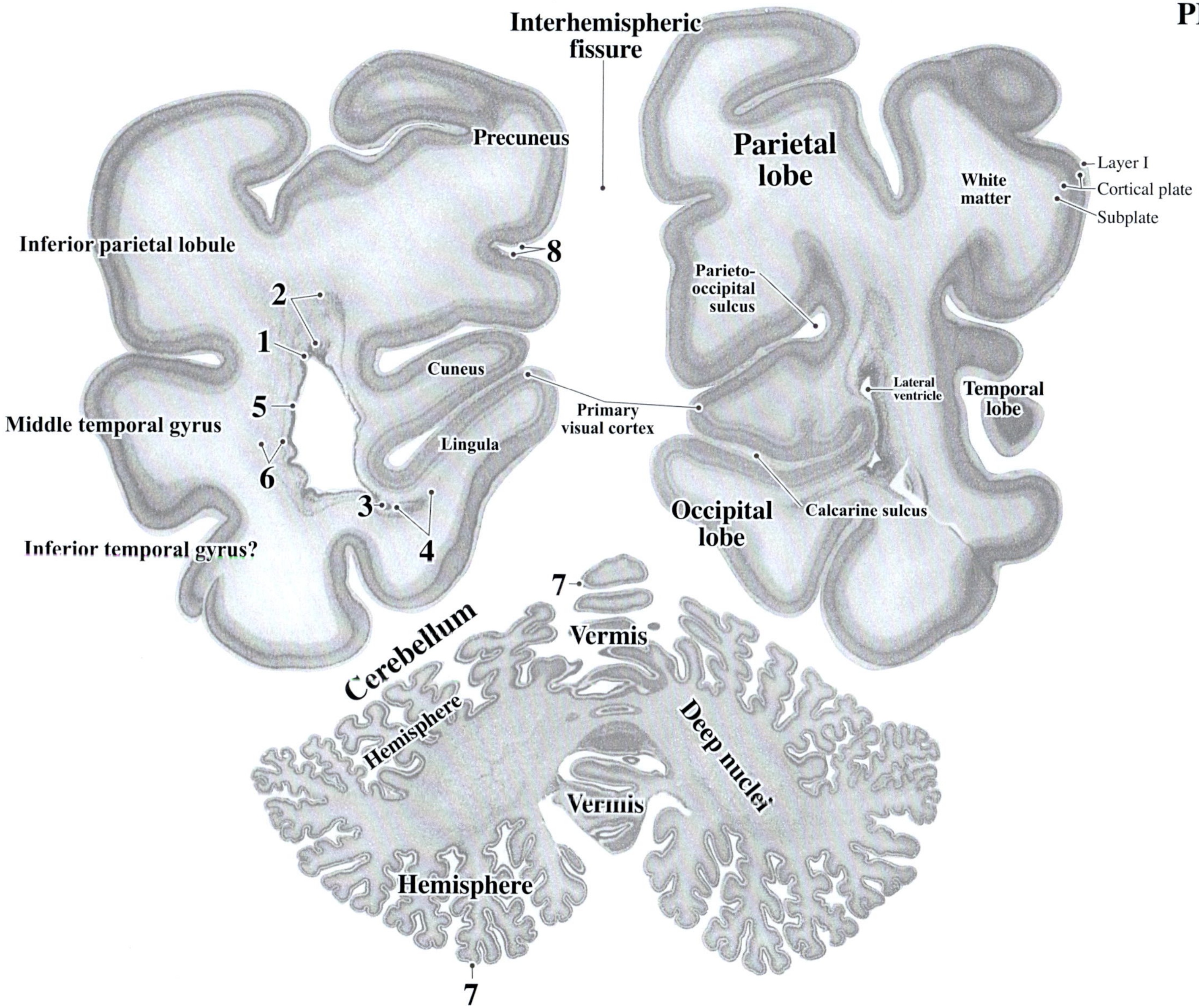
Interhemispheric fissure
Precuneus
Parietal lobe
Layer I
White matter
Cortical plate
Subplate
Inferior parietal lobule
8
2
Parieto-occipital sulcus
1
Cuneus
5
Lateral ventricle
Temporal lobe
Primary visual cortex
Middle temporal gyrus
Lingula
6
3
Calcarine sulcus
Occipital lobe
Inferior temporal gyrus?
4
7
Vermis
Cerebellum
Hemisphere
Deep nuclei
Vermis
Hemisphere
7

PLATE 69A
CR 350 mm
GW 37
Y217-65
Coronal
Section 1611

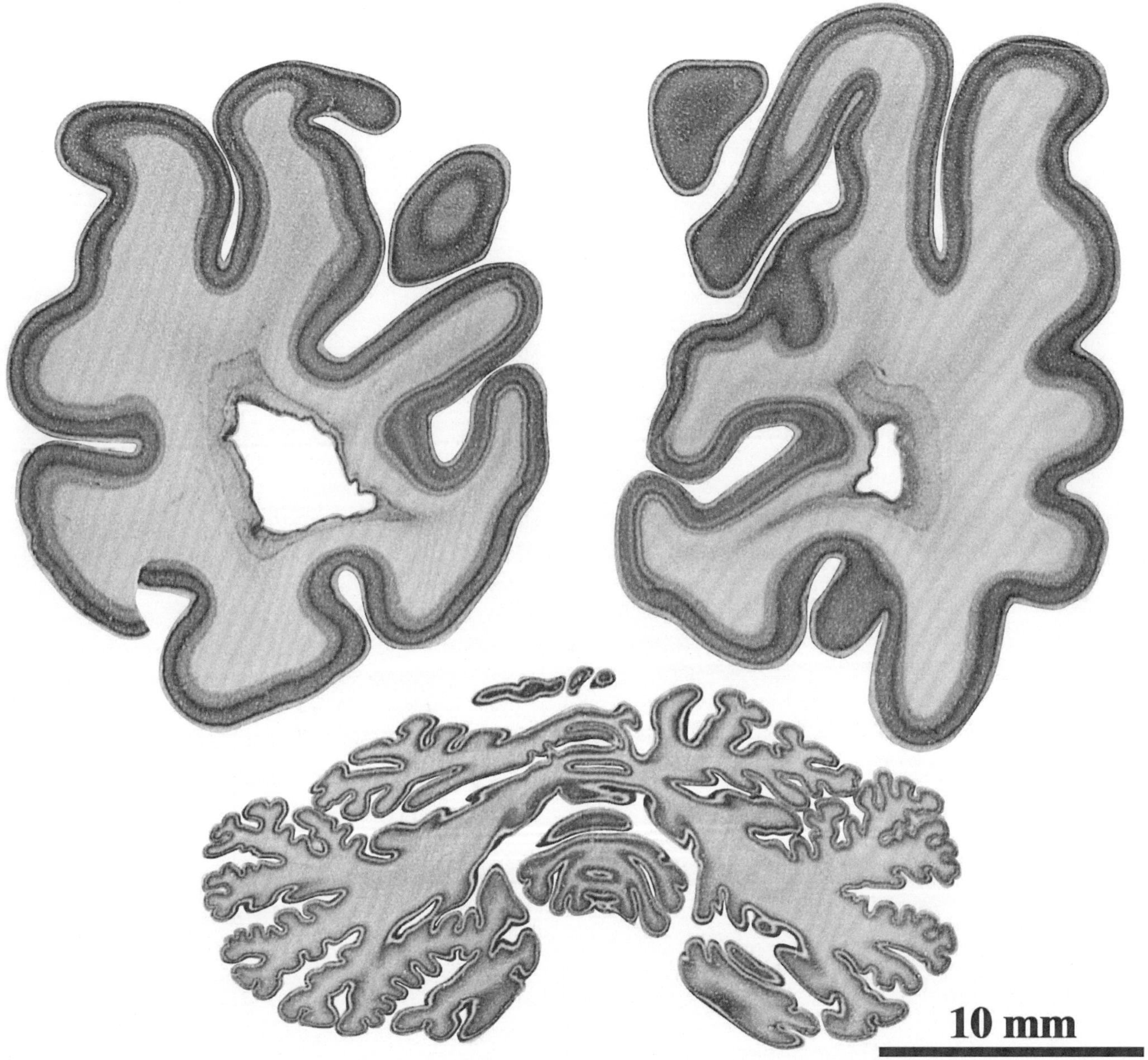

See detail of the cerebellum in Plates 95A and B.

Remnants of the germinal matrix, migratory streams, and transitional fields

1 ***Parietal NEP and SVZ***
2 ***Parietal STF***
3 ***Occipital NEP and SVZ***
4 ***Occipital STF***
5 ***Temporal NEP and SVZ***
6 ***Temporal STF***
7 ***External germinal layer (cerebellum)***
8 ***Subpial granular layer (cortical)***

NEP - Neuroepithelium
STF - Stratified transitional field
SVZ - Subventricular zone

PLATE 69B

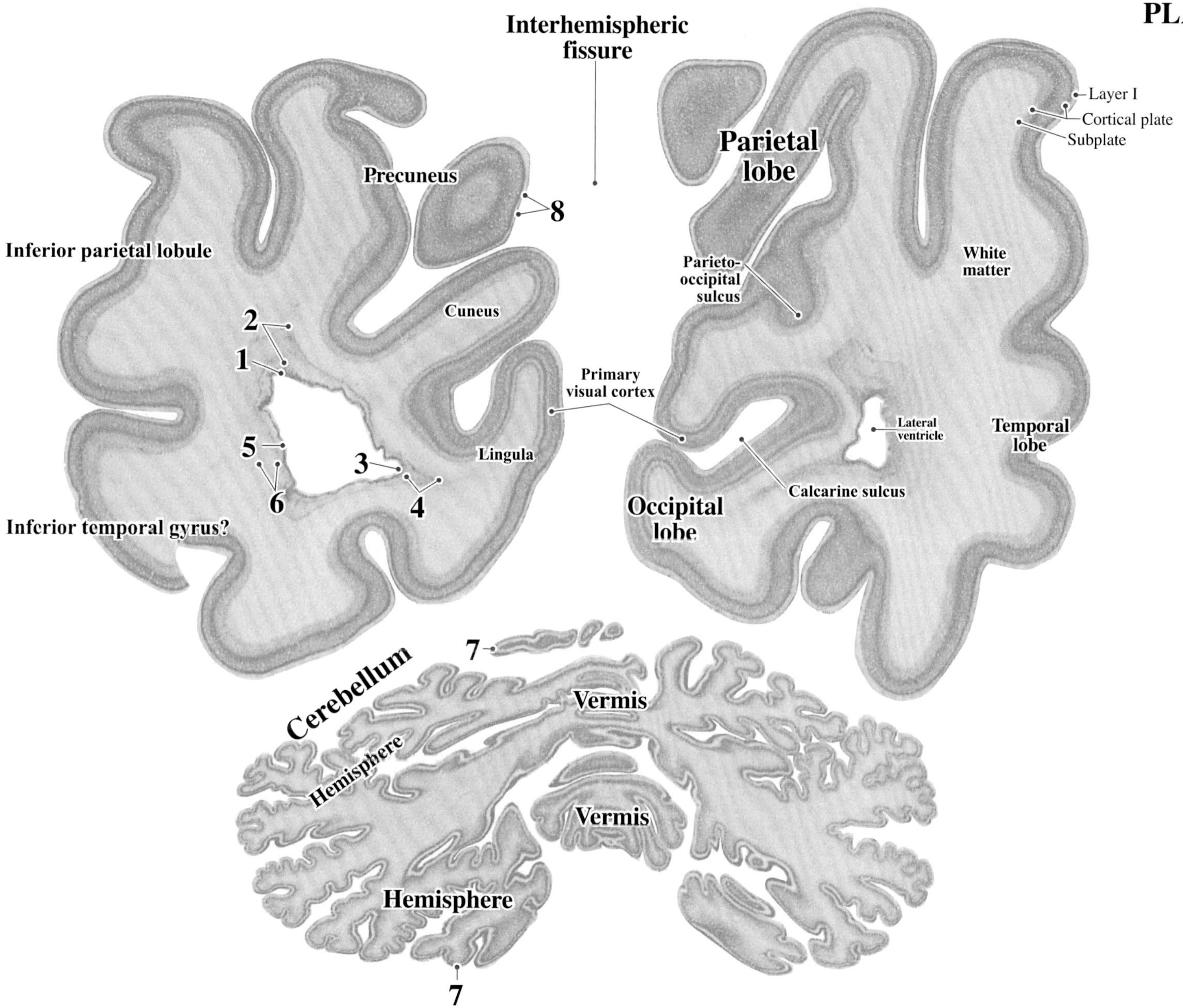
Interhemispheric fissure
Precuneus
8
Inferior parietal lobule
Cuneus
2
1
5
6
3
4
Lingula
Inferior temporal gyrus?
Primary visual cortex
Parietal lobe
Layer I
Cortical plate
Subplate
White matter
Parieto-occipital sulcus
Lateral ventricle
Temporal lobe
Calcarine sulcus
Occipital lobe
7
Cerebellum
Vermis
Hemisphere
Vermis
Hemisphere
7

PLATE 70A
CR 350 mm
GW 37
Y217-65
Coronal
Section 1711

See detail of the cerebellum in Plates 97A and B.

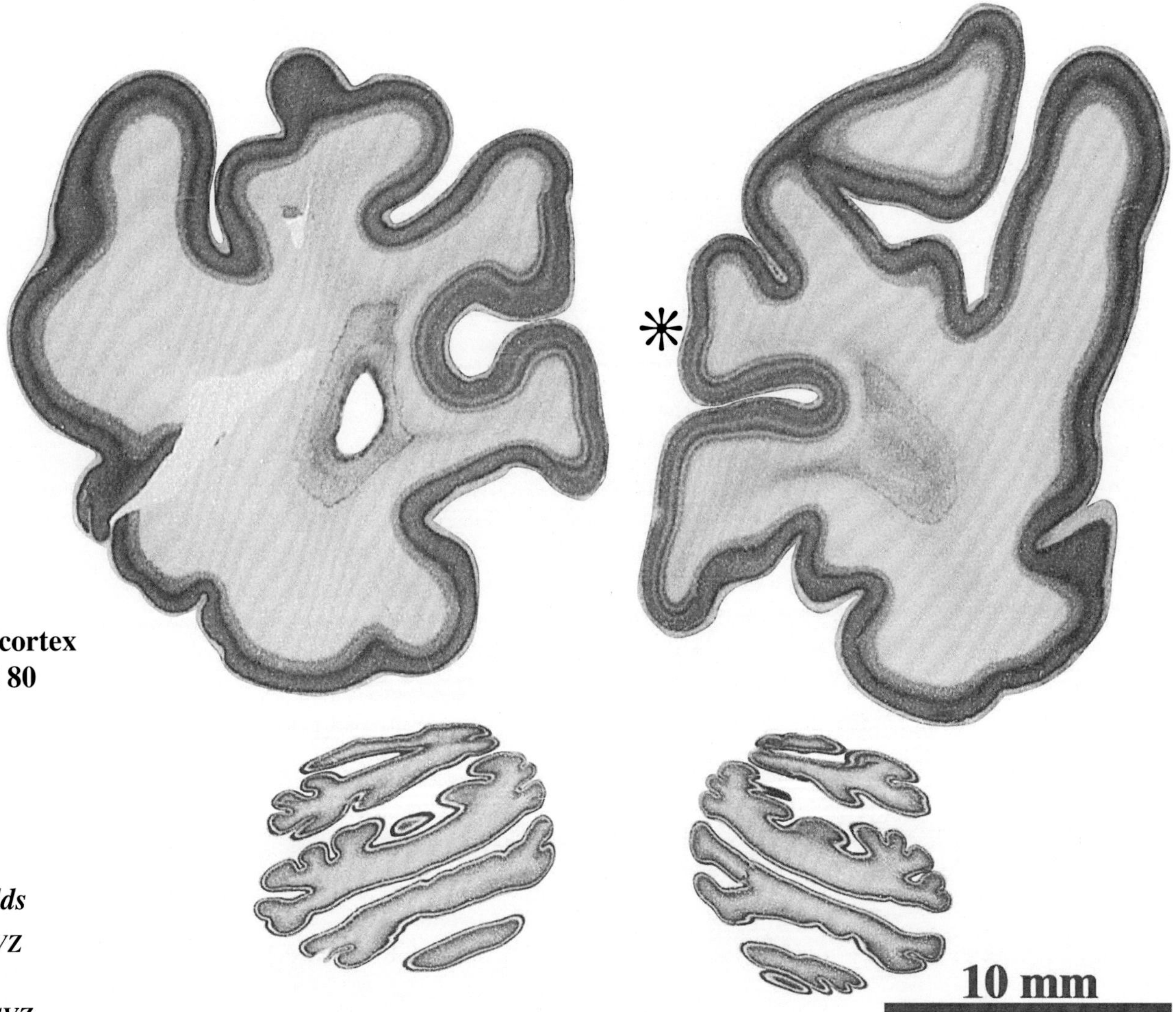

✳ See this area of cortex in Plates 79 and 80

Remnants of the germinal matrix, migratory streams, and transitional fields

1 ***Parietal NEP and SVZ***
2 ***Parietal STF***
3 ***Occipital NEP and SVZ***
4 ***Occipital STF***
5 ***External germinal layer (cerebellum)***
6 ***Subpial granular layer (cortical)***

NEP - Neuroepithelium
STF - Stratified transitional field
SVZ - Subventricular zone

PLATE 70B

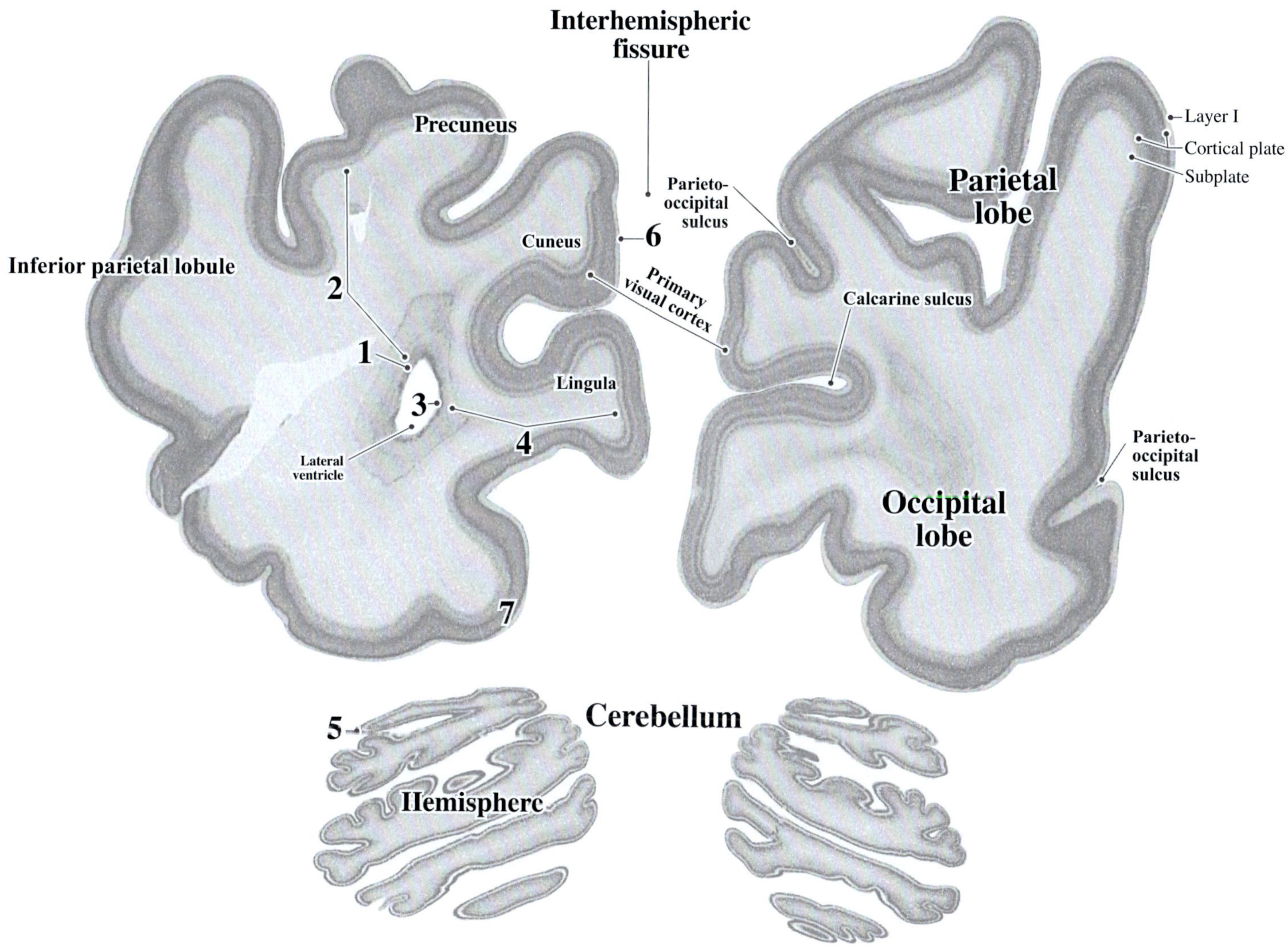
Interhemispheric fissure
Precuneus
Parieto-occipital sulcus
Layer I
Cortical plate
Subplate
Parietal lobe
Cuneus
6
Inferior parietal lobule
2
Primary visual cortex
Calcarine sulcus
1
Lingula
3
4
Lateral ventricle
Parieto-occipital sulcus
Occipital lobe
7
Cerebellum
5
Hemisphere

PLATE 71
CR 350 mm, GW 37, Y217-65
Coronal, Section 761
FRONTAL CORTEX

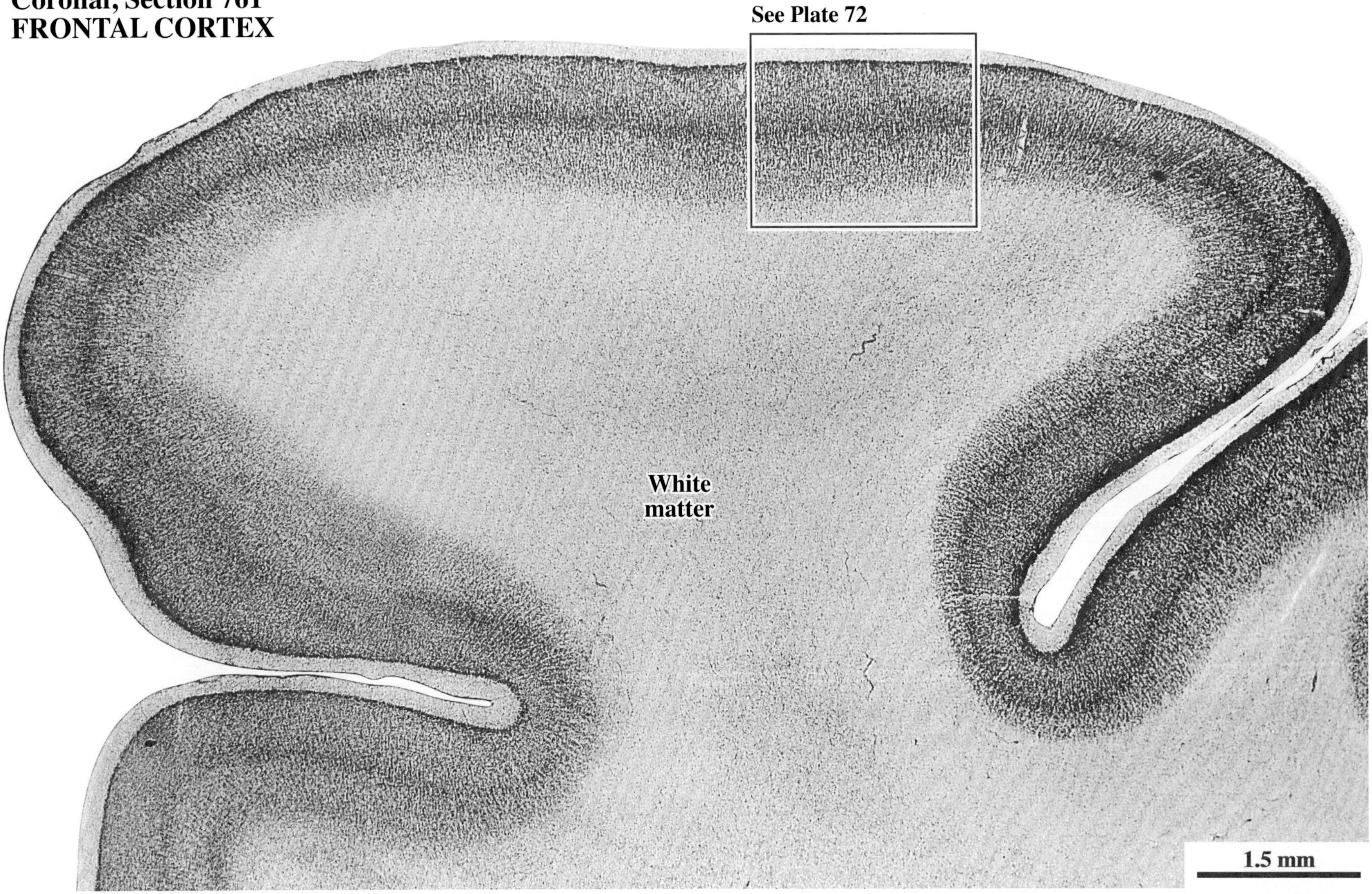

See the entire Section 761 in Plate 57.

PLATE 72
CR 350 mm, GW 37, Y217-65, Coronal, Section 761

FRONTAL CORTEX

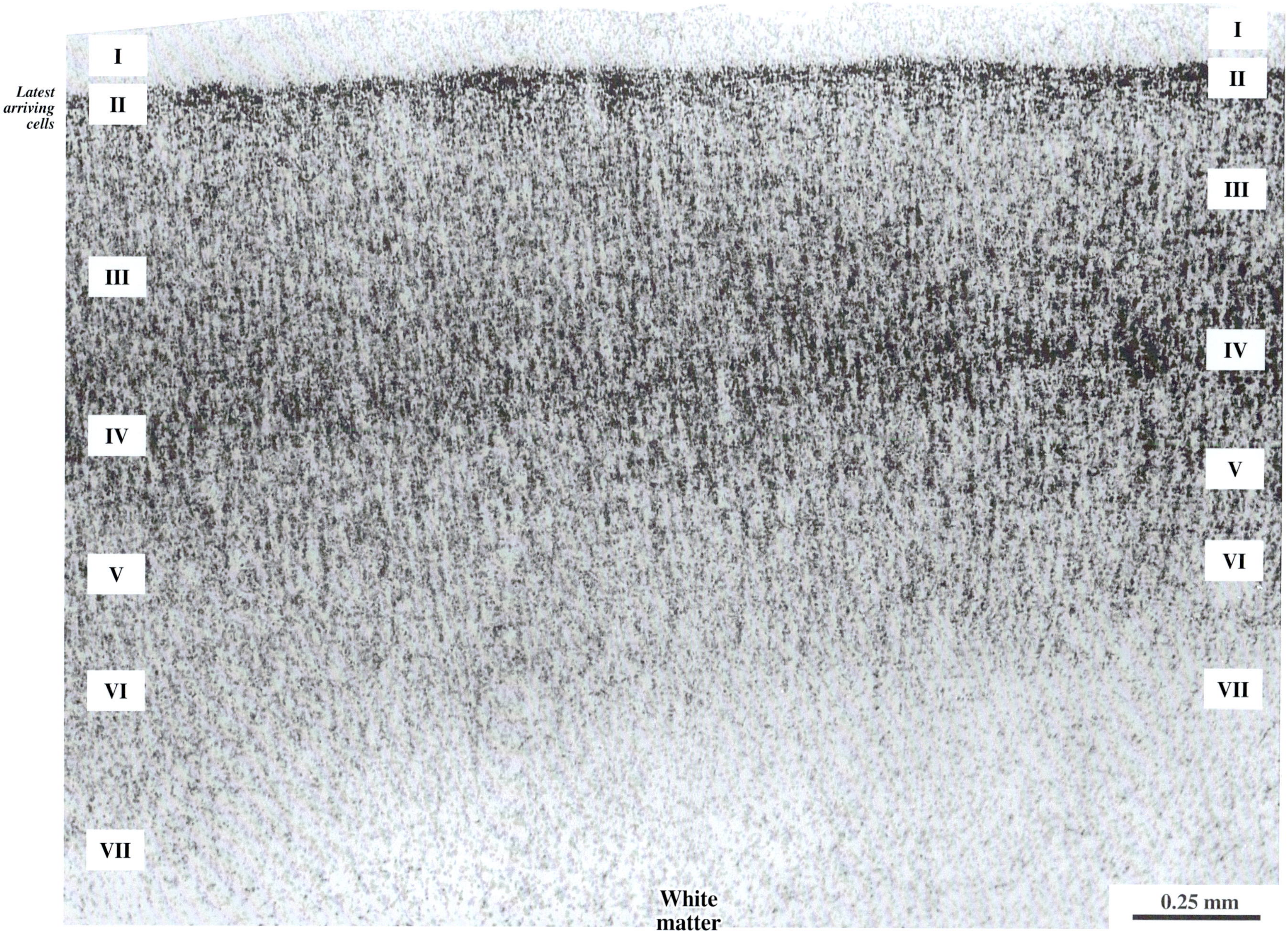

PLATE 73
CR 350 mm, GW 37, Y217-65
Coronal, Section 1081
CORTEX OF THE PRECENTRAL GYRUS
(Primary motor cortex)

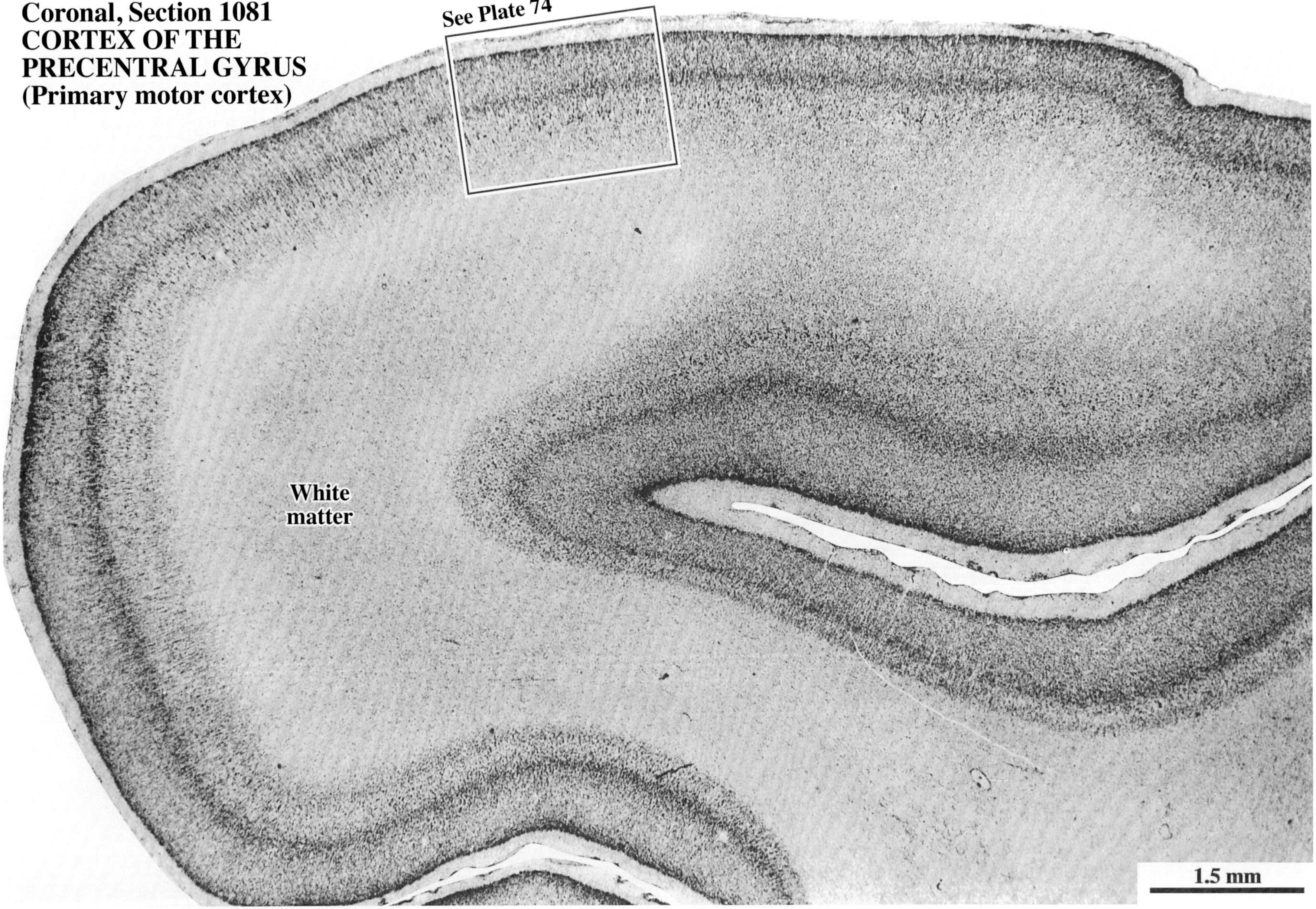

See the entire Section 1081 in Plate 62.

PLATE 74
CR 350 mm, GW 37, Y217-65, Coronal, Section 1081

CORTEX OF THE PRECENTRAL GYRUS

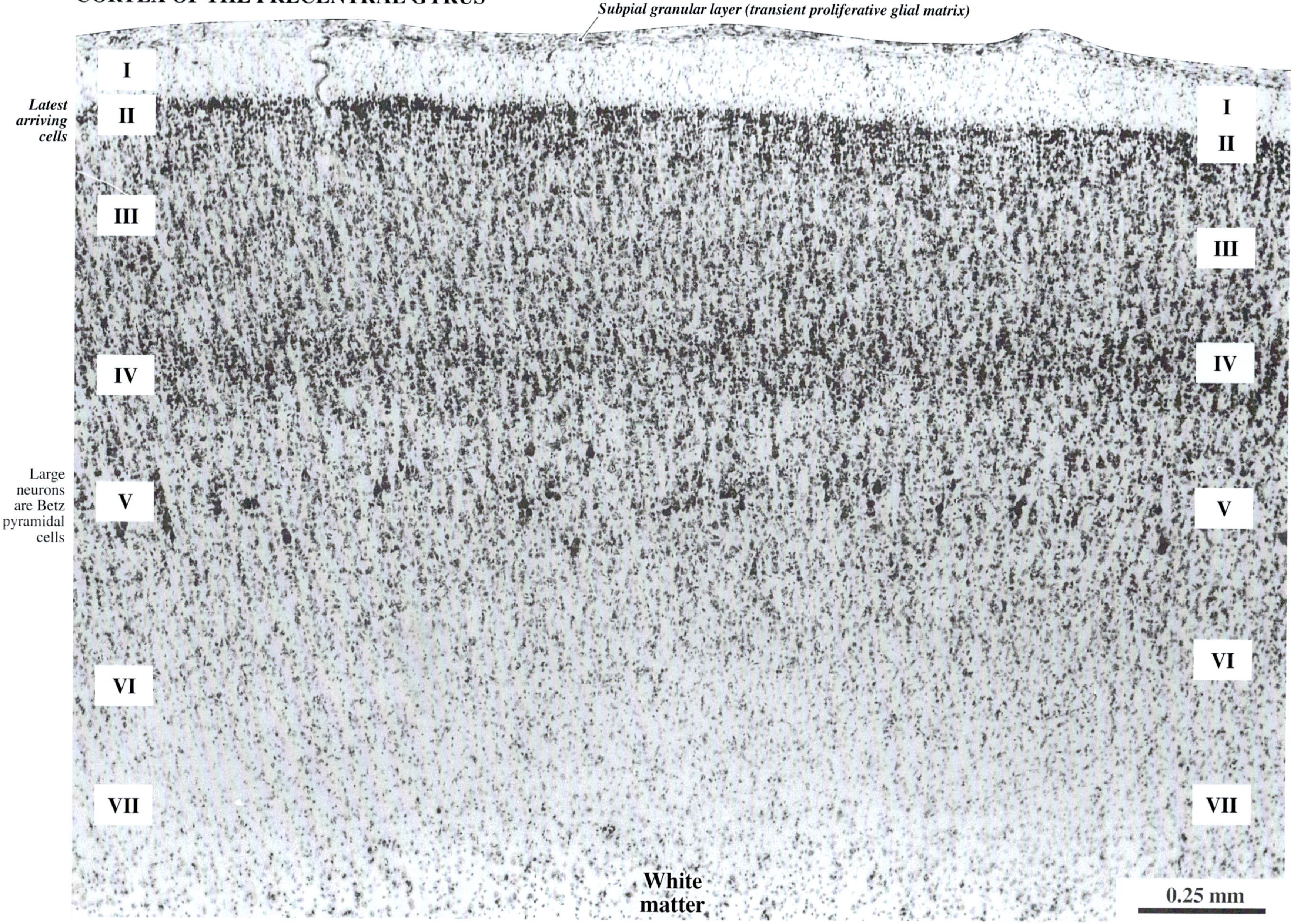

PLATE 75
CR 350 mm, GW 37, Y217-65
Coronal, Section 1161
CORTEX OF THE
POSTCENTRAL GYRUS
(Primary sensory cortex)

See Plate 76

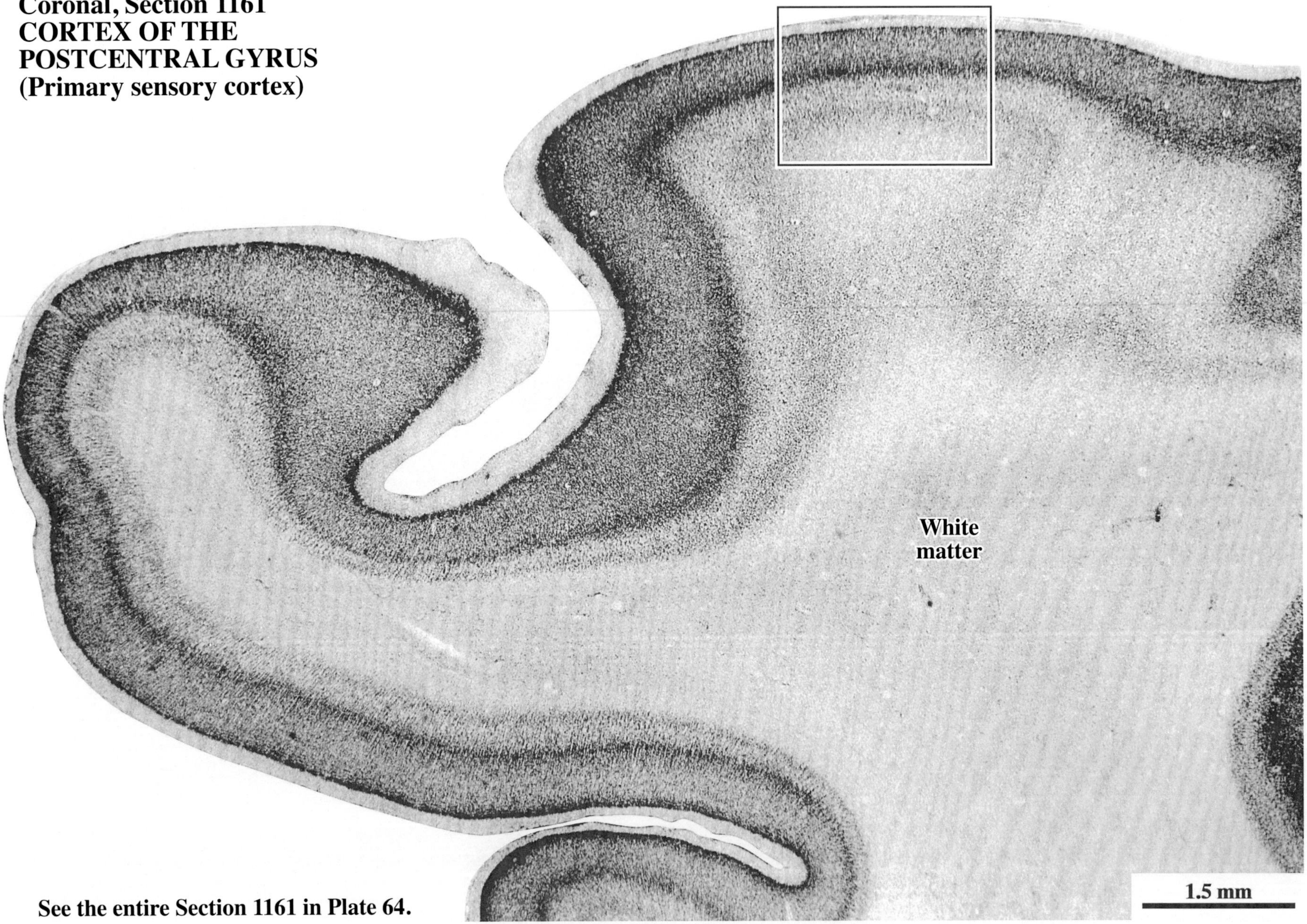

See the entire Section 1161 in Plate 64.

PLATE 76
CR 350 mm, GW 37, Y217-65, Coronal, Section 1161

CORTEX OF THE POSTCENTRAL GYRUS

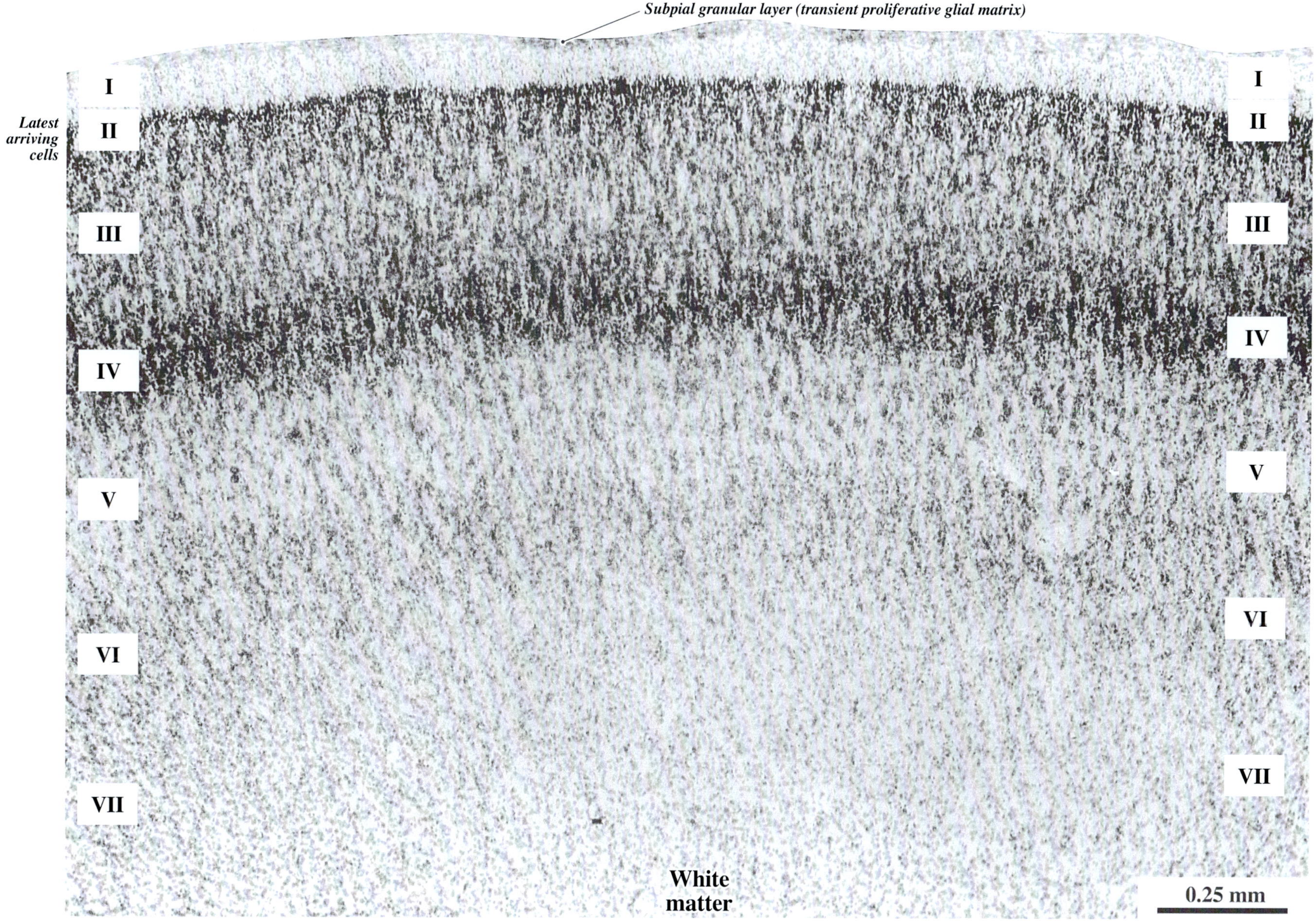

PLATE 77
CR 350 mm, GW 37, Y217-65
Coronal, Section 1371
PARIETAL CORTEX

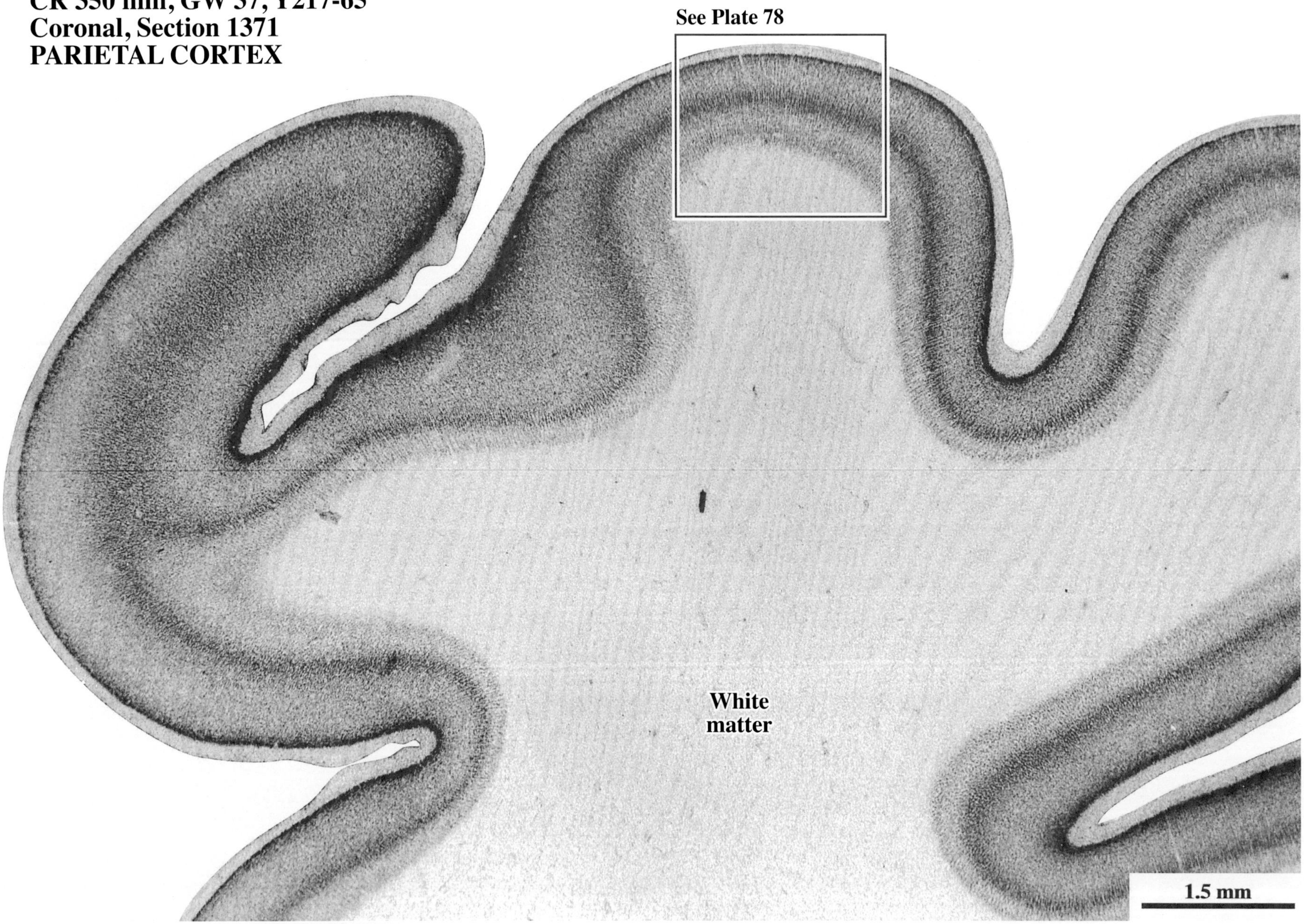

See the entire Section 1371 in Plate 67.

PLATE 78
CR 350 mm, GW 37, Y217-65, Coronal, Section 1371

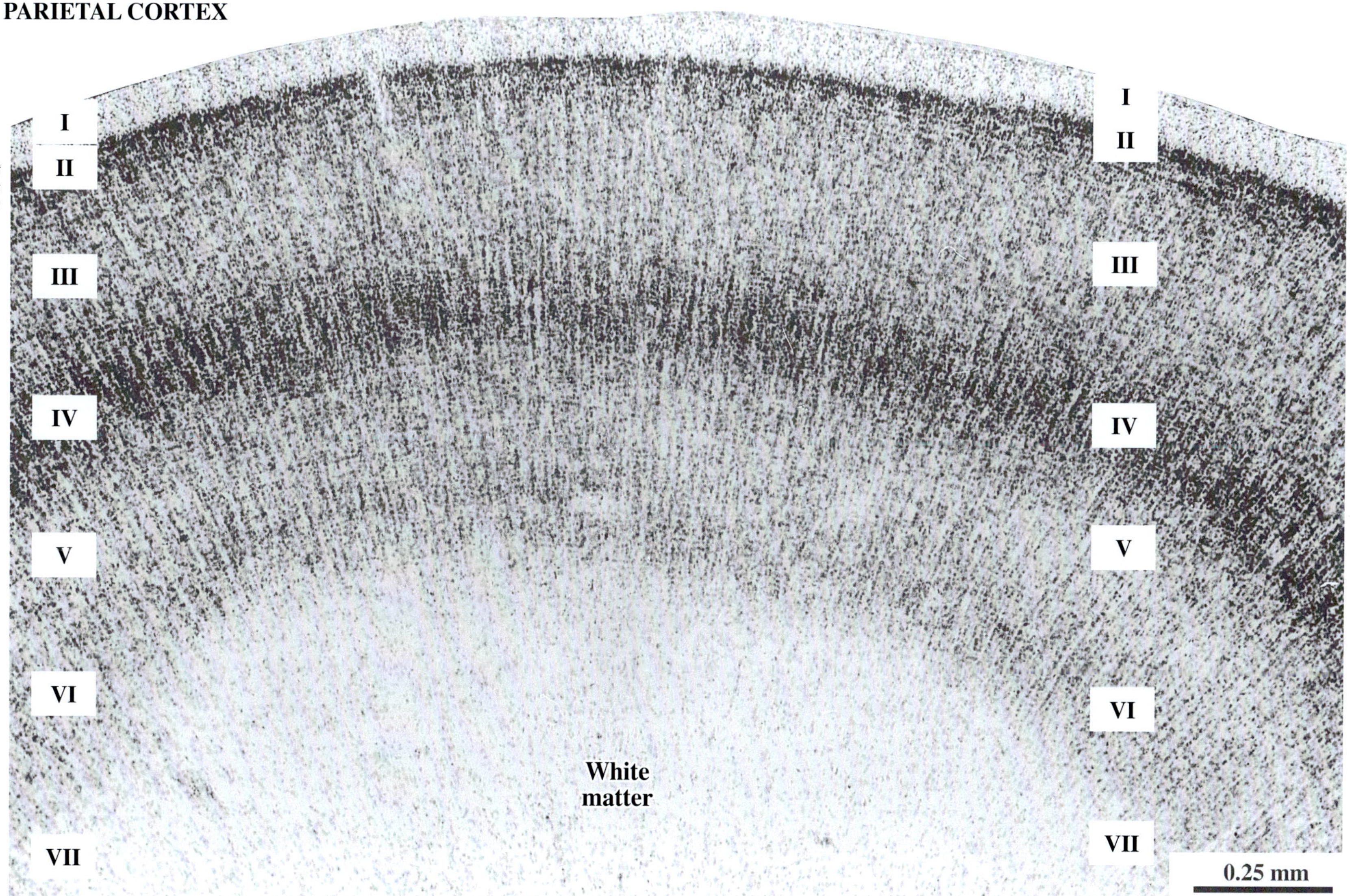

PLATE 79
CR 350 mm, GW 37, Y217-65
Coronal, Section 1751
STRIATE/PERISTRIATE CORTICAL TRANSITION AREA

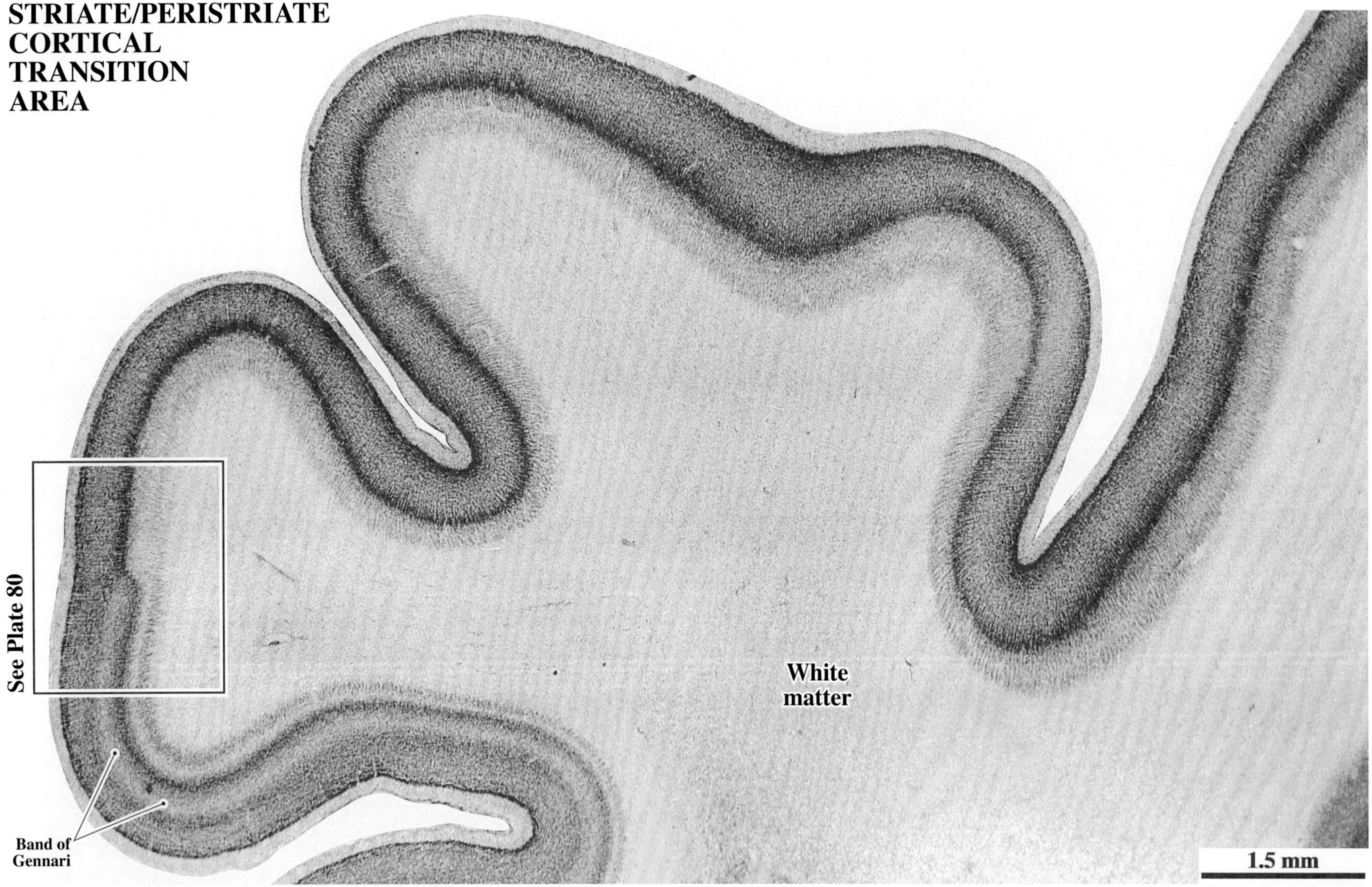

See the nearby entire Section 1711 in Plate 70.

PLATE 80
CR 350 mm, GW 37, Y217-65, Coronal, Section 1751

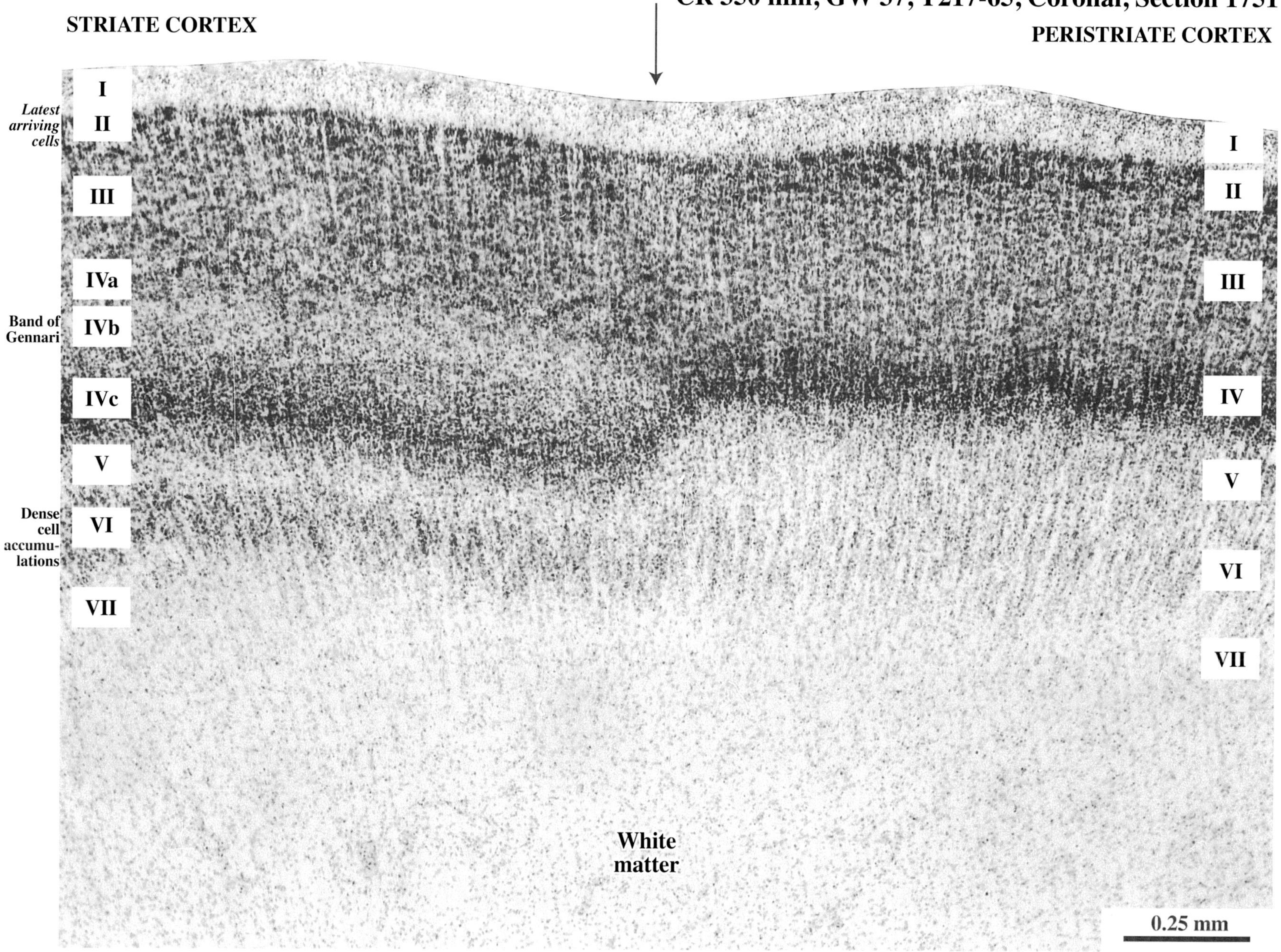

PLATE 81A
CR 350 mm
GW 37
Y217-65
Coronal
Section 611

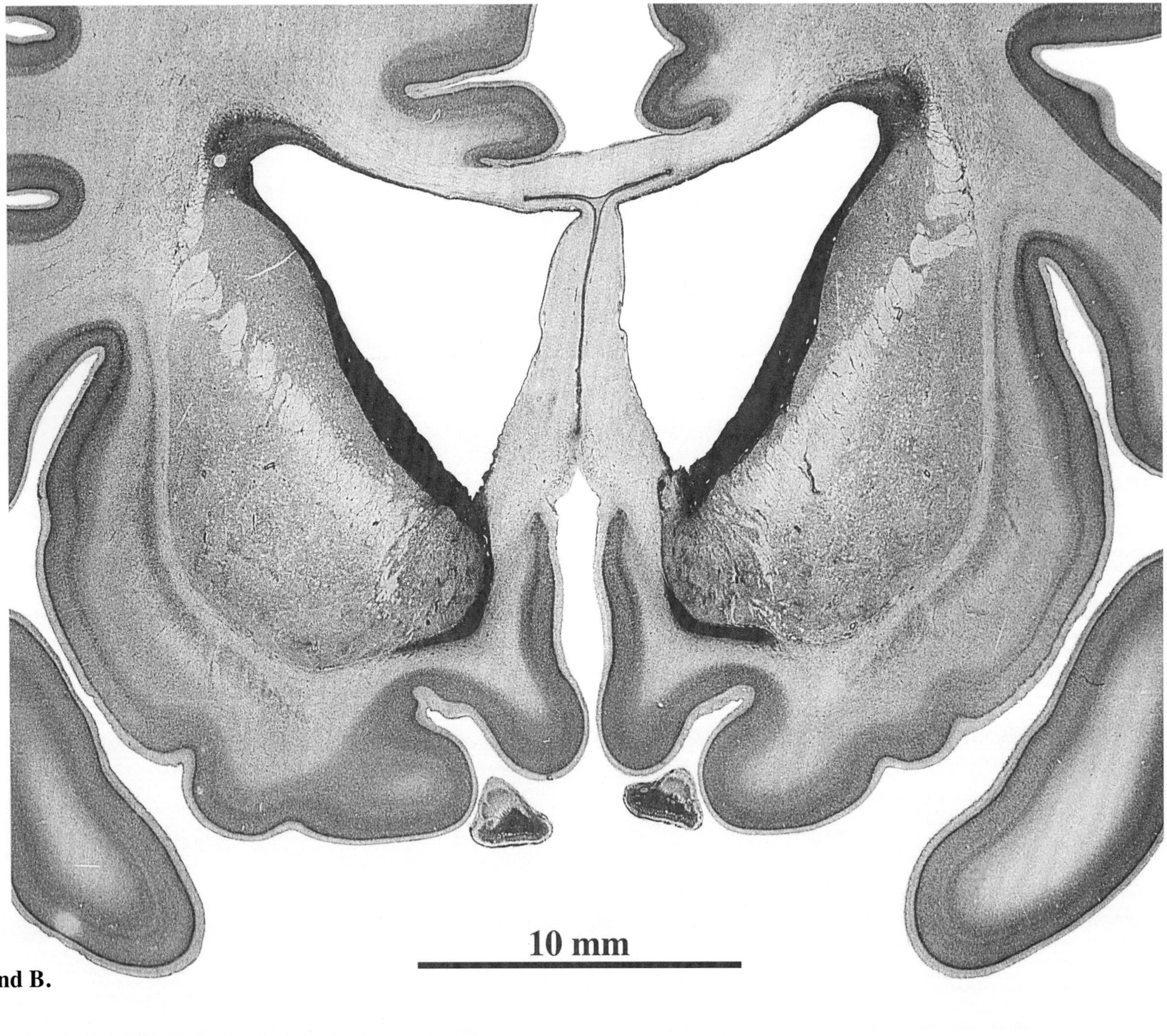

See the entire
Section 611
in Plates 54A and B.

PLATE 81B

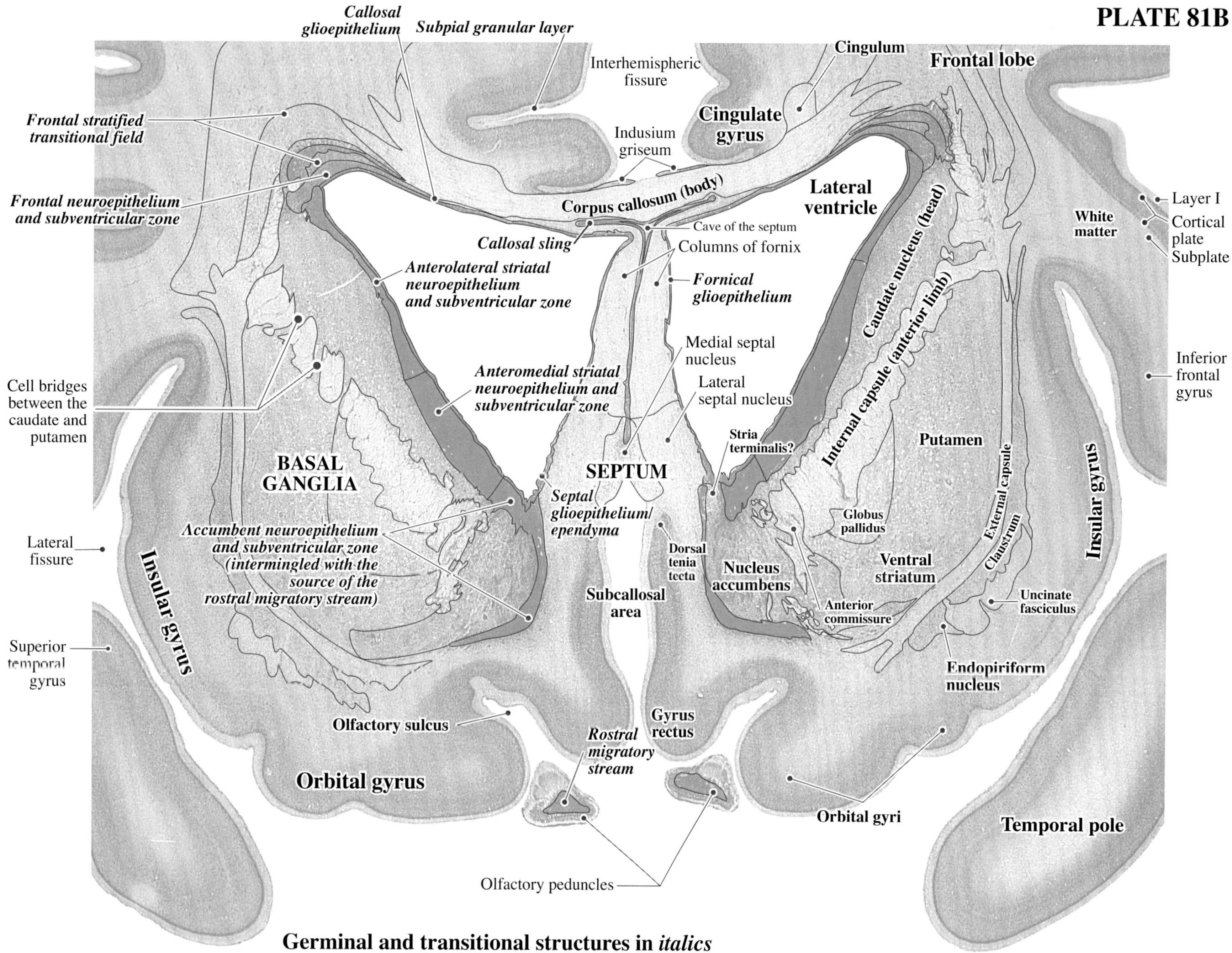

Germinal and transitional structures in *italics*

PLATE 82A
CR 350 mm, GW 37, Y217-65, Coronal, Section 691

See the entire Section 691 in Plate 55.

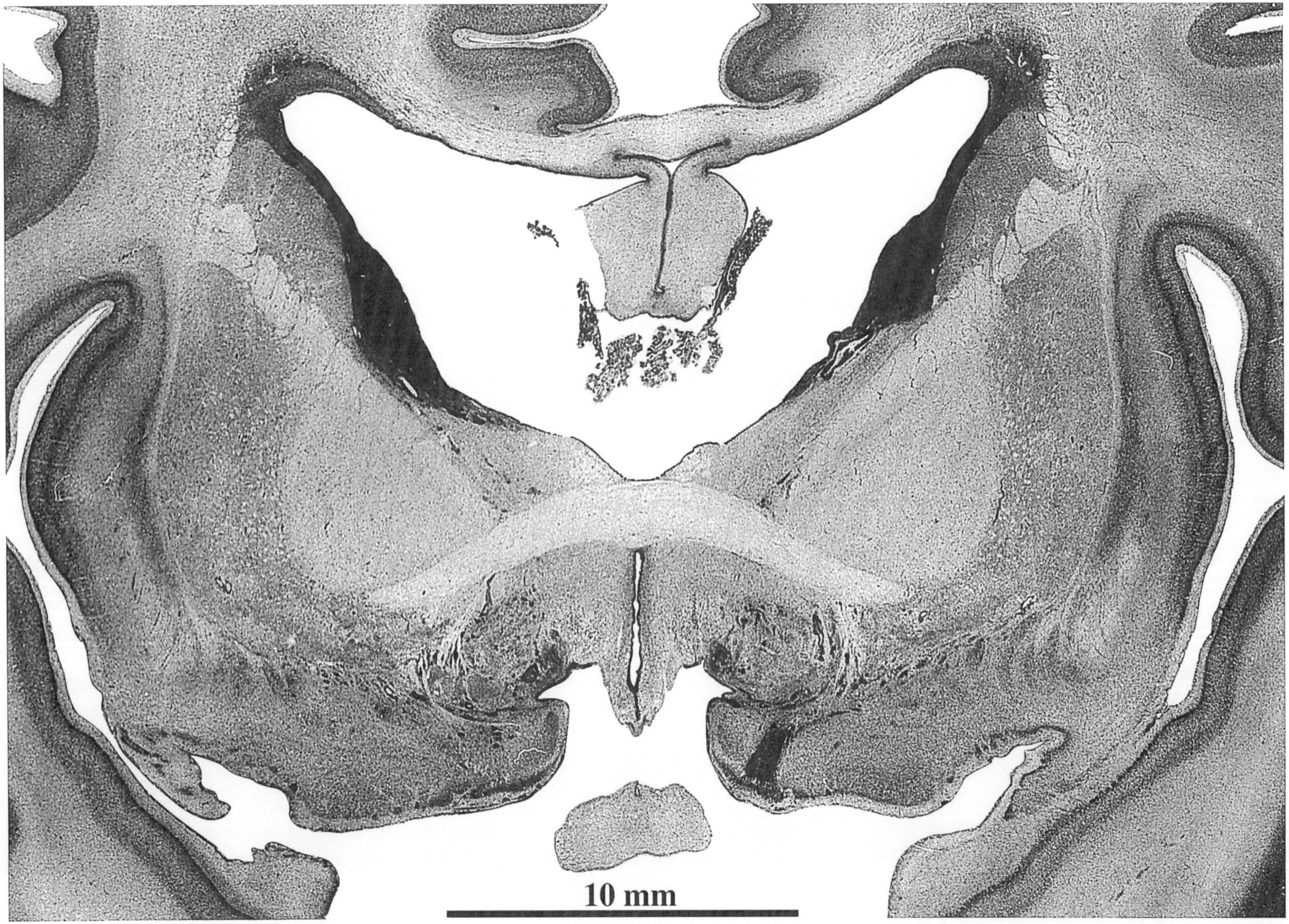

PLATE 82B

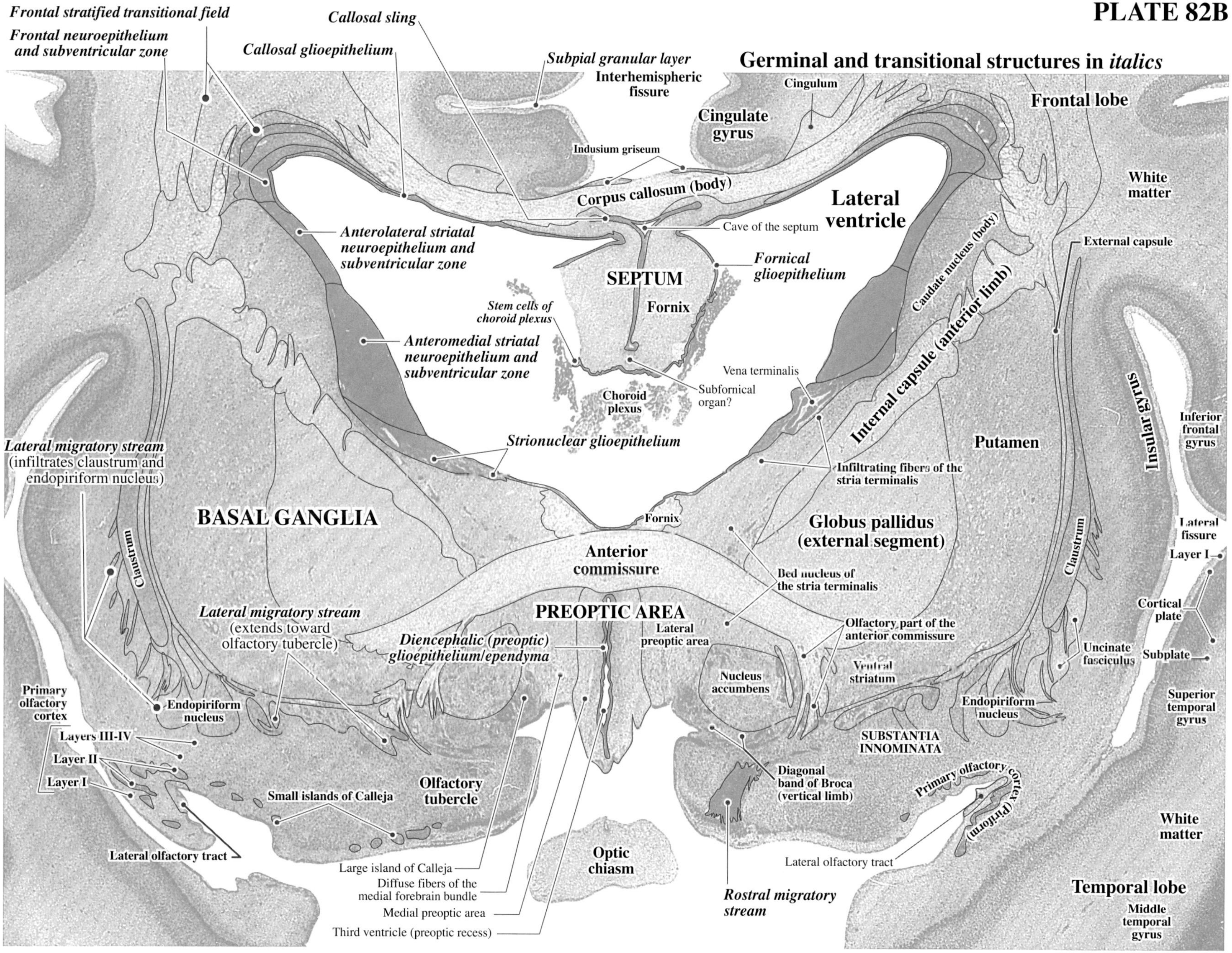
Frontal stratified transitional field
Frontal neuroepithelium and subventricular zone
Callosal sling
Callosal glioepithelium
Subpial granular layer
Interhemispheric fissure
Germinal and transitional structures in italics
Cingulum
Cingulate gyrus
Frontal lobe
Indusium griseum
Corpus callosum (body)
White matter
Lateral ventricle
Cave of the septum
Anterolateral striatal neuroepithelium and subventricular zone
SEPTUM
Fornical glioepithelium
External capsule
Caudate nucleus (body)
Internal capsule (anterior limb)
Stem cells of choroid plexus
Fornix
Anteromedial striatal neuroepithelium and subventricular zone
Vena terminalis
Subfornical organ?
Choroid plexus
Insular gyrus
Inferior frontal gyrus
Putamen
Strionuclear glioepithelium
Lateral migratory stream (infiltrates claustrum and endopiriform nucleus)
Infiltrating fibers of the stria terminalis
BASAL GANGLIA
Fornix
Globus pallidus (external segment)
Lateral fissure
Claustrum
Anterior commissure
Claustrum
Layer I
Bed nucleus of the stria terminalis
Cortical plate
PREOPTIC AREA
Lateral migratory stream (extends toward olfactory tubercle)
Lateral preoptic area
Olfactory part of the anterior commissure
Diencephalic (preoptic) glioepithelium/ependyma
Uncinate fasciculus
Subplate
Ventral striatum
Nucleus accumbens
Primary olfactory cortex
Endopiriform nucleus
Endopiriform nucleus
Superior temporal gyrus
Layers III-IV
SUBSTANTIA INNOMINATA
Layer II
Diagonal band of Broca (vertical limb)
Layer I
Small islands of Calleja
Olfactory tubercle
Primary olfactory cortex (Piriform)
White matter
Lateral olfactory tract
Optic chiasm
Lateral olfactory tract
Large island of Calleja
Temporal lobe
Diffuse fibers of the medial forebrain bundle
Rostral migratory stream
Middle temporal gyrus
Medial preoptic area
Third ventricle (preoptic recess)

PLATE 83A
CR 350 mm
GW 37
Y217-65
Coronal
Section 721

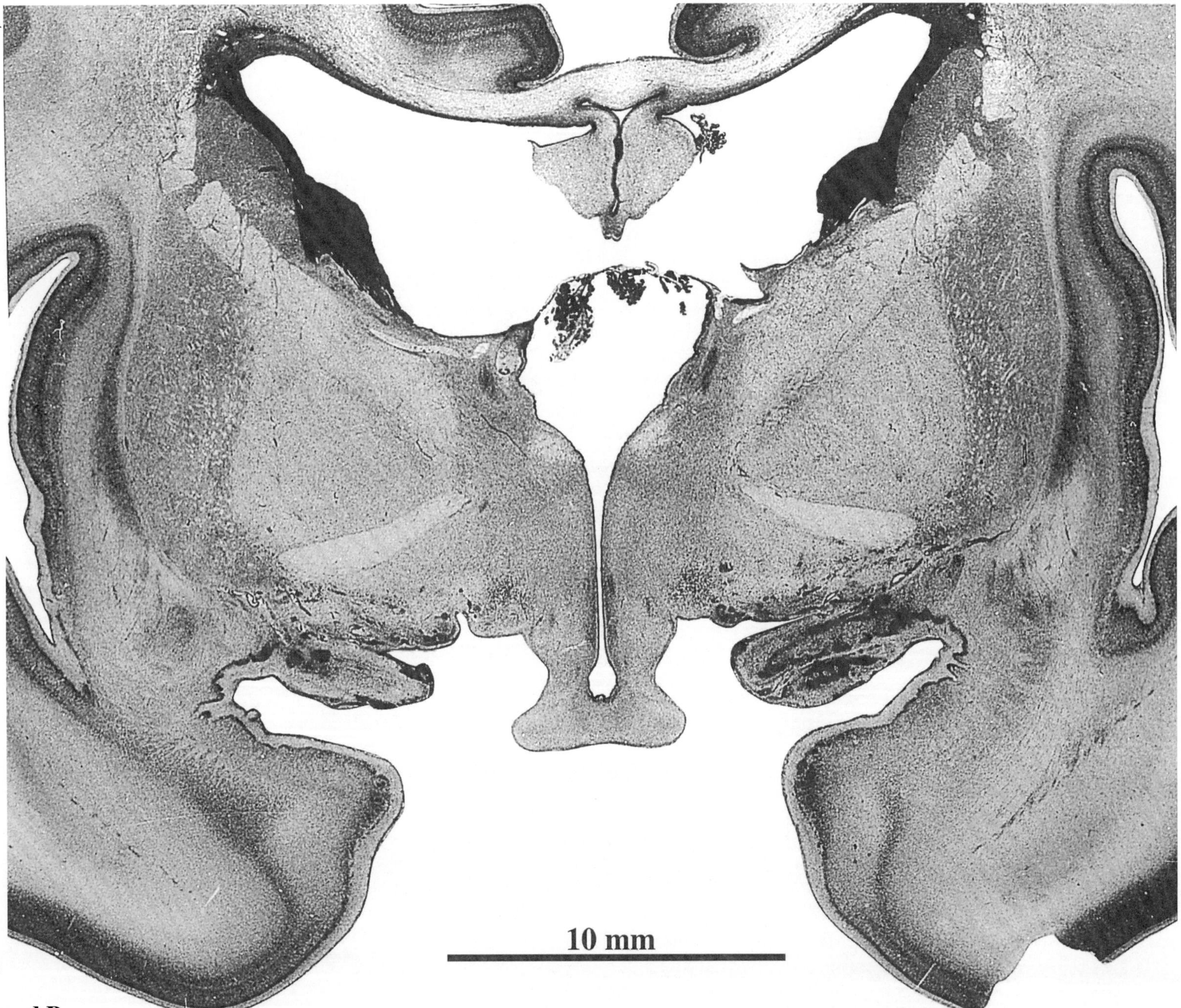

See the entire Section 721 in Plates 56A and B.

PLATE 83B

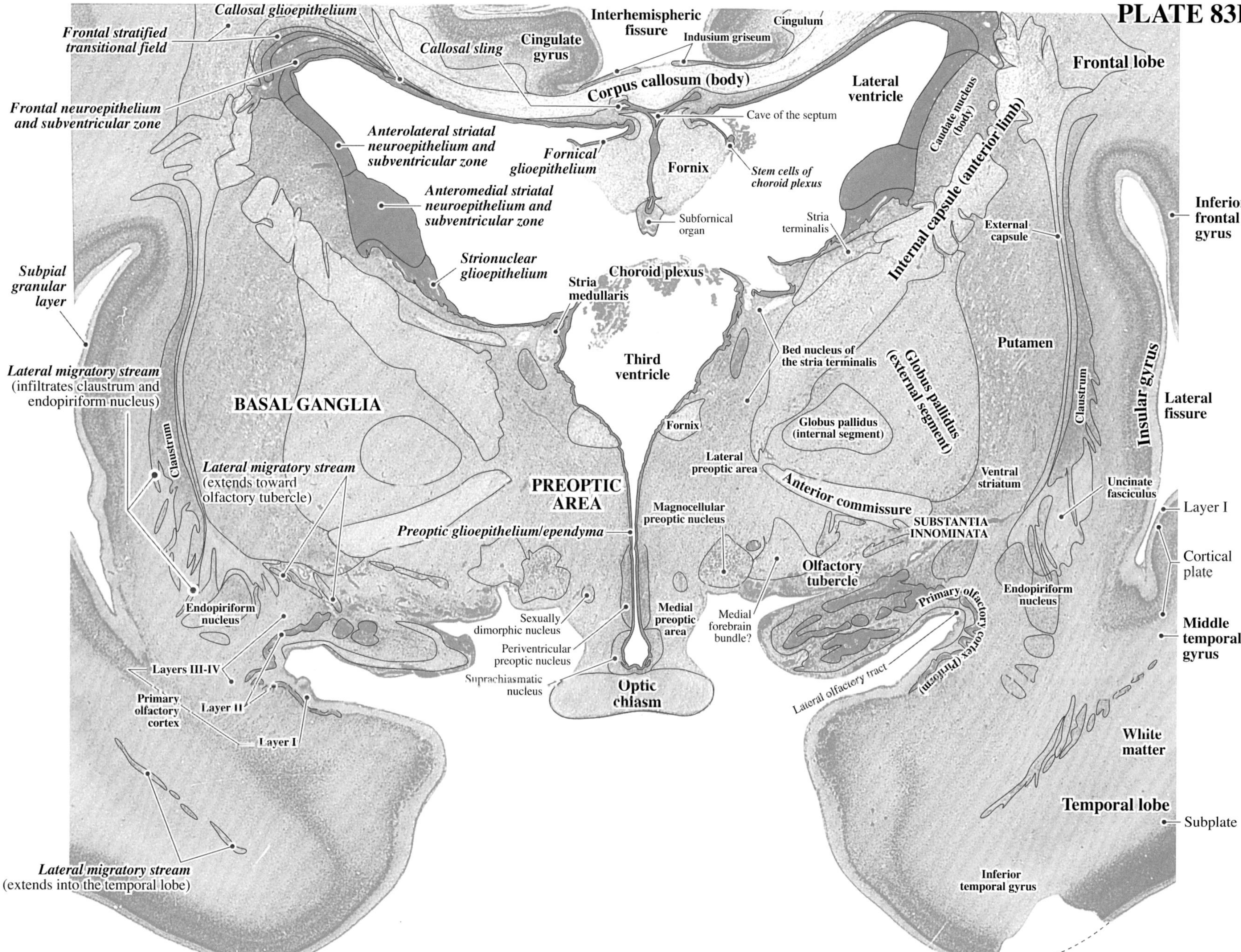

Germinal and transitional structures in *italics*

PLATE 84A
CR 350 mm
GW 37
Y217-65
Coronal
Section 761

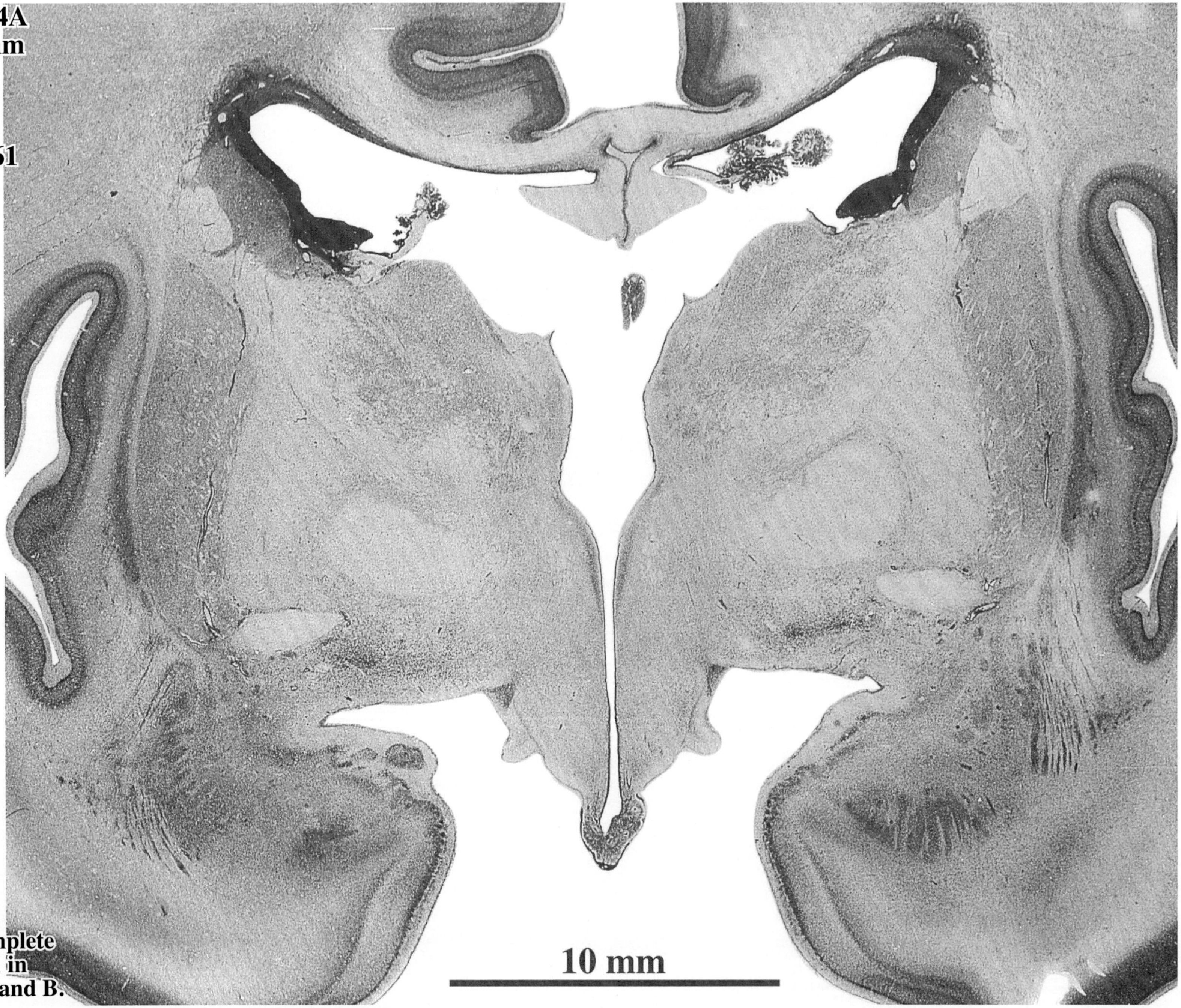

See the complete
Section 761 in
Plates 87A and B.

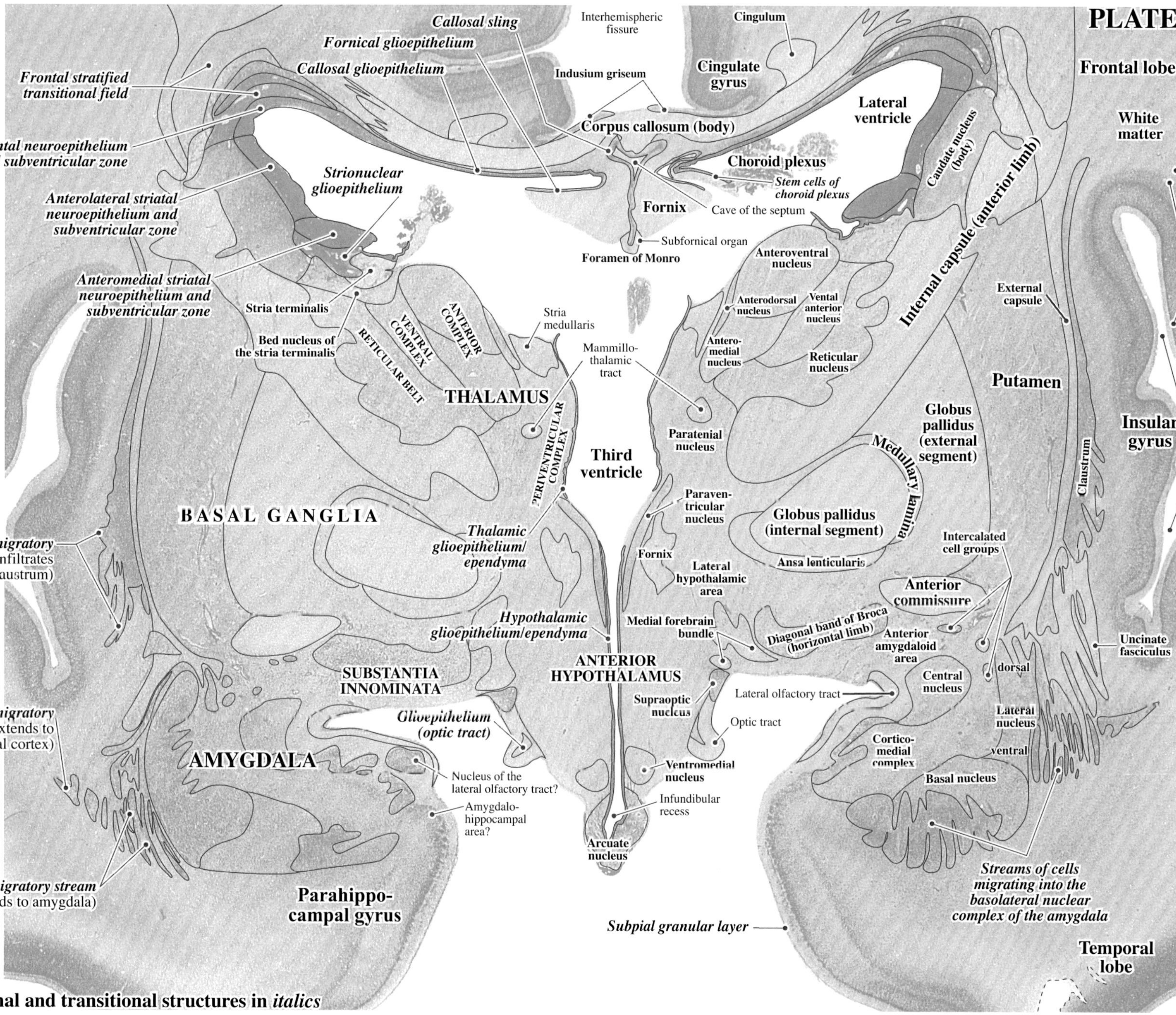

Germinal and transitional structures in ***italics***

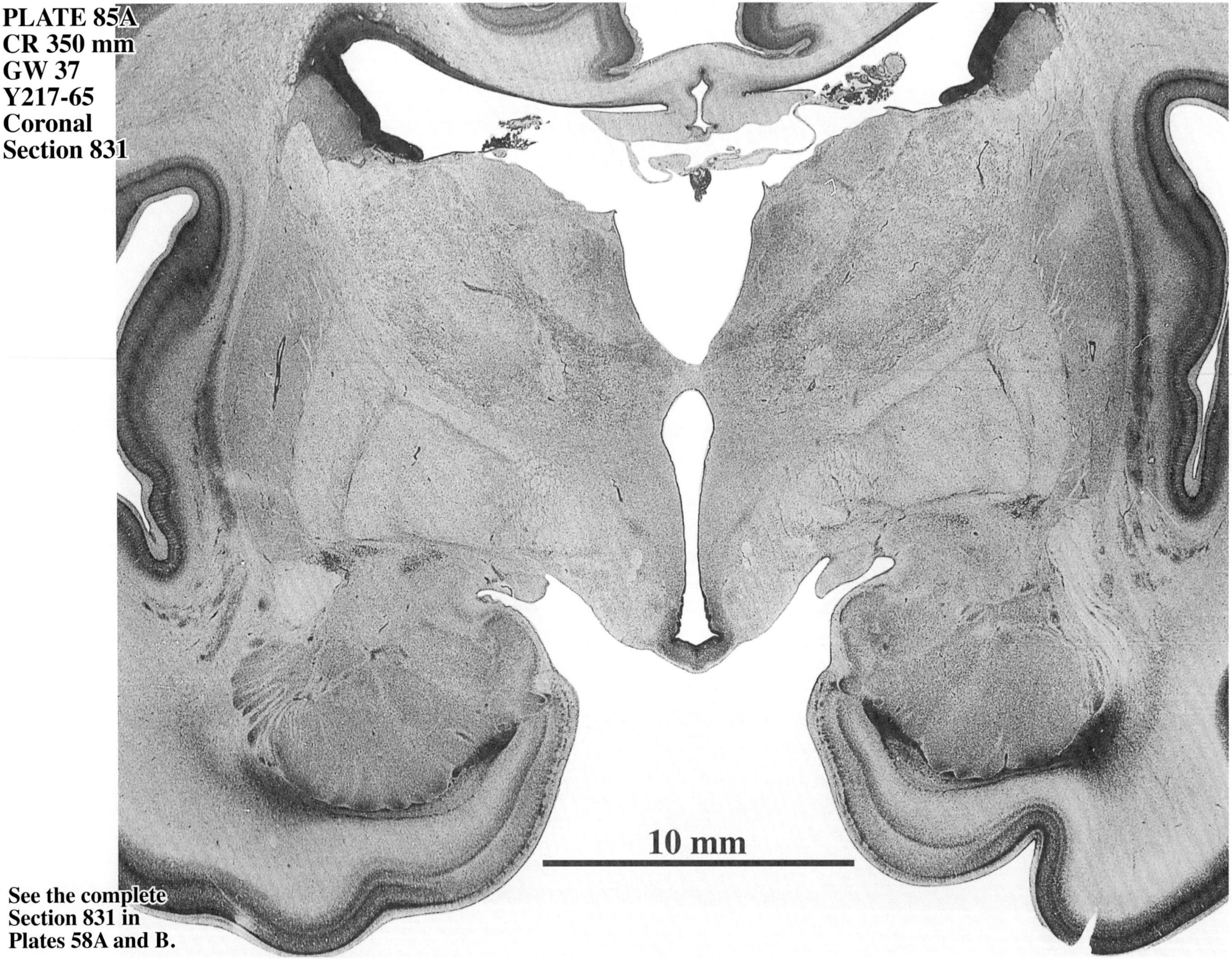

PLATE 85A
CR 350 mm
GW 37
Y217-65
Coronal
Section 831
10 mm
See the complete
Section 831 in
Plates 58A and B.

PLATE 85B

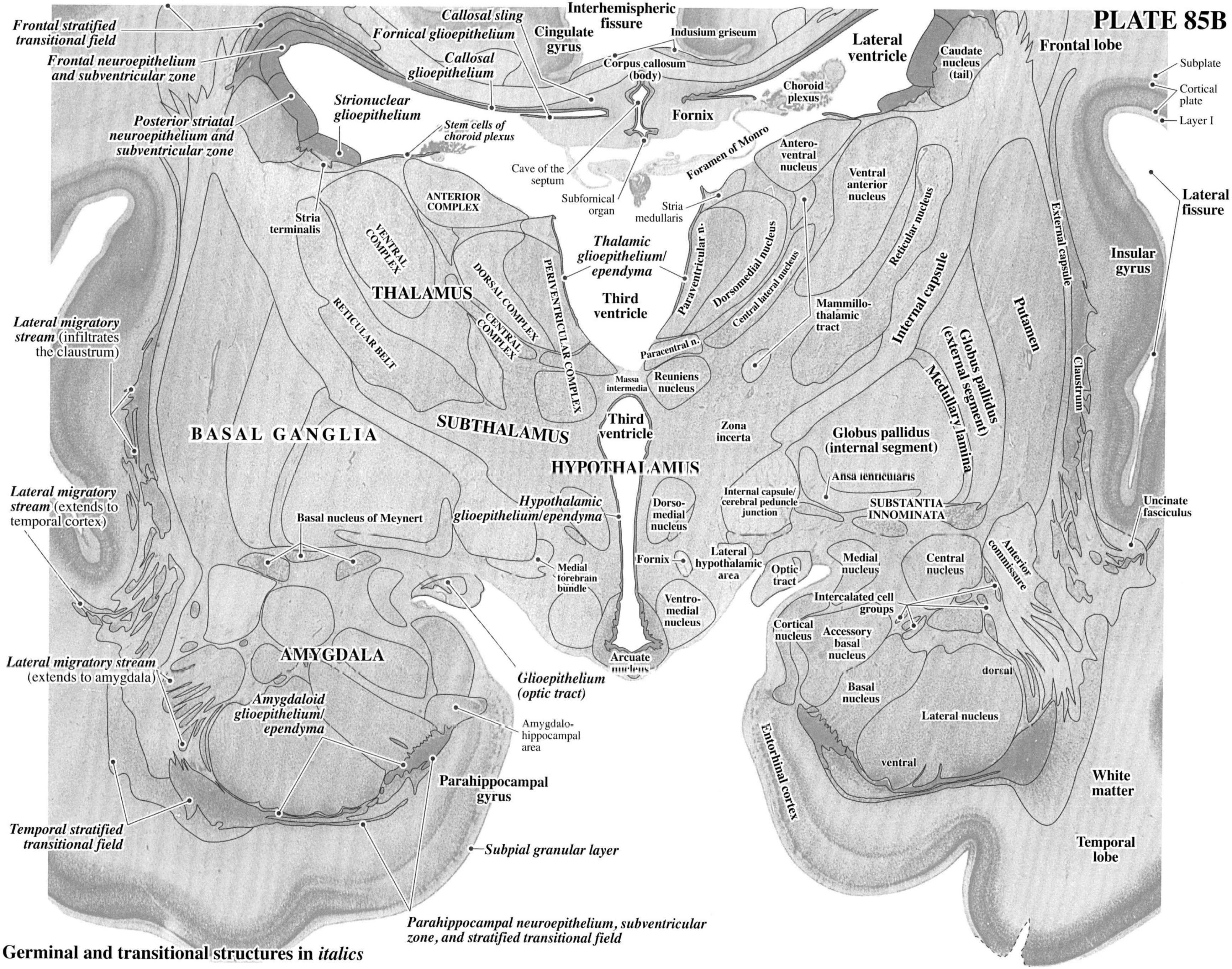

Germinal and transitional structures in *italics*

PLATE 86A
CR 350 mm
GW 37
Y217-65
Coronal
Section 881

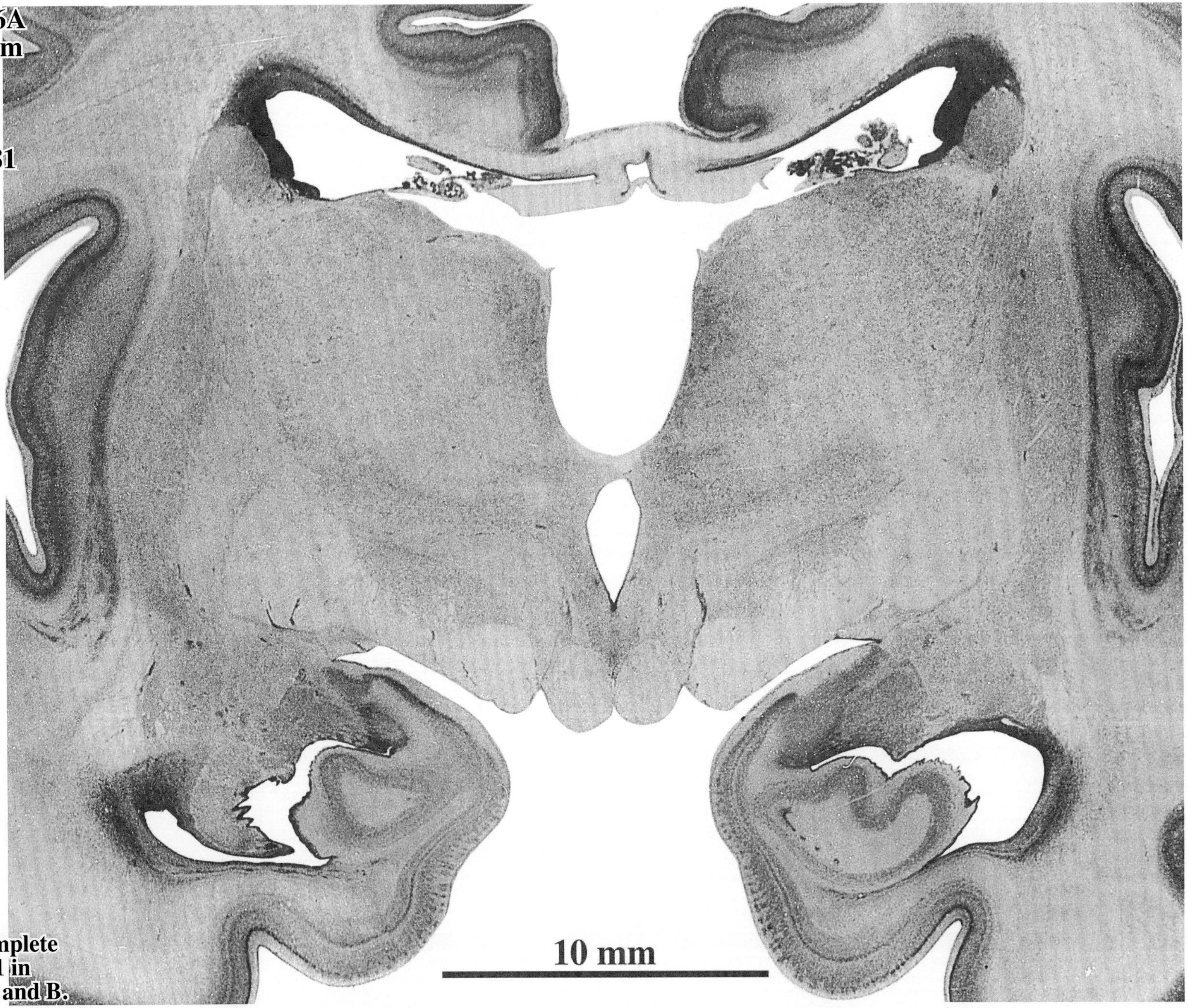

See the complete Section 881 in Plates 59A and B.

PLATE 86B

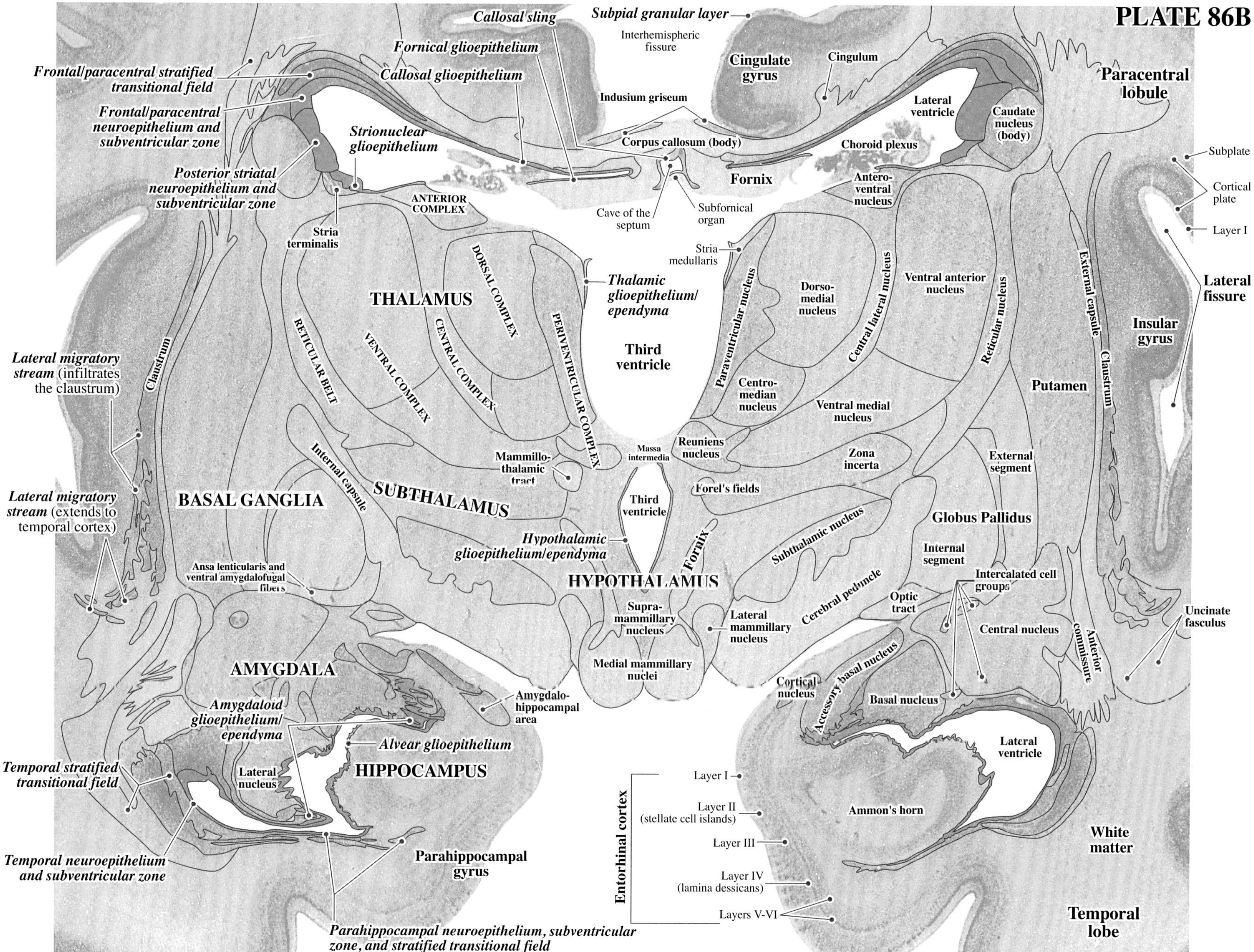

Germinal and transitional structures in *italics*

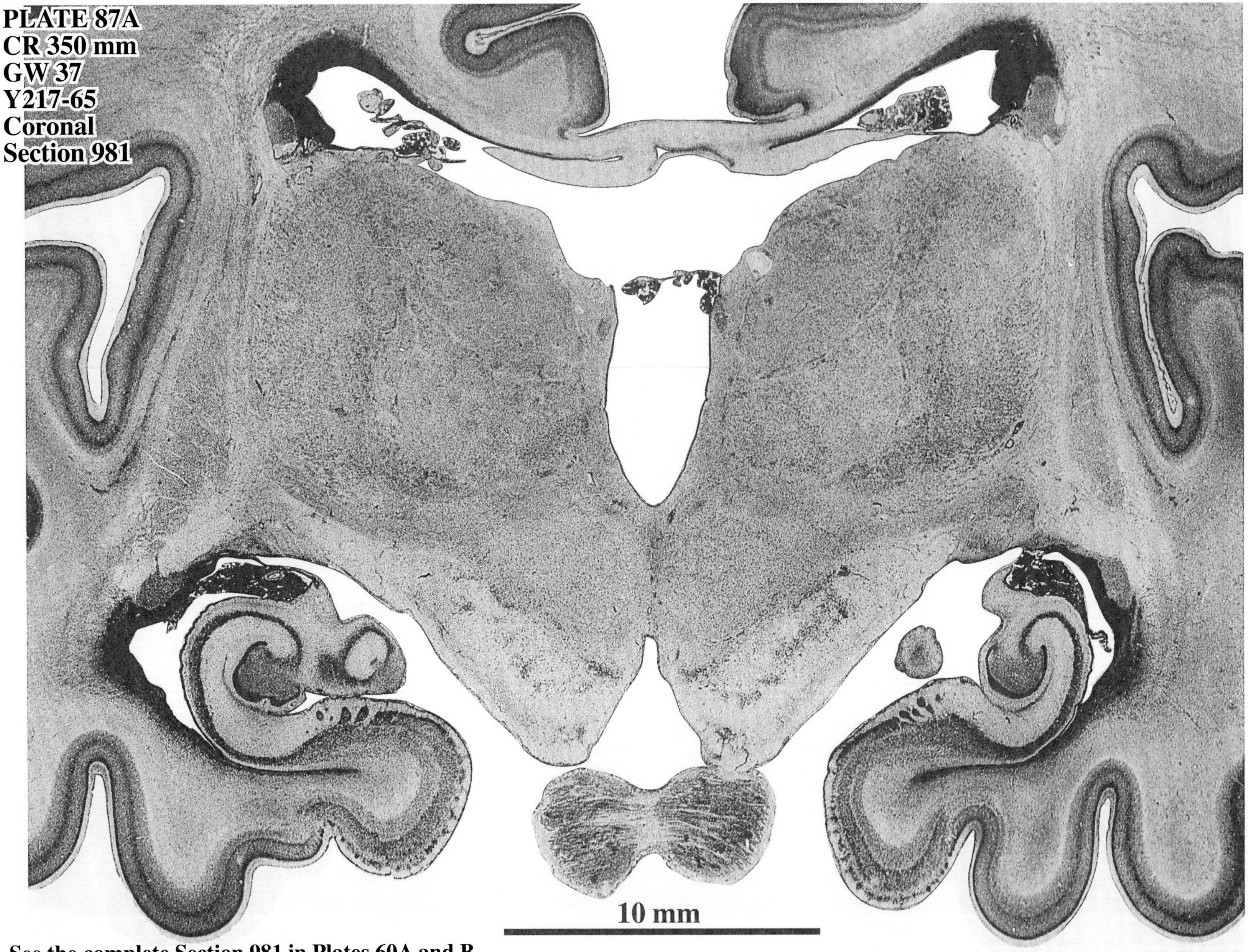

See the complete Section 981 in Plates 60A and B.

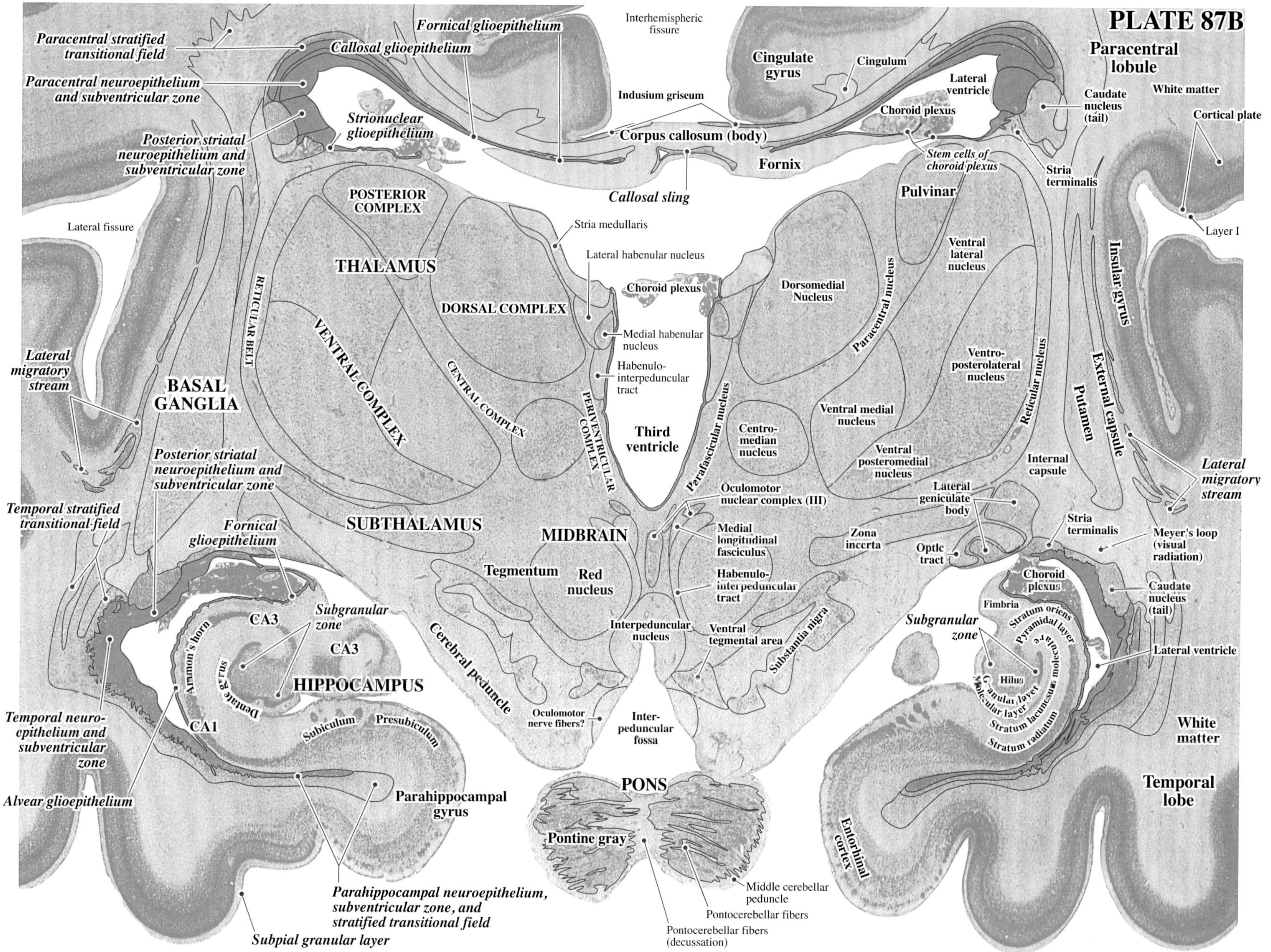

Germinal and transitional structures in *italics*

PLATE 88A
CR 350 mm
GW 37
Y217-65
Coronal
Section 1021

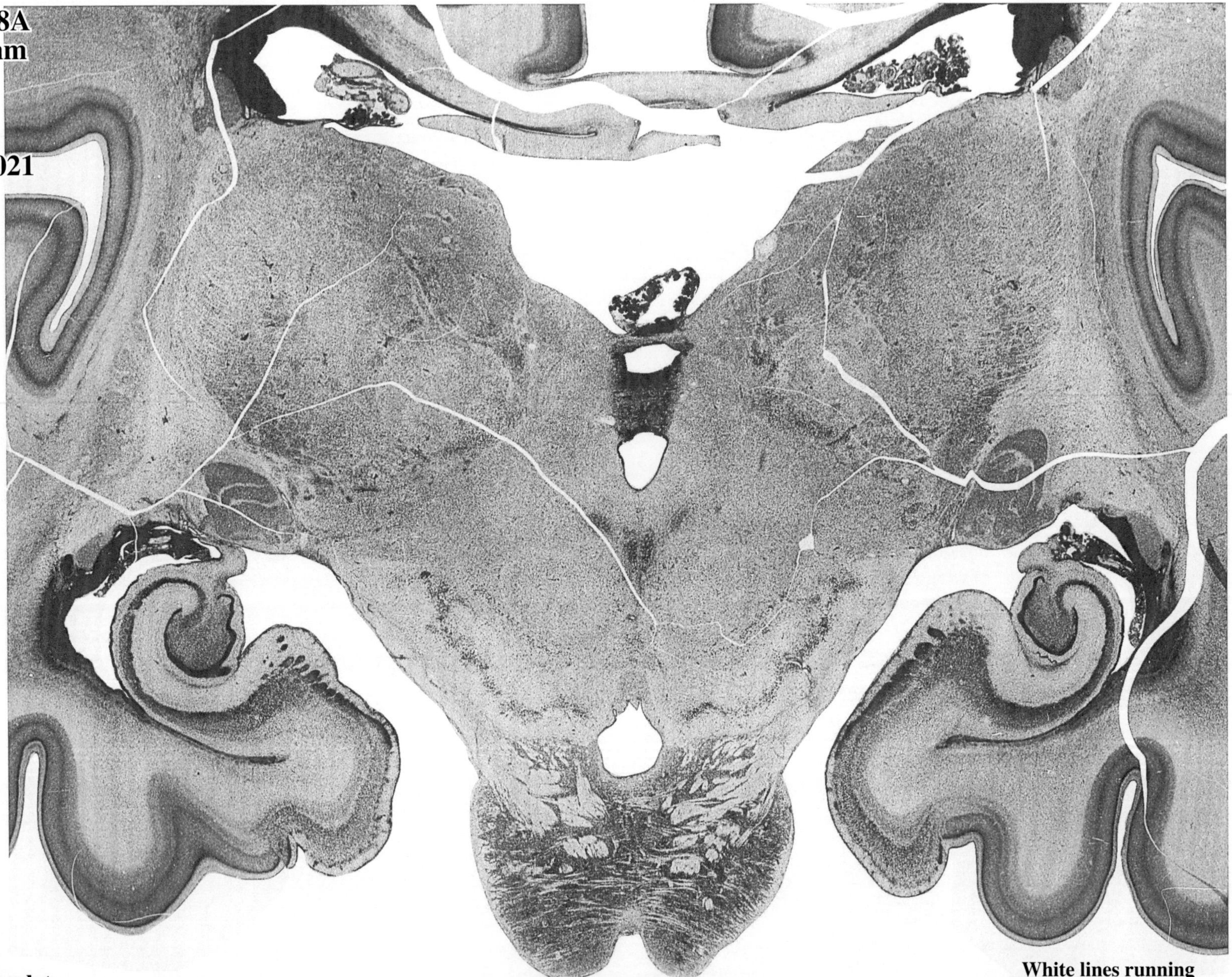

See the complete Section 1021 in Plates 61A and B.

10 mm

White lines running through this section are artifacts of histological processing.

PLATE 88B

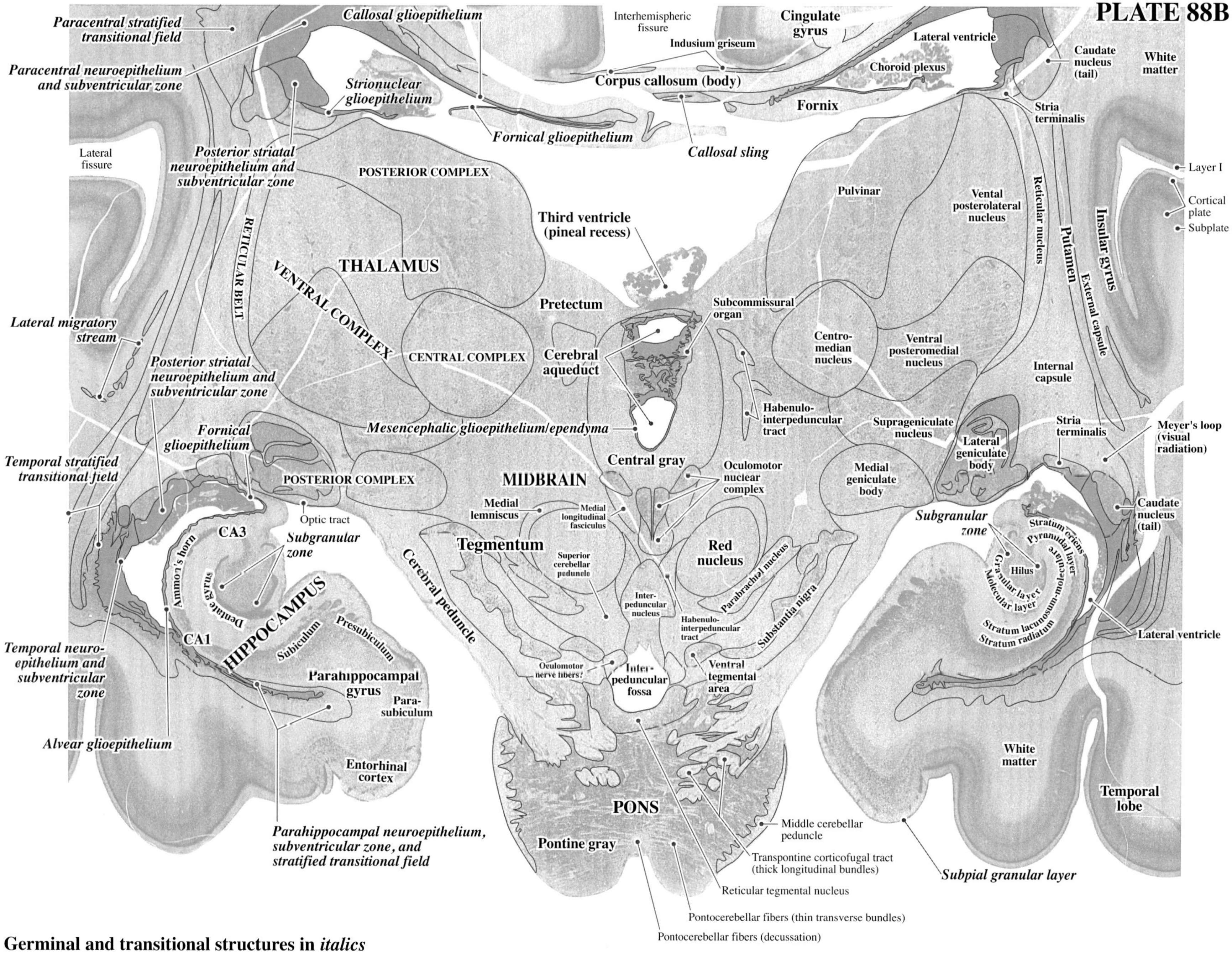

Germinal and transitional structures in *italics*

PLATE 89A
CR 350 mm
GW 37
Y217-65
Coronal
Section 1081

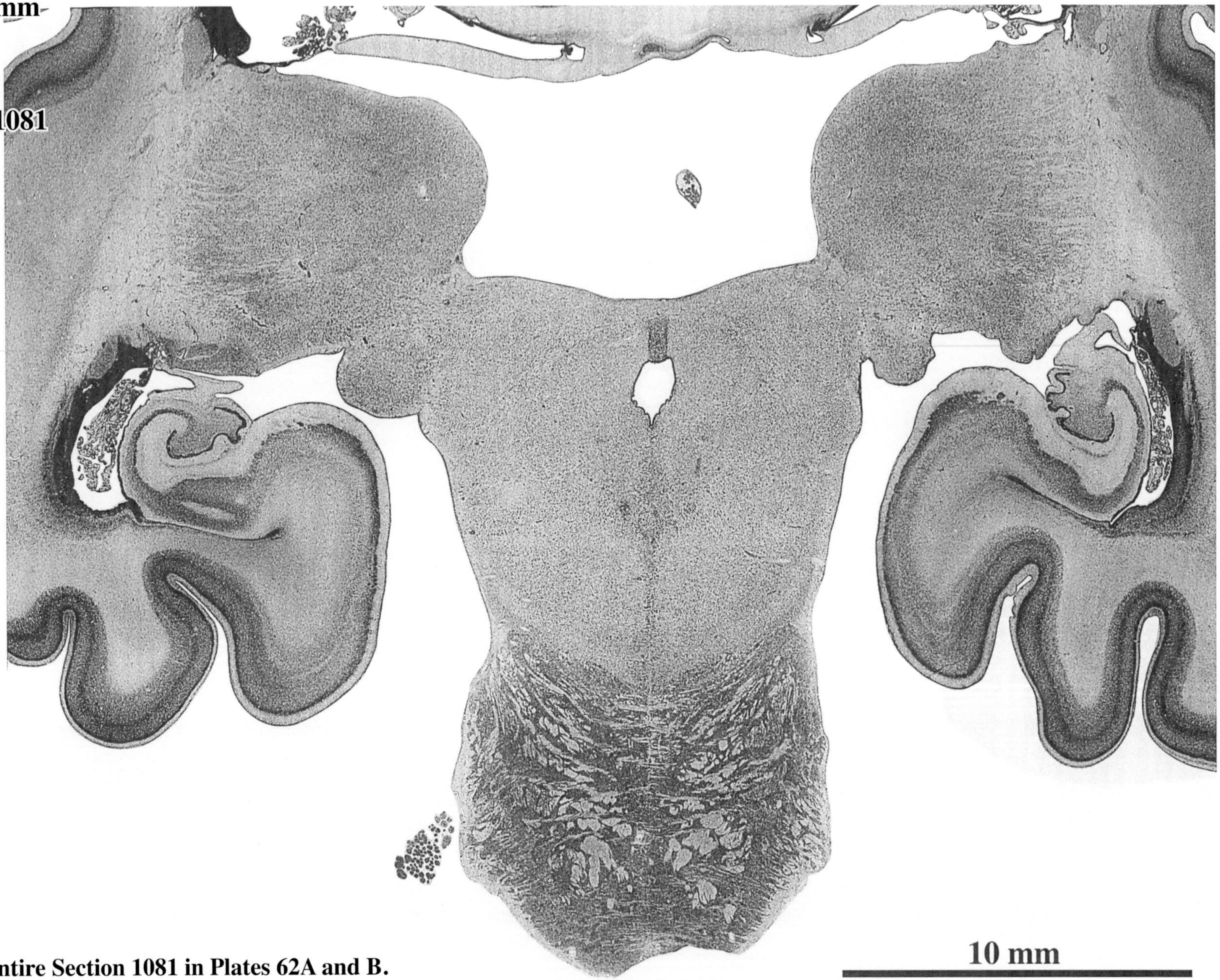

See the entire Section 1081 in Plates 62A and B.

PLATE 89B

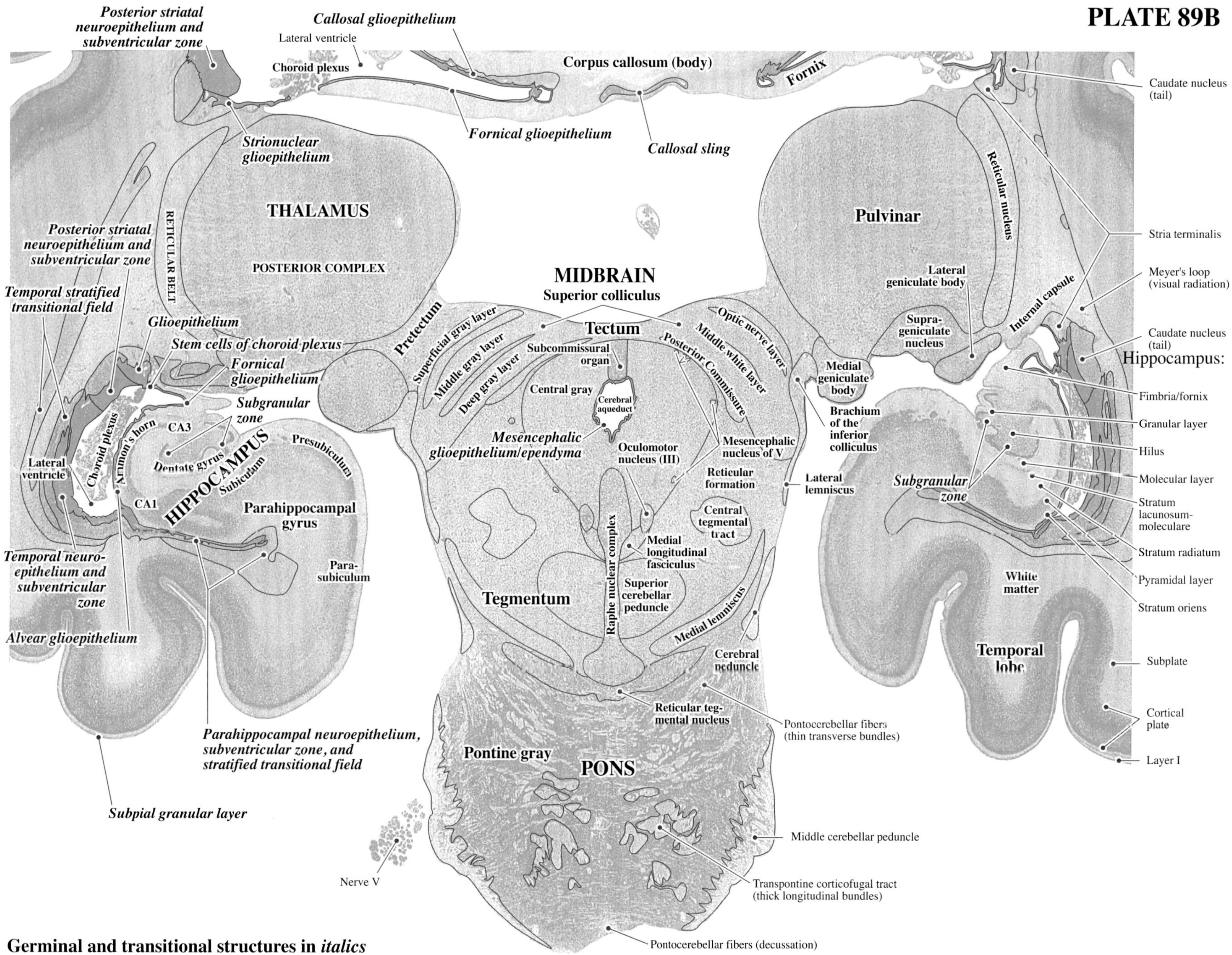

Germinal and transitional structures in *italics*

PLATE 90A
CR 350 mm, GW 37, Y217-65, Coronal, Section 1141

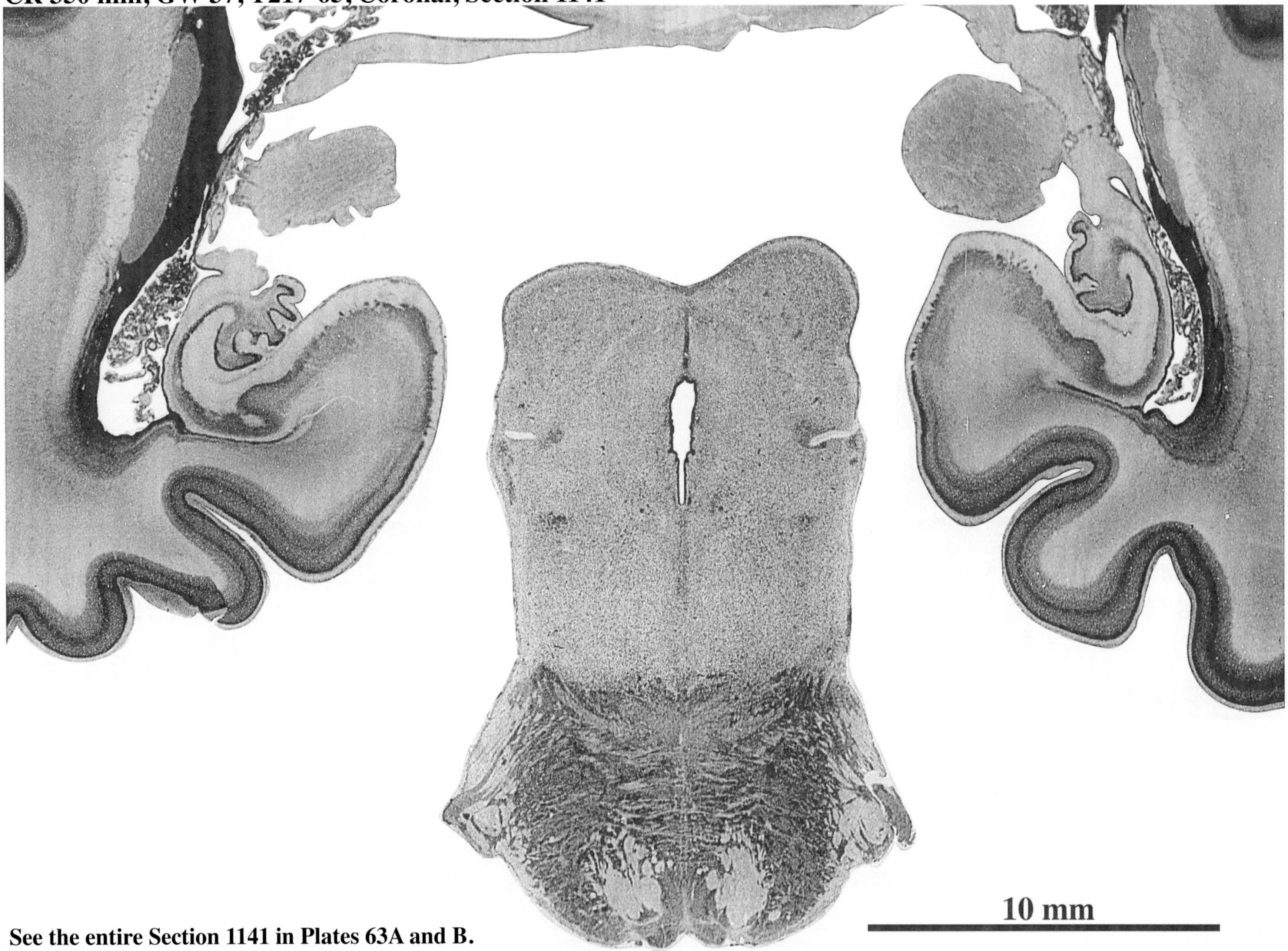

See the entire Section 1141 in Plates 63A and B.

PLATE 90B

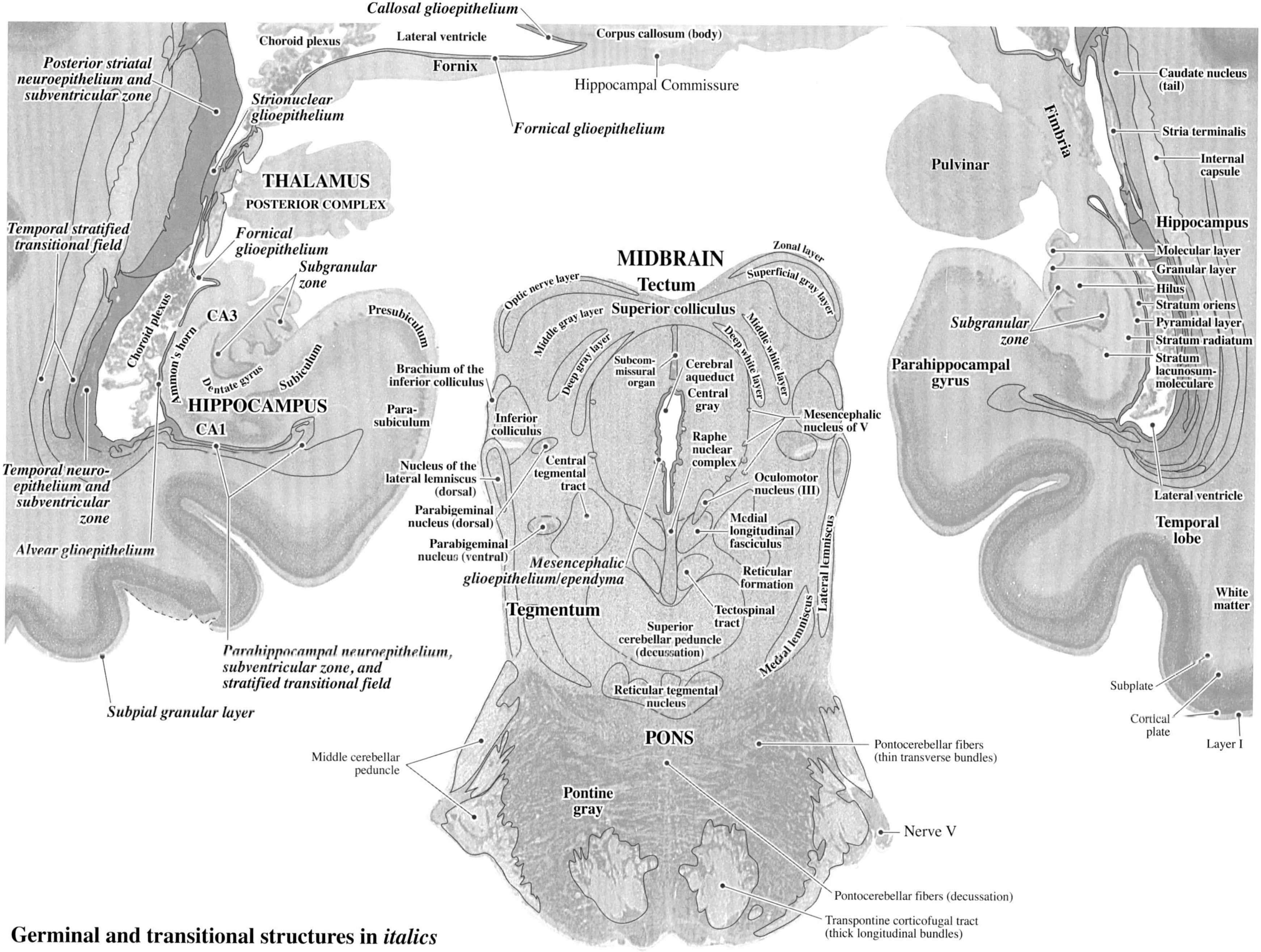

Germinal and transitional structures in *italics*

PLATE 91A
CR 350 mm, GW 37, Y217-65, Coronal, Section 1181

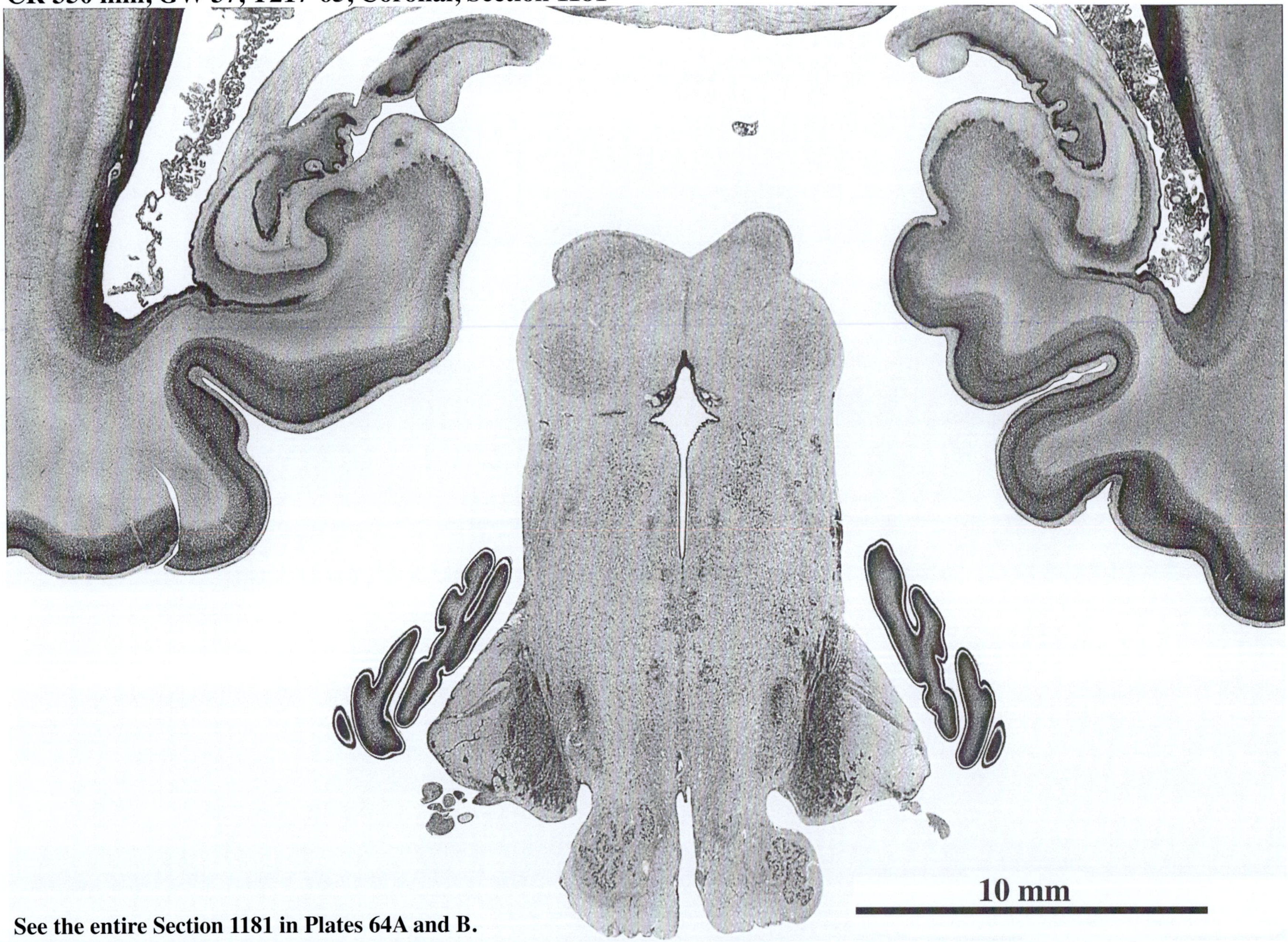

See the entire Section 1181 in Plates 64A and B.

PLATE 91B

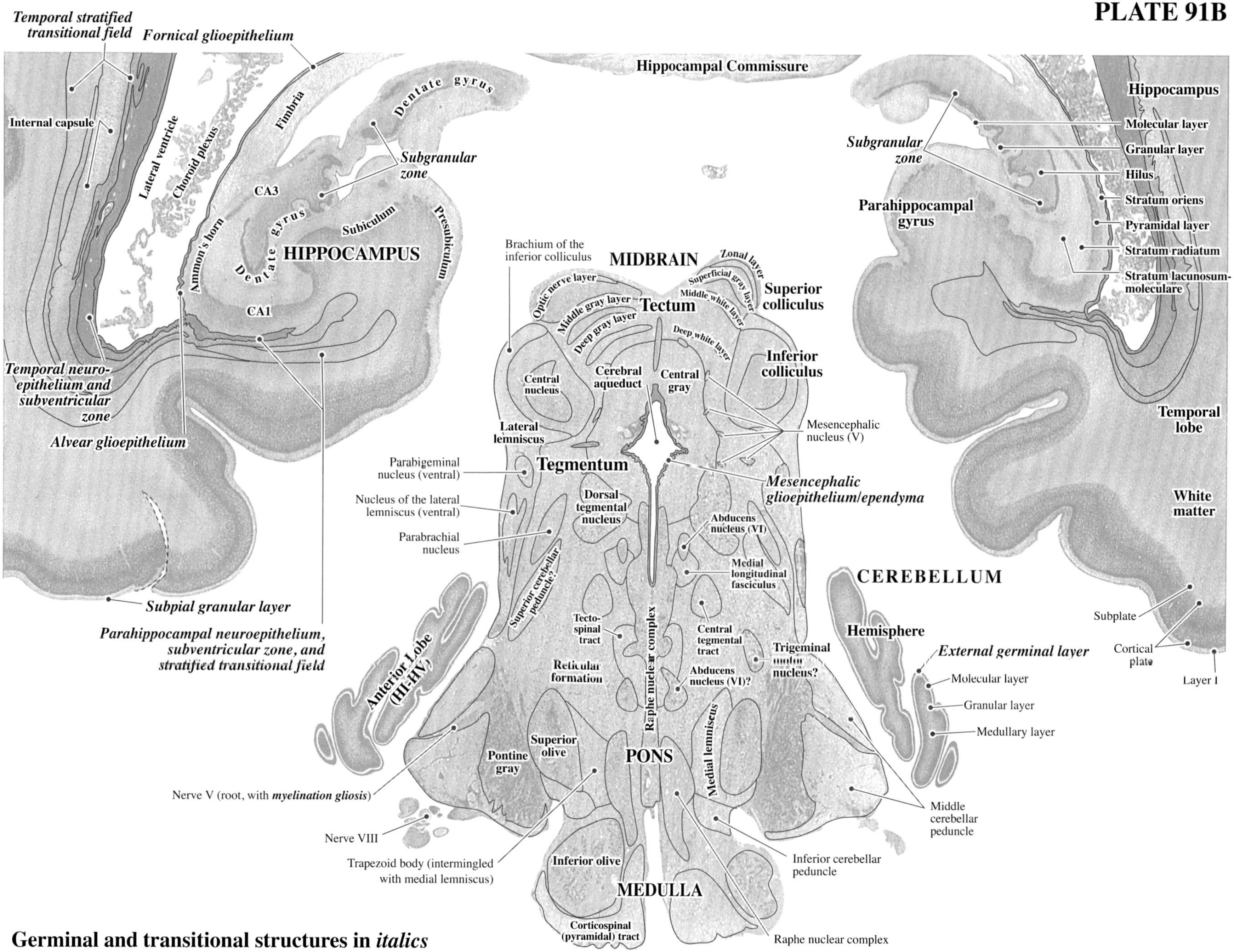

Germinal and transitional structures in *italics*

PLATE 92A
CR 350 mm
GW 37
Y217-65
Coronal
Section 1221

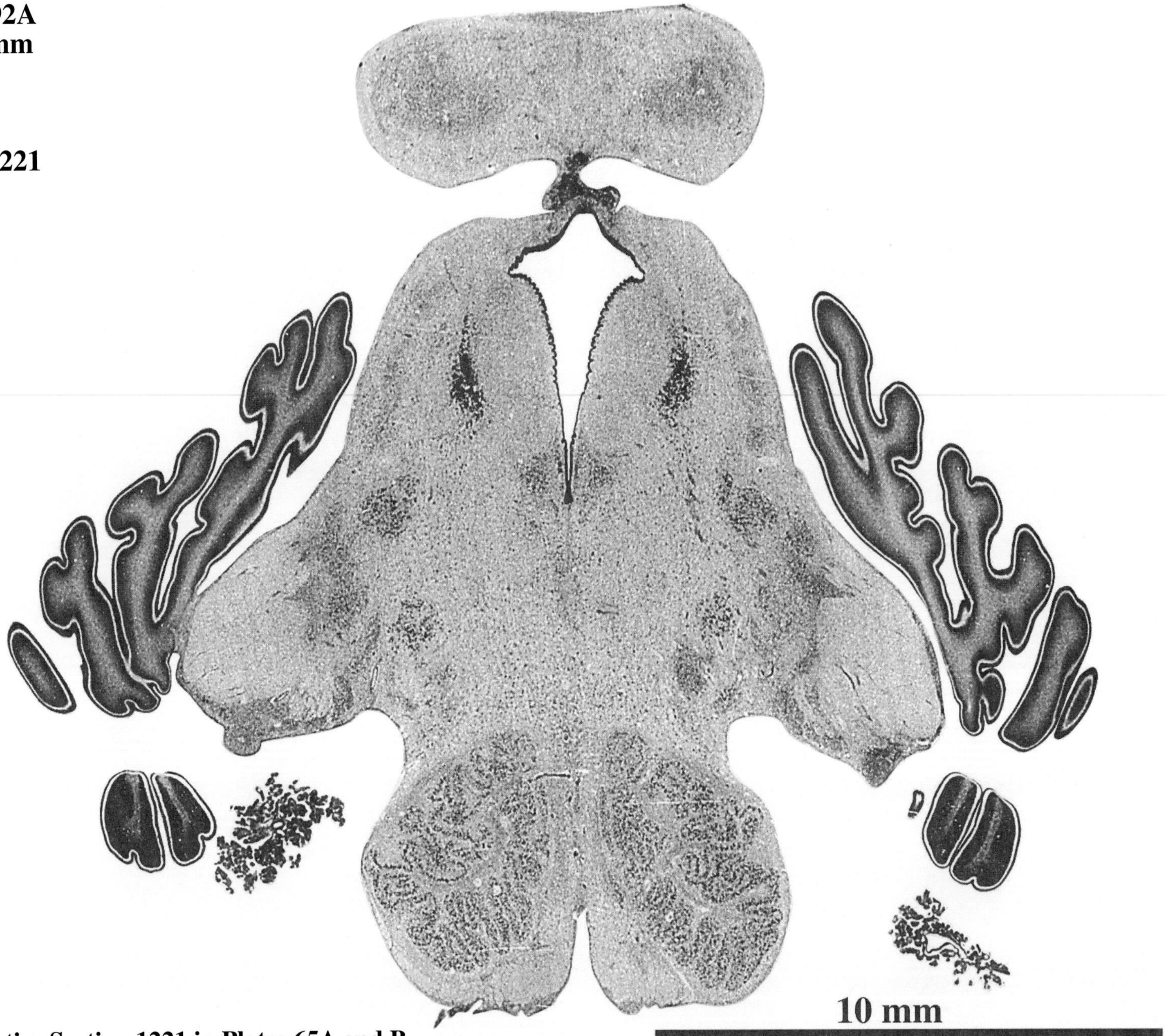

See the entire Section 1221 in Plates 65A and B.

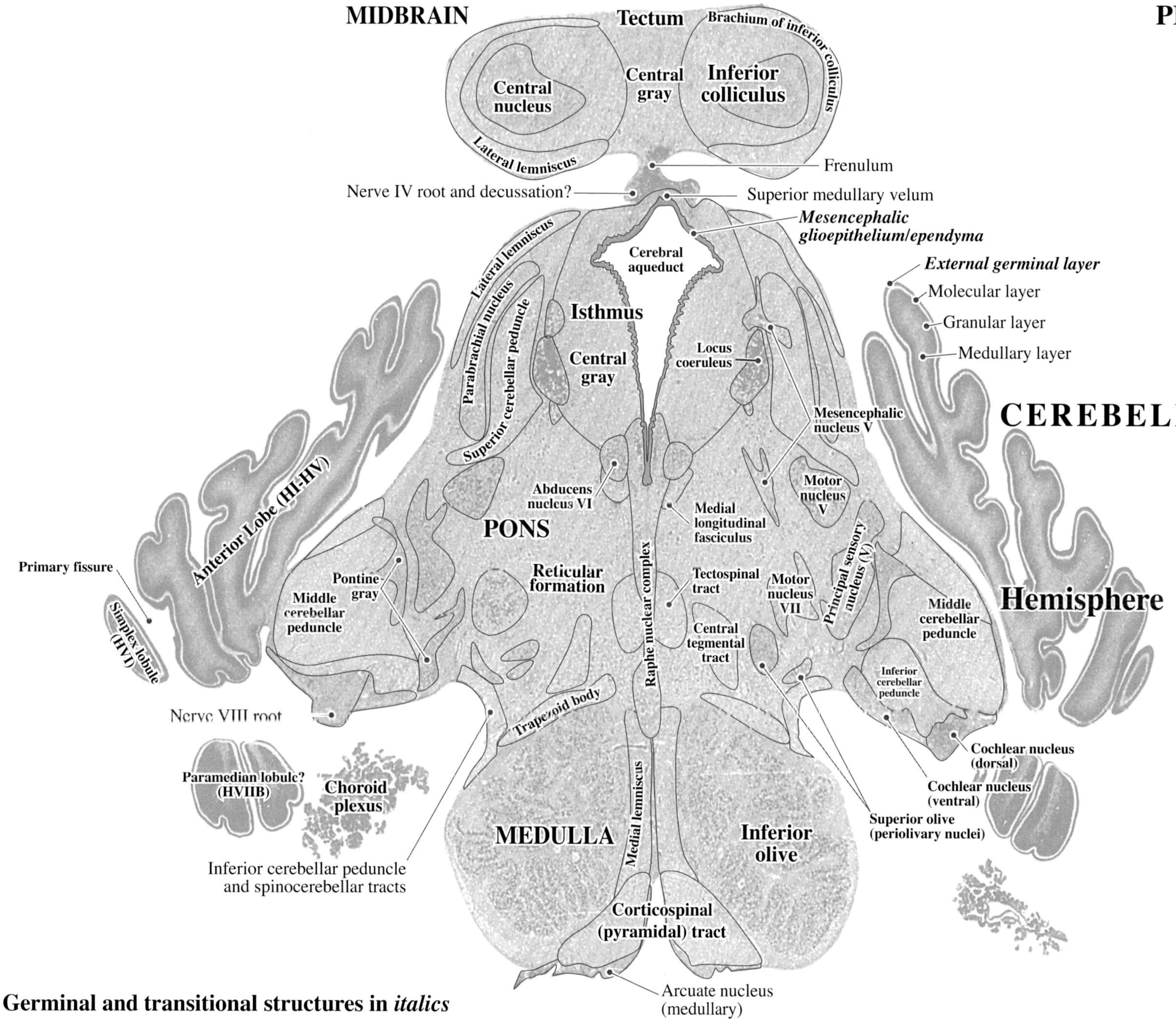

Germinal and transitional structures in *italics*

PLATE 93A
CR 350 mm
GW 37
Y217-65
Coronal
Section 1311

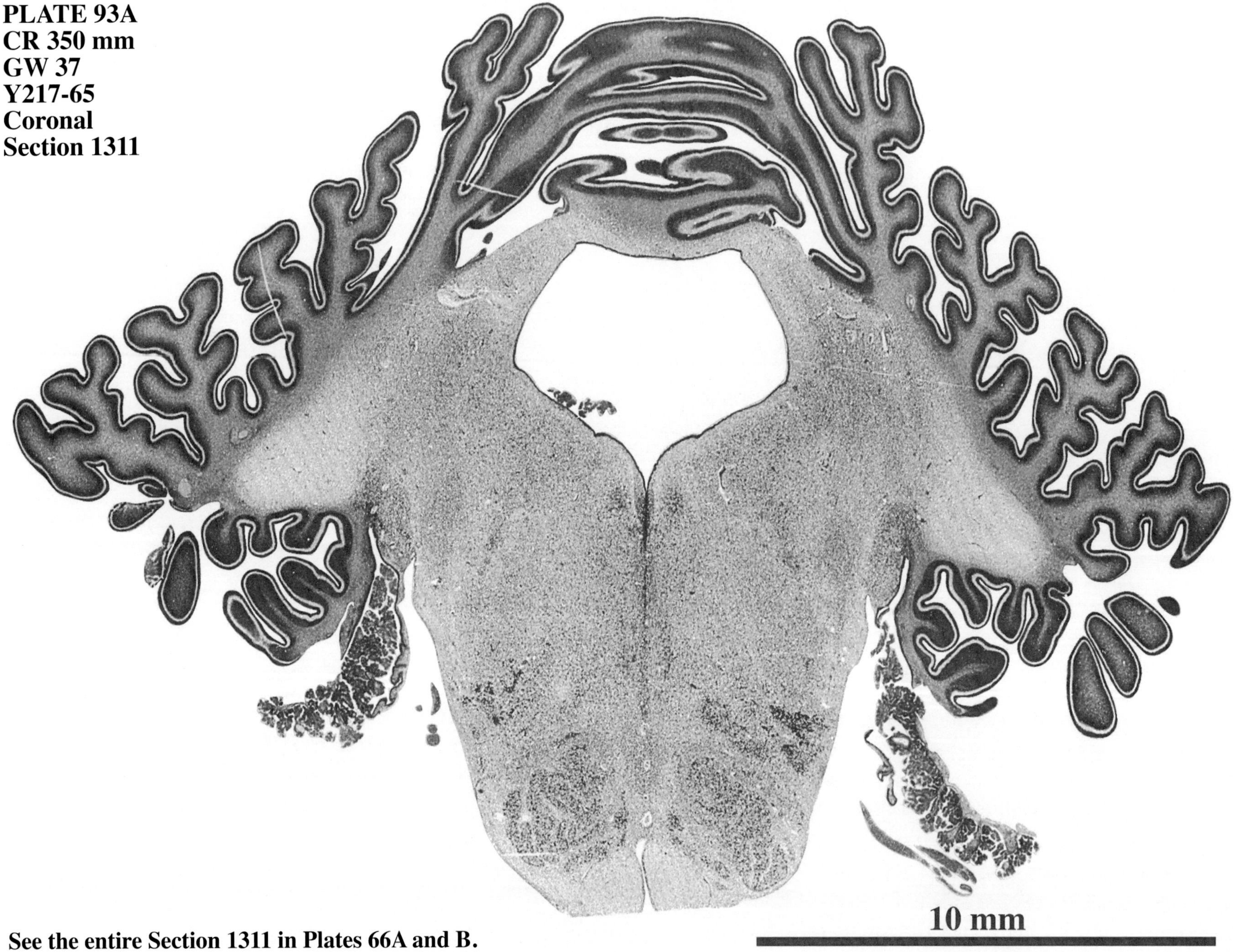

See the entire Section 1311 in Plates 66A and B.

PLATE 93B

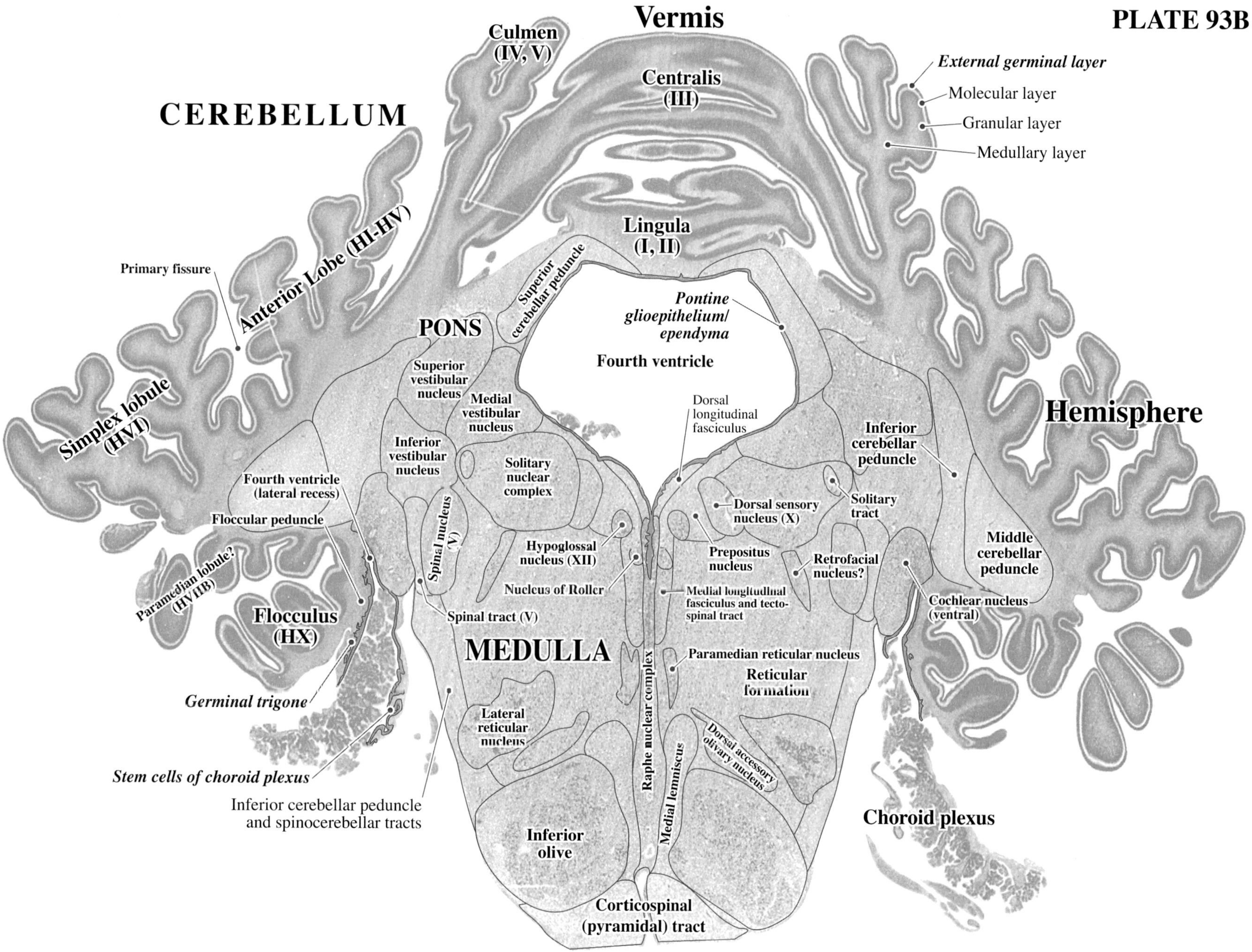

Germinal and transitional structures in *italics*

PLATE 94A
CR 350 mm
GW 37
Y217-65
Coronal
Section 1371

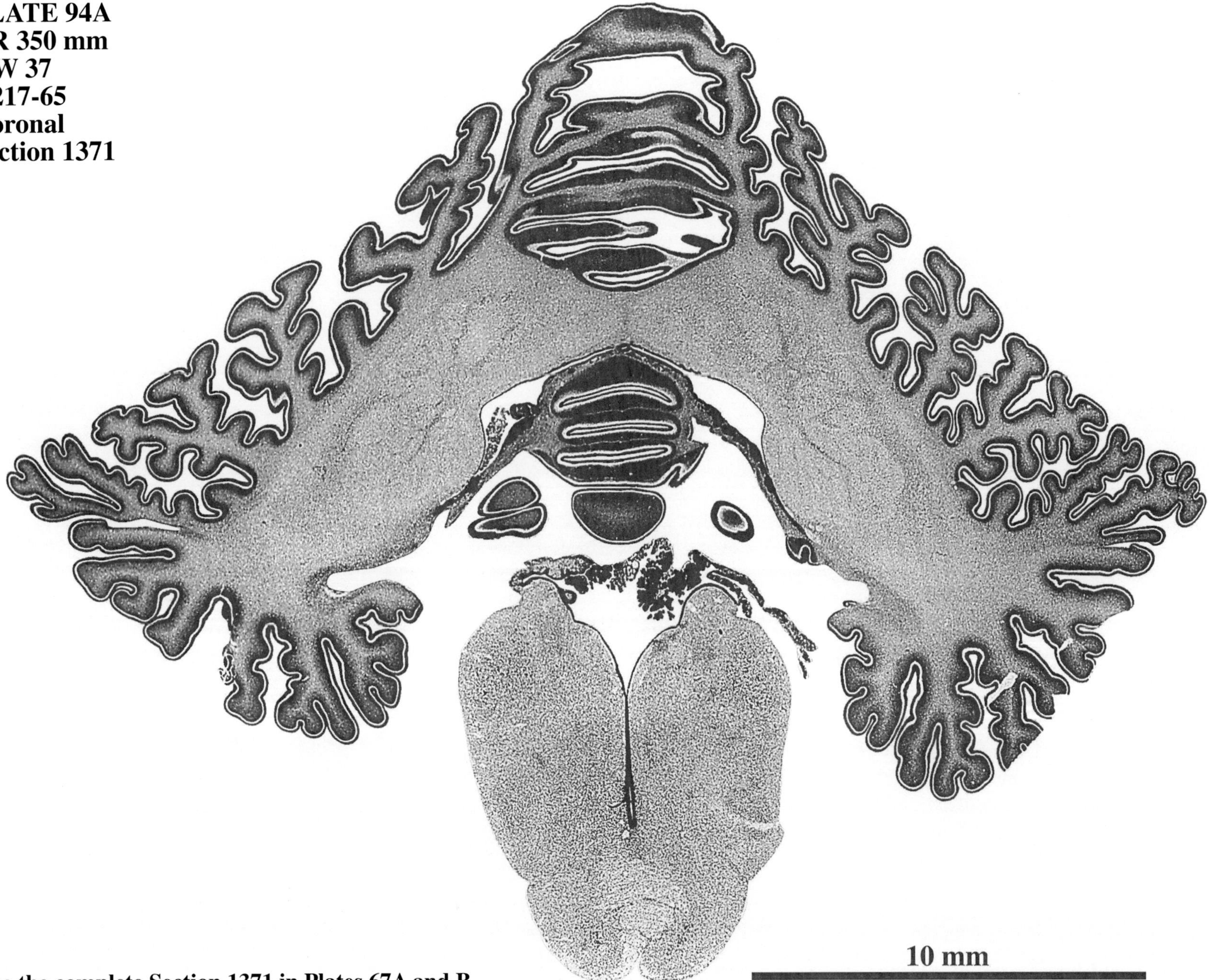

See the complete Section 1371 in Plates 67A and B.

PLATE 94B

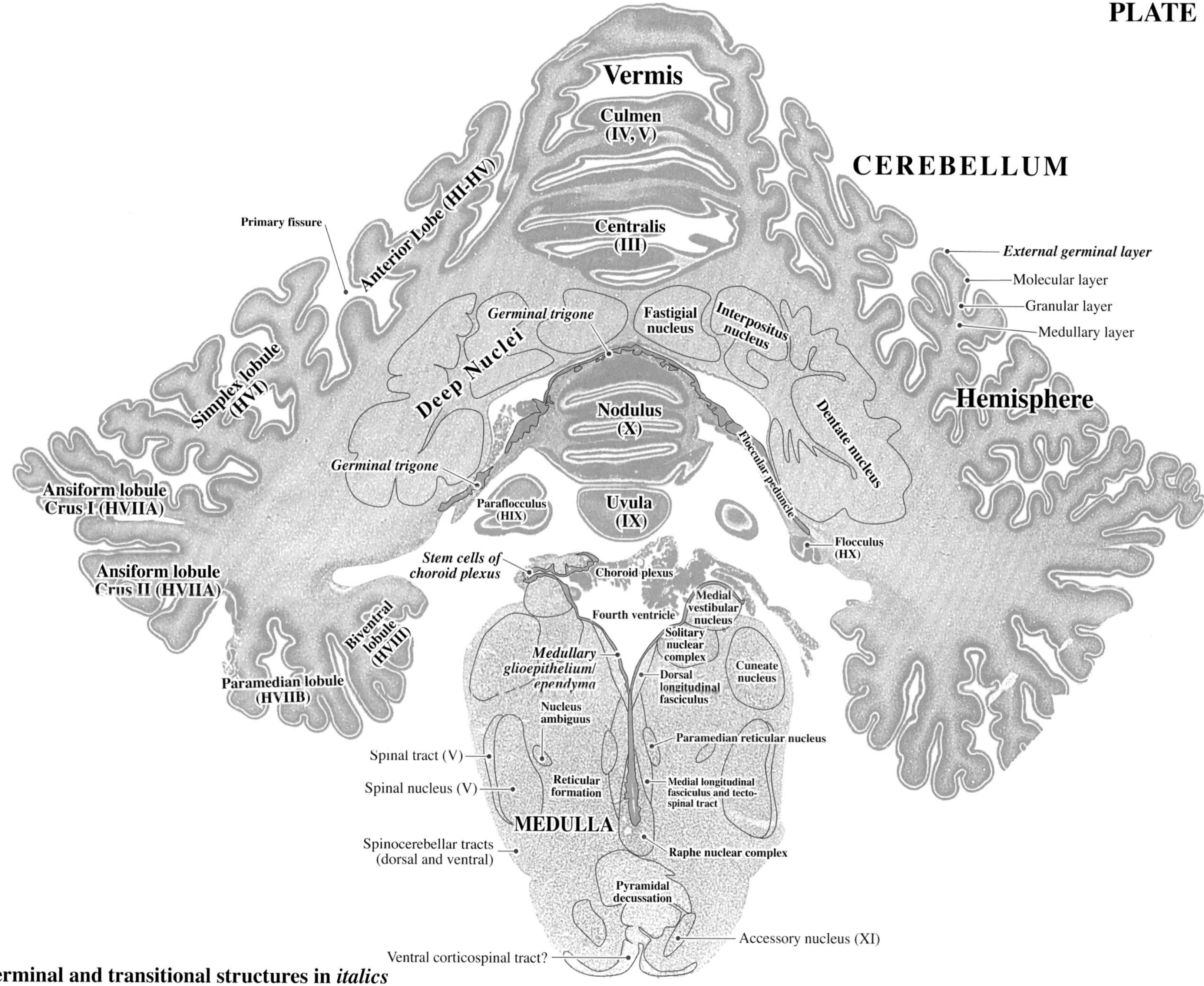

Germinal and transitional structures in *italics*

PLATE 95A
CR 350 mm
GW 37
Y217-65
Coronal
Section 1501

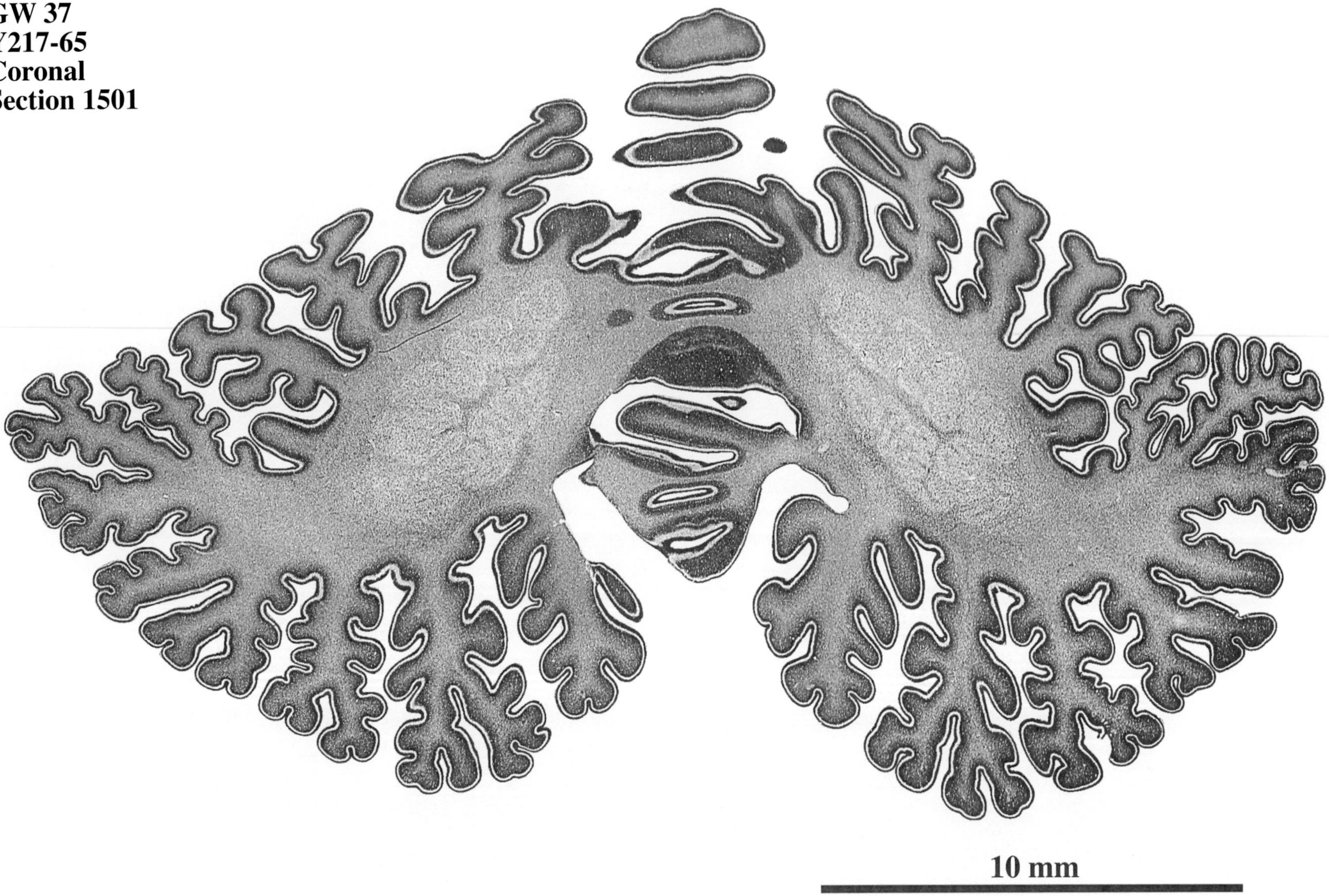

See the complete Section 1501 in Plates 68A and B.

PLATE 95B

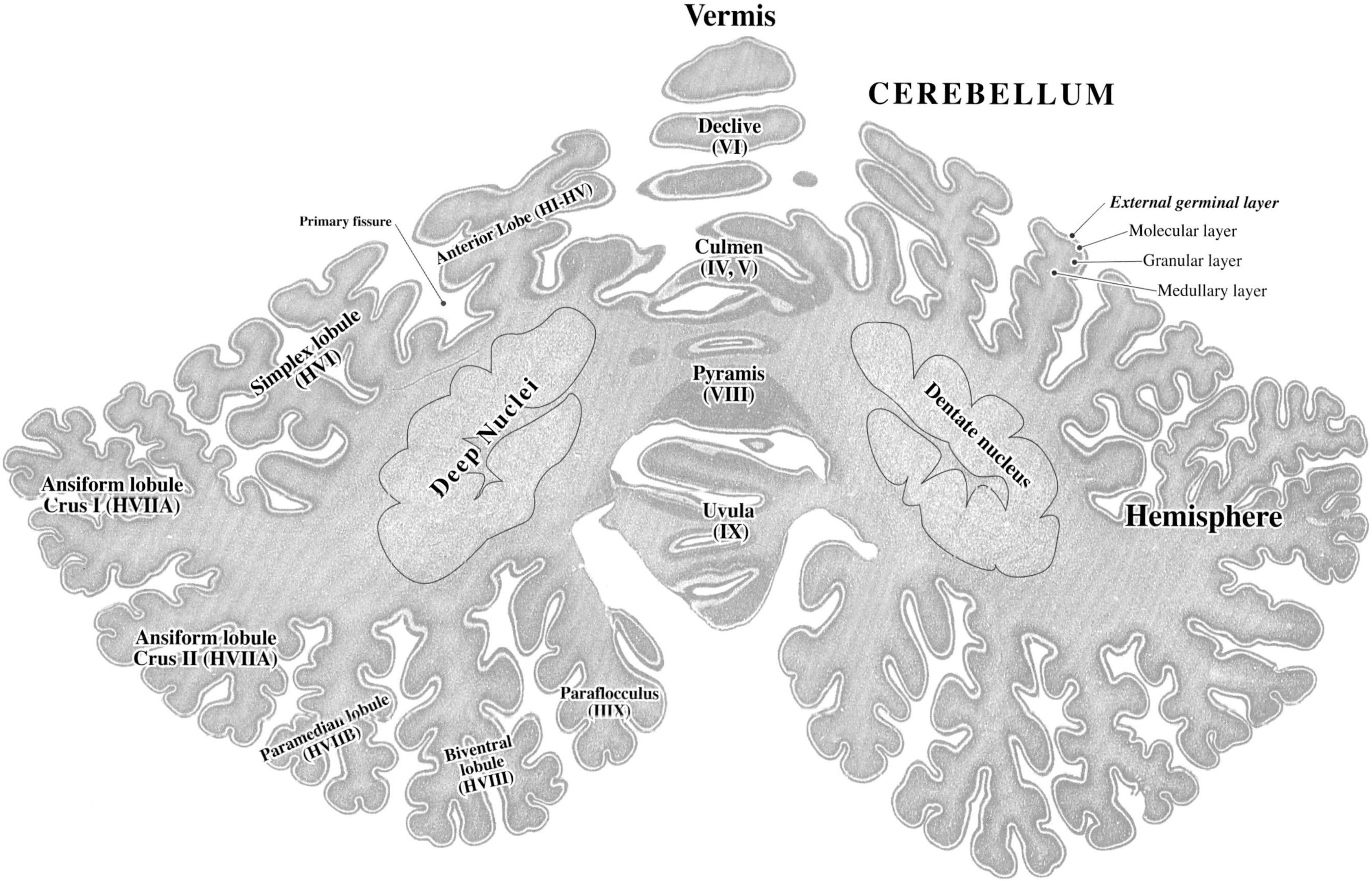

Germinal and transitional structures in *italics*

PLATE 96A
CR 350 mm
GW 37
Y217-65
Coronal
Section 1611

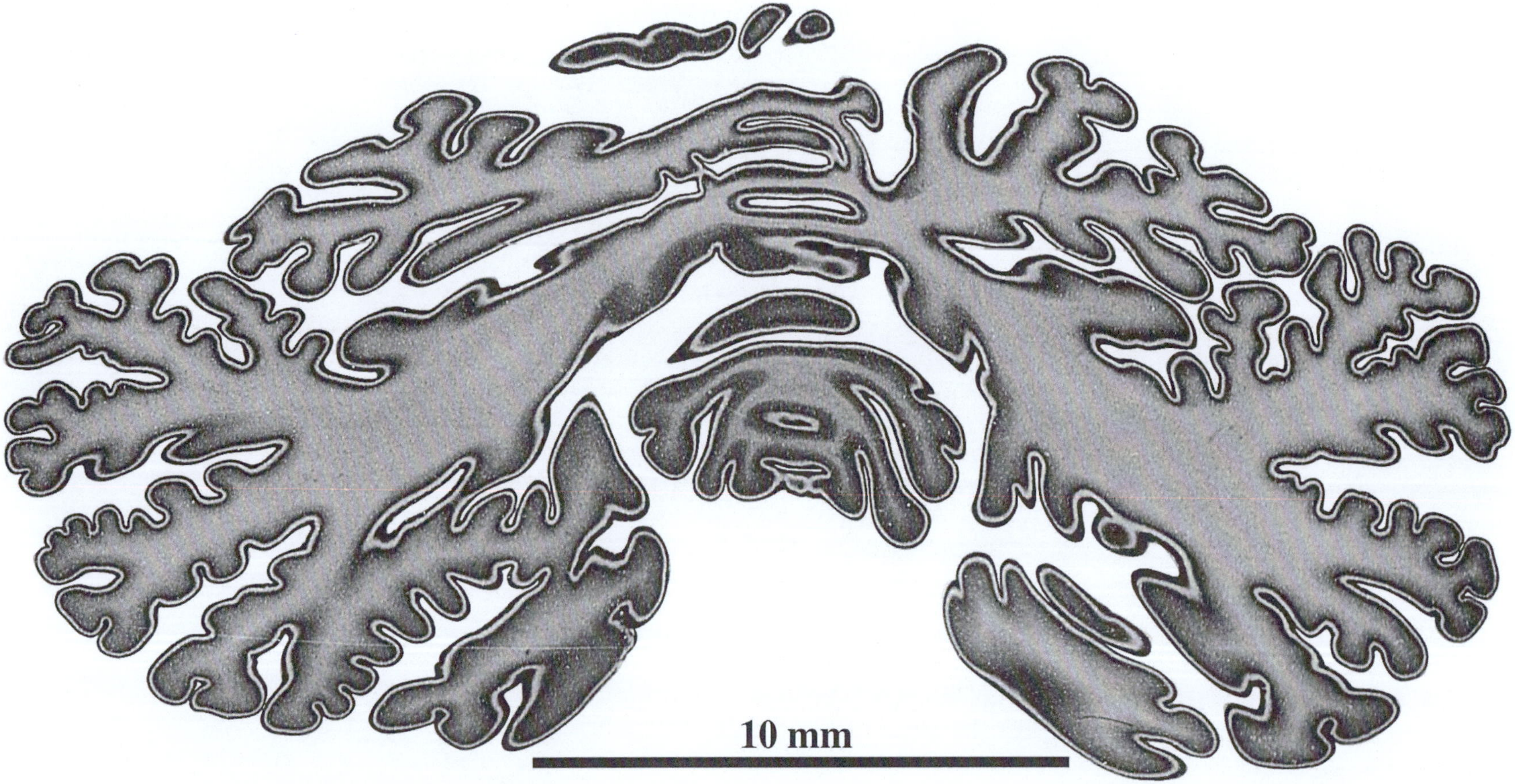

See the complete Section 1611 in Plates 69A and B.

PLATE 96B

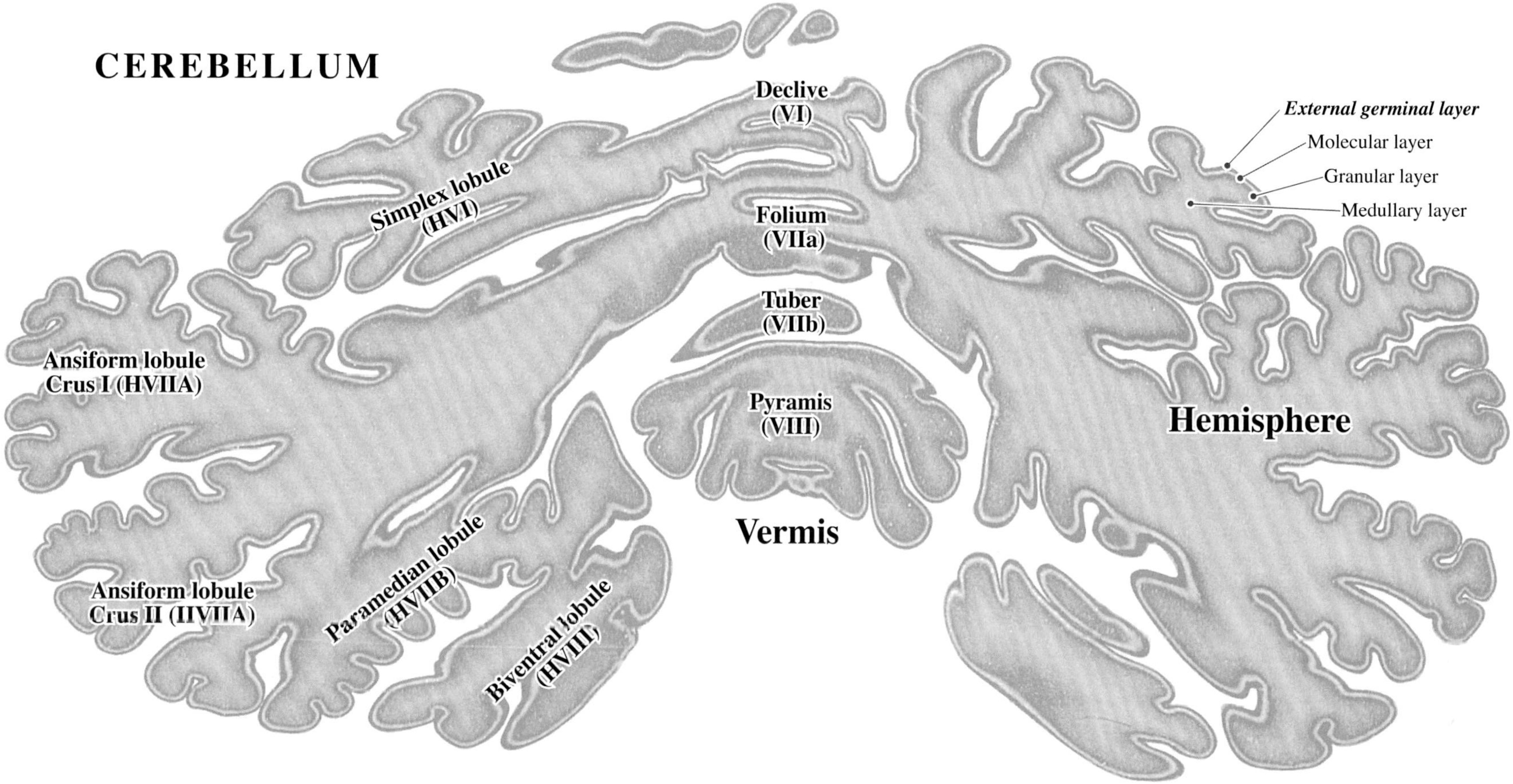

Germinal and transitional structures in *italics*

PLATE 97A
CR 350 mm
GW 37
Y217-65
Coronal
Section 1711

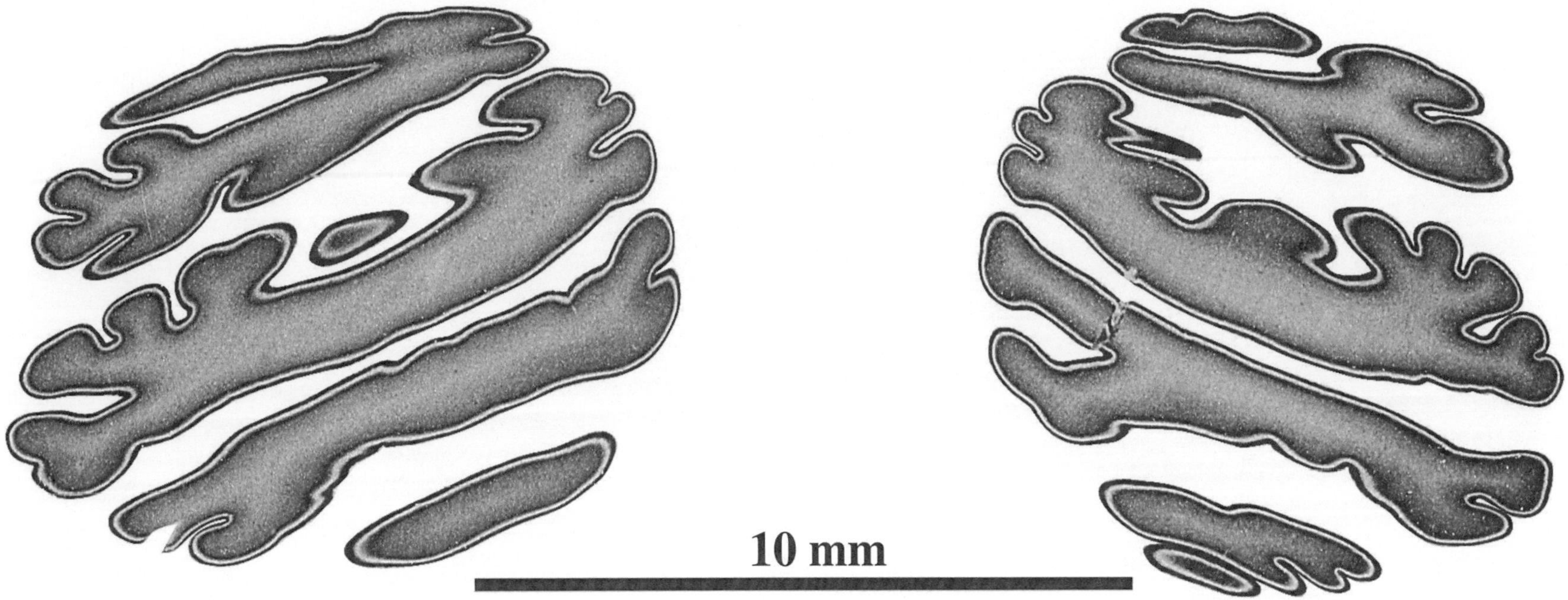

See the complete Section 1711 in Plates 70A and B.

PLATE 97B

CEREBELLUM

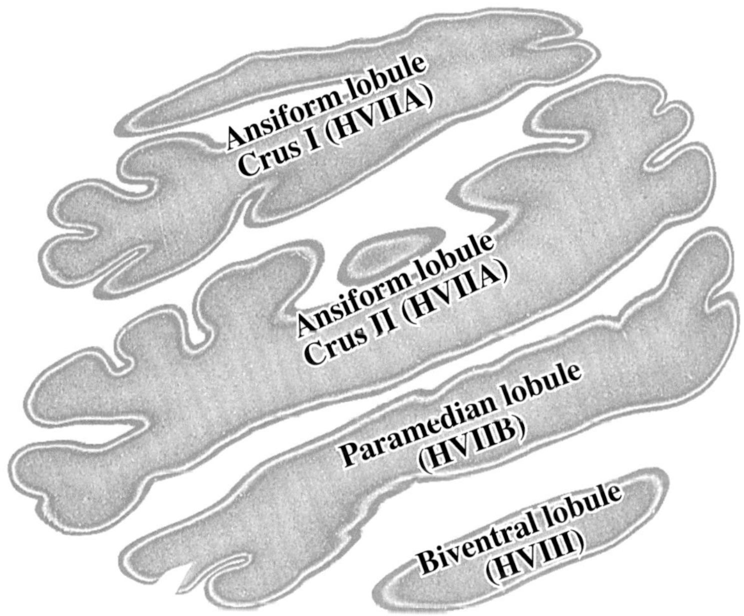

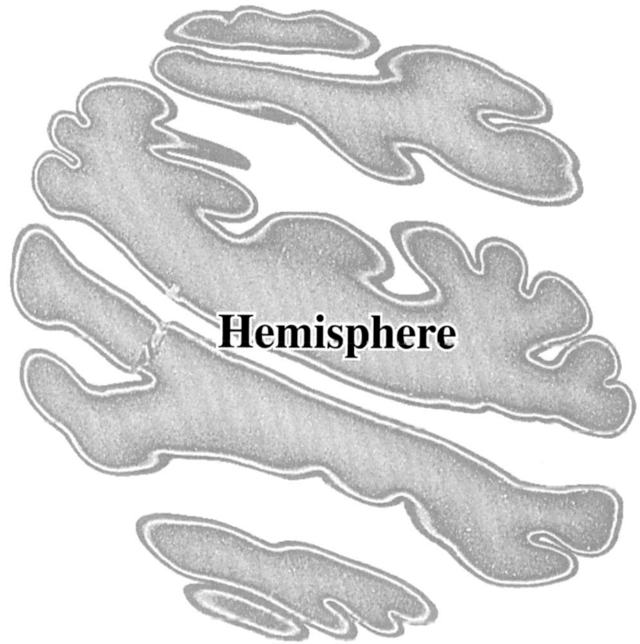

For Product Safety Concerns and Information please contact our EU representative GPSR@taylorandfrancis.com Taylor & Francis Verlag GmbH, Kaufingerstraße 24, 80331 München, Germany

Printed and bound by CPI Group (UK) Ltd, Croydon, CR0 4YY
14/09/2026
02216264-0002